Remote Sensing of Evapotranspiration (ET)

Remote Sensing of Evapotranspiration (ET)

Special Issue Editors

Pradeep Wagle
Prasanna H. Gowda

MDPI • Basel • Beijing • Wuhan • Barcelona • Belgrade

Special Issue Editors

Pradeep Wagle
USDA-ARS, Grazinglands Research
Laboratory
USA

Prasanna H. Gowda
USDA-ARS, Southeast Area
USA

Editorial Office
MDPI
St. Alban-Anlage 66
4052 Basel, Switzerland

This is a reprint of articles from the Special Issue published online in the open access journal *Remote Sensing* (ISSN 2072-4292) from 2018 to 2019 (available at: https://www.mdpi.com/journal/remotesensing/special_issues/rs-ET)

For citation purposes, cite each article independently as indicated on the article page online and as indicated below:

LastName, A.A.; LastName, B.B.; LastName, C.C. Article Title. *Journal Name* **Year**, *Article Number*, Page Range.

ISBN 978-3-03921-602-4 (Pbk)
ISBN 978-3-03921-603-1 (PDF)

Contents

About the Special Issue Editors

Pradeep Wagle is an internationally recognized scientist in biosphere-atmosphere interactions, carbon and water cycles of terrestrial ecosystems, and remote sensing of vegetation as documented in nearly 50 SCI indexed peer-reviewed articles in various journals, such as Remote Sensing of Environment; ISPRS Journal of Photogrammetry and Remote Sensing; Remote Sensing; Global Change Biology—Bioenergy; Agricultural and Forest Meteorology; Agricultural Water Management; Agriculture, Ecosystems and Environment; Ecological Applications; Agronomy; Atmospheric Environment; Science of the Total Environment; Ecohydrology. He received his Ph.D. in Crop Science from Oklahoma State University (OSU), USA in 2013. From 2013 to 2015, Dr. Wagle worked as a postdoctoral research associate and research scientist at the University of Oklahoma. Since 2016, he has worked as a Research Ecologist at the United States Department of Agriculture (USDA), Grazinglands Research Laboratory, El Reno, Oklahoma. He serves as the handling editor and secretary on the editorial board for the Global Journal of Agricultural and Allied Sciences, and as an Editorial Board Member for the Remote Sensing Journal. He worked as a Guest Editor for "Sustainability in the Mountains Region" for Sustainability Journal from 2016 to 2017 and "Remote Sensing of Evapotranspiration" for Remote Sensing Journal from 2018 and 2019. Dr. Wagle is a reviewer for myriad professional journals of meteorology, ecosystems/ecology, climate change, and remote sensing. For his exceptional research and professional achievements, Dr. Wagle received a Certificate of Merit/Outstanding Performance Awards from the USDA for FYs 2017 and 2018. He was included in the 71st edition of Who is Who in America by Marquis and featured as a Marquis Who's Who Top Scientist in August 2019. As a graduate faculty member, he also supervises/mentors graduate students at OSU.

Prasanna H. Gowda obtained his B.S at Bangalore University and M.S. at University of Mysore, India and Ph.D. in Agricultural Engineering at Ohio State University, USA. After his postdoctoral assignment with U.S. Geological Survey in Onalaska, Wisconsin, Dr. Gowda joined the University of Minnesota as the Senior Research Associate where he conducted tillage mapping and watershed modeling research for developing TMDLs for the Minnesota River Basin. In 2005, Dr. Gowda joined the USDA-ARS Conservation and Production Research Laboratory in Texas as an agricultural engineer. His research focused on evapotranspiration mapping using surface energy balance models, irrigation management, and calibration and validation of hydrologic and crop simulation models using lysimetric data. In 2015, Dr. Gowda joined the USDA-ARS Grazinglands Research Laboratory in Oklahoma as a Research Leader of the Forage and Livestock Production Research Unit where he conducted research on forage and rangeland management, water conservation in dryland agricultural systems, and greenhouse gas emissions in the Southern Great Plains. He was also a co-lead of Southern Great Plains Long Term Agricultural Research Project. Dr. Gowda is an internationally recognized authority on irrigation management, tillage mapping, evapotranspiration mapping, and hydrologic modeling as documented in more than 525 publications, including more than 200 peer-reviewed publications. He served on the editorial boards of Transactions of ASABE, Agronomy Journal, Remote Sensing, and Vadose Zone Journal. Considering his stature in the scientific community, Dr. Gowda was invited to serve as a co-lead author on the Agriculture and Rural Communities Chapter of the 4th National Climate Assessment. He has received a Fellow Award from the American Society of Agronomy and Soil Science Society of America in addition to a Laj Ahuja Agricultural Systems Modeling Award and Crops, Soils, and an Environmental Sciences Mentoring Award.

Editorial

Editorial for the Special Issue "Remote Sensing of Evapotranspiration (ET)"

Pradeep Wagle [1,*] and Prasanna H. Gowda [2]

[1] Grazinglands Research Laboratory, USDA Agricultural Research Service, El Reno, OK 73036, USA
[2] Southeast Area, USDA Agricultural Research Service, Stoneville, MS 38776, USA;
    prasanna.gowda@ars.usda.gov
* Correspondence: pradeep.wagle@usda.gov

Received: 11 September 2019; Accepted: 15 September 2019; Published: 15 September 2019

Abstract: Evapotranspiration (ET) is a critical component of the water and energy balances, and the number of remote sensing-based ET products and estimation methods has increased in recent years. Various aspects of remote sensing of ET are reported in 11 papers published in this special issue. The major research topics covered by this special issue include inter-comparison and performance evaluation of widely used one- and two-source energy balance models, a new dual-source model (Soil Plant Atmosphere and Remote Sensing Evapotranspiration, SPARSE), and a process-based model (ETMonitor); assessment of multi-source (e.g., remote sensing, reanalysis, and land surface model) ET products; development or improvement of data fusion frameworks to provide continuous daily ET at a high spatial resolution (field-scale or 30 m) by fusing the advanced space-borne thermal emission reflectance radiometer (ASTER), the moderate resolution imaging spectroradiometer (MODIS), and Landsat data; and investigating uncertainties in ET estimates using an ET ensemble composed of 36 land surface models and four diagnostic datasets. The effects of the differences among ET products on water resources and ecosystem management were also investigated. More accurate ET estimates and improved understanding of remotely sensed ET products can help maximize crop productivity while minimizing water loses and management costs.

Keywords: data fusion; evapotranspiration partitioning; land surface model; process-based model; water stress

---

1. Introduction

Evapotranspiration (ET), a critical and major component of the water and energy balances, is a key variable for linking ecosystem functions and climate feedbacks [1], determination of crop water or irrigation requirements and crop coefficients [2], and estimation of productivity and water use efficiency of ecosystems [3,4]. Although the eddy covariance (EC) technique has been widely used for continuous measurements of ET in recent decades [5,6], it is not possible to measure ET by the EC technique at all places all the time and especially over heterogeneous landscapes. Thus, a wide range of remote sensing-based ET products at the global and regional scales has been developed in recent decades to complement the limited land surface coverage of the ground-based ET measurements [7–9]. These ET products include numerous remote sensing reanalysis-based [10–12], land surface model (LSM)-based [13,14], surface energy balance (SEB)-based [15–17], and empirical up-scaling of in situ ET observations [18,19]. The SEB-based models are gaining increased popularity because remote sensing in the thermal infrared provides information not only on the partitioning of the available energy to sensible and latent heat fluxes, but also on the predicting water stress levels [17,20]. However, a major shortcoming of SEB-based models is that they rely on available land surface temperature (LST) data from satellite observations. Consequently, SEB modeling estimates are not available for cloudy days. Thus, the process-based ET models are gaining more acceptance to generate continuous ET estimates by utilizing a variety of biophysical parameters derived from microwave and optical remote sensing observations [21,22]. It is also recognized that there are large differences among a wide range of ET products. Validations and inter-comparisons of various ET models or ET products under diverse ecosystems and agrometeorological conditions are needed due to different levels of uncertainties and accuracies that vary over space and time [23,24].

Although several remote sensing-based ET products are available, these datasets cannot generally provide ET data at both higher spatial and temporal resolutions to derive field-scale ET estimates over heterogeneous landscapes due to satellite orbital dynamics and physical limitations of the satellite sensors. Thus, downscaling and data fusion approaches have been employed to improve the higher spatial and temporal resolutions of remote sensing-based ET products [25–28].

Accurate ET estimates are crucial to manage water resources and to assess the impacts of climate on agriculture and food security [29]. High uncertainty in ET estimates is a major obstacle to examine spatial and temporal variability in regional hydrology [30]. Thus, understanding the uncertainty of ET estimates can help to better determine water availability for agriculture and livelihoods.

This special issue compiles contributions on research related to the above-mentioned various aspects of remote sensing of ET. The major topics covered by the 11 papers in this special issue include inter-comparison and performance evaluation of several ET models or products, data fusion approach to generate higher spatial and temporal resolution ET products, model development and/or improvement, and investigating uncertainties in ET estimates. A short summary of the varied contributions to this special issue is presented in the next section.

2. Overview of Contributions

2.1. Inter-Comparison and Performance Evaluation of Several ET Models or Products

Yang et al. [31] compared three Two-Source Energy Balance (TSEB) models for estimating ET and its components (evaporation, E and transpiration, T) in semiarid climates of China. Those three TSEB models were: TSEB model with the Priestley–Taylor equation (TSEB-PT), TSEB model with the Penman–Monteith equation (TSEB-PM), and TSEB model using component temperatures derived from vegetation fractional cover and land surface temperature (VFC/LST) space (TSEB-TC-TS). The study provided valuable insights into understanding the performances of TSEB models with different temperature decomposition methods since they were responsible for the observed discrepancies in the partitioned E and T fluxes. Based on the soil wetness isoline in the VFC/LST space, the VFC/LST-based temperature decomposition method can add a further constraint on vegetation T. This could also be used as a substitution for the interactive procedure adopted in the TSEB model.

Grosso et al. [32] employed the Surface Energy Balance Algorithm for Land (SEBAL) in a salt-affected and water-stressed maize field using Landsat images to map the spatial structure of water fluxes and crop yield. The SEBAL results were compared with ET estimates of the Food and Agriculture Organization (FAO) method and three-dimensional soil–plant simulations. The study highlighted that the integration of SEBAL with field observations and soil–plant simulations could be beneficial for precision agriculture practices (e.g., precision irrigation).

Li et al. [33] evaluated four popular global ET products: Global Land Evaporation Amsterdam Model version 3.0a (GLEAM3.0a), Modern Era Retrospective-Analysis for Research and Applications-Land (MERRA-Land), Global Land Data Assimilation System version 2.0 with the Noah model (GLDAS2.0-Noah), and EartH2Observe ensemble (EartH2Observe-En) over China using a stratification method, six validation criteria, and EC measurements at 12 sites. The model performances were evaluated by biome, elevation, and climate regime as well. The study recommended the use of multi-source ET datasets since no ET product consistently performed best for the selected validation criterion.

Delogu et al. [34] assessed the model predictions of water stress and ET components for the two proposed versions (the "patch" and "layer" resistances network) of the new dual-source Soil

Plant Atmosphere and Remote Sensing Evapotranspiration (SPARSE) model over 20 in situ datasets encompassing diverse vegetation and climate conditions. The SPARSE model showed good estimates of latent and sensible heat fluxes and water stress over a large range of leaf area indexes and contrasting water stress levels.

Zheng et al. [22] used ETMonitor, a process-based model, with satellite earth observation datasets as main inputs to derive daily ET by utilizing surface soil moisture from microwave remote sensing and LST from thermal remote sensing. Estimated daily ET showed good agreement with EC-measured ET in Northeastern Thailand.

Khand et al. [35] developed an automated modeling framework to construct daily time series of ET maps, addressing the challenges related to processing and gap filling of non-continuous satellite data using the moderate resolution imaging spectroradiometer (MODIS) imagery and the Surface Energy Balance System (SEBS) model. The daily ET maps generated by this modeling framework captured the spatial and temporal variations (2001–2014) of ET across Oklahoma, USA. The proposed ET modeling framework provided a pathway to construct daily time series of ET maps at a regional scale and highlighted a range of potential applications for making informed decision and policies.

Lu et al. [36] evaluated the effects of differences among five representative ET products (Australian Water Availability Project (AWAP) as a reference, ET product developed by Commonwealth Scientific and Industrial Research Organization (CSIRO), LSM-based ET product from GLDAS, remote sensing-based ET product from MODIS, and water budget-based ET product from TerraClimate) on water resources and ecosystem management in the Murrumbidgee River catchment in Australia. Large differences in ET budgets among these five ET products propagated into the estimates of mean annual runoff, soil water storage, and irrigation demands.

2.2. Data Fusion Approach to Generate Higher Spatial and Temporal Resolution ET Products

Considering the lack of concurrent higher spatial and temporal resolution ET products, Yi et al. [37] employed a data fusion framework for predicting continuous daily ET at the field-scale over heterogeneous agricultural areas of Northwest China by fusing the advanced space-borne thermal emission reflectance radiometer (ASTER) and the MODIS data. Through a combination with the linear unmixing-based method, the spatial and temporal adaptive reflectance fusion model (STARFM) was modified to generate high-resolution ET estimates over heterogeneous areas. As compared with the original STARFM, the modified STARFM showed a significant improvement in daily ET estimation, preserved more spatial details for heterogeneous agricultural fields, and provided field-to-field variability in water use.

Wang et al. [38] proposed an improved ET fusion method— the Spatio-temporal Adaptive Data Fusion Algorithm for EvapoTranspiration mapping (SADFAET)—by introducing critical surface temperature (the corresponding temperature to determine soil moisture), importing the weights of surface ET-indicative similarity (the influencing factor of ET), and modifying the spectral similarity (the differences in spectral characteristics of different spatial resolution images) for the Enhanced Spatial and Temporal Adaptive Reflectance Fusion Model (ESTARFM). The study successfully fused daily MODIS and periodic Landsat 8 ET data in the SADFAET for producing ET at high spatial (30 m) and temporal (daily) resolutions.

2.3. Model Development and/or Improvement

Considering the knowledge gaps in differences among final ET estimates resulting from subjectivity in selecting "hot" and "cold" pixel pair, Dhungel and Barber [39] tested the assumption of low variability of surface properties by first applying an automated calibration pixel selection process for a SEB model—Mapping EvapoTranspiration at high Resolution with Internalized Calibration (METRIC). Consequently, they computed vertical near-surface temperature differences (dT) vs. surface temperature (Ts) relationships at all pixels, which could potentially be used for model calibration to explore ET variance among the outcomes from multiple calibration schemes where normalized difference vegetation index (NDVI) and Ts variability are intrinsically negligible. Significant variability in ET (ranging from 5% to 20%) and a high and statistically consistent variability in dT suggested that additional surface properties, which were not captured when using only NDVI and Ts, affected the calibration process. This approach of quantifying ET variability based on candidate pixel selection helps to quantify the biases inadvertently introduced by user subjectivity as well as to improve the model's usability and performance.

Zheng et al. [22] developed and applied a new scheme in ETMonitor, a process-based model, to take advantage of thermal remote sensing. In the improved scheme, the evaporation fraction was obtained by LST-vegetation index triangle method to estimate ET in clear days. The soil moisture stress index (SMSI) was defined to express the impact of soil moisture on ET. Clear sky SMSI, retrieved according to the estimated clear sky ET, was interpolated to cloudy days to obtain the SMSI for all sky conditions. Finally, interpolated spatio-temporal continuous SMSI was used to derive daily time-series ET.

Wang et al. [38] developed an improved ET fusion method (SADFAET) based on ESTARFM. The improvements in SADFAET were as follows: consideration of soil moisture by introducing the critical surface temperature while selecting similar pixels, use of multiple spectral bands, and introduction of the surface ET-indicative similarity to calculate the weights of similar pixels. This new method can effectively fuse ET at high and low spatial resolutions.

2.4. Investigating Uncertainties in ET Estimates

Jung et al. [40] investigated uncertainties in ET estimates over five different climatic regions in West Africa using an ET ensemble composed of 36 LSM experiments and four diagnostic datasets (GLEAM, ALEXI, MOD16, and FLUXNET). The LSM-based ET values had greater uncertainty estimates and larger seasonal variations than the diagnostic ET datasets. The LSM formulations and parameters had the largest impact on ET in humid regions (contributing to 90% of the ET uncertainty estimates), while precipitation contributed to the ET uncertainty primarily in arid regions. The results indicated that assimilating diagnostic ET datasets into LSMs or hydrological models could improve the accuracy of ET estimates.

3. Conclusions

The 11 papers published in this special issue highlight a variety of topics related to remote sensing of ET. This special issue provides valuable insights into understanding the performances of different ET models and products under diverse ecosystems and agrometeorological conditions. In addition, improvements on the ET models have also been proposed. Proposed ET data fusion approaches provide unique means of monitoring continuous daily ET at higher spatial resolutions (e.g., field-scale or less) over heterogeneous landscapes. More accurate ET estimates and improved understanding of remotely sensed ET products are crucial to maximize crop productivity while minimizing water losses and management costs.

Author Contributions: P.W. wrote the editorial and P.G. revised and contributed for intellectual contents.

Acknowledgments: We would like to thank all the authors who contributed to the special issue and the staff in the editorial office.

Conflicts of Interest: The authors declare no conflict of interest.

References

1. Fisher, J.B.; Melton, F.; Middleton, E.; Hain, C.; Anderson, M.; Allen, R.; McCabe, M.F.; Hook, S.; Baldocchi, D.; Townsend, P.A.; et al. The future of evapotranspiration: Global requirements for ecosystem functioning, carbon and climate feedbacks, agricultural management, and water resources. *Water Resour. Res.* **2017**, *53*, 2618–2626. [CrossRef]
2. Marek, T.; Piccinni, G.; Schneider, A.; Howell, T.; Jett, M.; Dusek, D. Weighing lysimeters for the determination of crop water requirements and crop coefficients. *Appl. Eng. Agric.* **2006**, *22*, 851–856. [CrossRef]
3. Law, B.E.; Falge, E.; Gu, L.V.; Baldocchi, D.D.; Bakwin, P.; Berbigier, P.; Davis, K.; Dolman, A.J.; Falk, M.; Fuentes, J.D.; et al. Environmental controls over carbon dioxide and water vapor exchange of terrestrial vegetation. *Agric. For. Meteorol.* **2002**, *113*, 97–120. [CrossRef]
4. Wagle, P.; Gowda, P.H.; Xiao, X.; Anup, K.C. Parameterizing ecosystem light use efficiency and water use efficiency to estimate maize gross primary production and evapotranspiration using MODIS EVI. *Agric. For. Meteorol.* **2016**, *222*, 87–97. [CrossRef]
5. Baldocchi, D.D.; Hincks, B.B.; Meyers, T.P. Measuring Biosphere-Atmosphere Exchanges of Biologically Related Gases with Micrometeorological Methods. *Ecology* **1988**, *69*, 1331–1340. [CrossRef]
6. Rana, G.; Katerji, N. Measurement and estimation of actual evapotranspiration in the field under Mediterranean climate: a review. *Eur. J. Agron.* **2000**, *13*, 125–153. [CrossRef]
7. Mu, Q.; Heinsch, F.A.; Zhao, M.; Running, S.W. Development of a global evapotranspiration algorithm based on MODIS and global meteorology data. *Remote Sens. Environ.* **2007**, *111*, 519–536. [CrossRef]
8. Mueller, B.; Seneviratne, S.I.; Jimenez, C.; Corti, T.; Hirschi, M.; Balsamo, G.; Ciais, P.; Dirmeyer, P.; Fisher, J.B.; Guo, Z.; et al. Evaluation of global observations-based evapotranspiration datasets and IPCC AR4 simulations. *Geophys. Res. Lett.* **2011**, *38*. [CrossRef]
9. Yuan, W.; Liu, S.; Yu, G.; Bonnefond, J.M.; Chen, J.; Davis, K.; Desai, A.R.; Goldstein, A.H.; Gianelle, D.; Rossi, F.; et al. Global estimates of evapotranspiration and gross primary production based on MODIS and global meteorology data. *Remote Sens. Environ.* **2010**, *114*, 1416–1431. [CrossRef]
10. Onogi, K.; Tsutsui, J.; Koide, H.; Sakamoto, M.; Kobayashi, S.; Hatsushika, H.; Matsumoto, T.; Yamazaki, N.; Kamahori, H.; Takahashi, K.; et al. The JRA-25 reanalysis. *J. Meteorol. Soc. Jpn.* **2007**, *85*, 369–432. [CrossRef]
11. Dee, D.P.; Uppala, S.M.; Simmons, A.J.; Berrisford, P.; Poli, P.; Kobayashi, S.; Andrae, U.; Balmaseda, M.A.; Balsamo, G.; Bauer, D.P.; et al. The ERA-Interim reanalysis: Configuration and performance of the data assimilation system. *Q. J. R. Meteorol. Soc.* **2011**, *137*, 553–597. [CrossRef]
12. Bosilovich, M.G.; Robertson, F.R.; Takacs, L.; Molod, A.; Mocko, D. Atmospheric water balance and variability in the MERRA-2 reanalysis. *J. Clim.* **2017**, *30*, 1177–1196. [CrossRef]
13. Rodell, M.; Houser, P.R.; Jambor, U.E.A.; Gottschalck, J.; Mitchell, K.; Meng, C.J.; Arsenault, K.; Cosgrove, B.; Radakovich, J.; Bosilovich, M.; et al. The global land data assimilation system. *Bull. Am. Meteorol. Soc.* **2004**, *85*, 381–394. [CrossRef]
14. Schellekens, J.; Dutra, E.; la Torre, A.M.D.; Balsamo, G.; van Dijk, A.; Weiland, F.S.; Minvielle, M.; Calvet, J.C.; Decharme, B.; Eisner, S.; et al. A global water resources ensemble of hydrological models: the eartH2Observe Tier-1 dataset. *Earth Syst. Sci. Data* **2017**, *9*, 389–413. [CrossRef]
15. Bastiaanssen, W.G.; Menenti, M.; Feddes, R.A.; Holtslag, A.A.M. A remote sensing surface energy balance algorithm for land (SEBAL). 1. Formulation. *J. Hydrol.* **1998**, *212*, 198–212. [CrossRef]
16. Roerink, G.J.; Su, Z.; Menenti, M. S-SEBI: A simple remote sensing algorithm to estimate the surface energy balance. *Phys. Chem. Earth* **2000**, *25*, 147–157. [CrossRef]
17. Allen, R.G.; Tasumi, M.; Trezza, R. Satellite-based energy balance for mapping evapotranspiration with internalized calibration (METRIC)—Model. *J. Irrig. Drain. Eng.* **2007**, *133*, 380–394. [CrossRef]
18. Jung, M.; Reichstein, M.; Bondeau, A. Towards global empirical upscaling of FLUXNET eddy covariance observations: validation of a model tree ensemble approach using a biosphere model. *Biogeosciences* **2009**, *6*, 2001–2013.

19. Wagle, P.; Xiao, X.; Gowda, P.; Basara, J.; Brunsell, N.; Steiner, J.; Anup, K.C. Analysis and estimation of tallgrass prairie evapotranspiration in the central United States. *Agric. For. Meteorol.* **2017**, *232*, 35–47. [CrossRef]

20. Bhattarai, N.; Wagle, P.; Gowda, P.H.; Kakani, V.G. Utility of remote sensing-based surface energy balance models to track water stress in rain-fed switchgrass under dry and wet conditions. *ISPRS J. Photogramm. Remote Sens.* **2017**, *133*, 128–141. [CrossRef]

21. Hu, G.; Jia, L. Monitoring of evapotranspiration in a semi-arid inland river basin by combining microwave and optical remote sensing observations. *Remote Sens.* **2015**, *7*, 3056–3087. [CrossRef]

22. Zheng, C.; Jia, L.; Hu, G.; Lu, J. Earth Observations-Based Evapotranspiration in Northeastern Thailand. *Remote Sens.* **2019**, *11*, 138. [CrossRef]

23. Ershadi, A.; McCabe, M.F.; Evans, J.P.; Chaney, N.W.; Wood, E.F. Multi-site evaluation of terrestrial evaporation models using FLUXNET data. *Agric. For. Meteorol.* **2014**, *187*, 46–61. [CrossRef]

24. Wagle, P.; Bhattarai, N.; Gowda, P.H.; Kakani, V.G. Performance of five surface energy balance models for estimating daily evapotranspiration in high biomass sorghum. *ISPRS J. Photogramm. Remote Sens.* **2017**, *128*, 192–203. [CrossRef]

25. Singh, R.; Senay, G.; Velpuri, N.; Bohms, S.; Verdin, J. On the downscaling of actual evapotranspiration maps based on combination of MODIS and Landsat-based actual evapotranspiration estimates. *Remote Sens.* **2014**, *6*, 10483–10509. [CrossRef]

26. Ke, Y.; Im, J.; Park, S.; Gong, H. Downscaling of MODIS One kilometer evapotranspiration using Landsat-8 data and machine learning approaches. *Remote Sens.* **2016**, *8*, 215. [CrossRef]

27. Cammalleri, C.; Anderson, M.C.; Gao, F.; Hain, C.R.; Kustas, W.P. A data fusion approach for mapping daily evapotranspiration at field scale. *Water Resour. Res.* **2013**, *49*, 4672–4686. [CrossRef]

28. Cammalleri, C.; Anderson, M.C.; Gao, F.; Hain, C.R.; Kustas, W.P. Mapping daily evapotranspiration at field scales over rainfed and irrigated agricultural areas using remote sensing data fusion. *Agric. For. Meteorol.* **2014**, *186*, 1–11. [CrossRef]

29. Lei, F.; Crow, W.T.; Holmes, T.R.; Hain, C.; Anderson, M.C. Global investigation of soil moisture and latent heat flux coupling strength. *Water Resour. Res.* **2018**, *54*, 8196–8215. [CrossRef]

30. Li, K.Y.; Coe, M.T.; Ramankutty, N. Investigation of hydrological variability in West Africa using land surface models. *J. Clim.* **2005**, *18*, 3173–3188. [CrossRef]

31. Yang, Y.; Qiu, J.; Zhang, R.; Huang, S.; Chen, S.; Wang, H.; Luo, J.; Fan, Y. Intercomparison of three two-source energy balance models for partitioning evaporation and transpiration in semiarid climates. *Remote Sens.* **2018**, *10*, 1149. [CrossRef]

32. Grosso, C.; Manoli, G.; Martello, M.; Chemin, Y.; Pons, D.; Teatini, P.; Piccoli, I.; Morari, F. Mapping maize evapotranspiration at field scale using SEBAL: A comparison with the FAO method and soil-plant model simulations. *Remote Sens.* **2018**, *10*, 1452. [CrossRef]

33. Li, S.; Wang, G.; Sun, S.; Chen, H.; Bai, P.; Zhou, S.; Huang, Y.; Wang, J.; Deng, P. Assessment of Multi-Source Evapotranspiration Products over China Using Eddy Covariance Observations. *Remote Sens.* **2018**, *10*, 1692. [CrossRef]

34. Delogu, E.; Boulet, G.; Olioso, A.; Garrigues, S.; Brut, A.; Tallec, T.; Demarty, J.; Soudani, K.; Lagouarde, J.P. Evaluation of the SPARSE Dual-Source Model for Predicting Water Stress and Evapotranspiration from Thermal Infrared Data over Multiple Crops and Climates. *Remote Sens.* **2018**, *10*, 1806. [CrossRef]

35. Khand, K.; Taghvaeian, S.; Gowda, P.; Paul, G. A modeling framework for deriving daily time series of evapotranspiration maps using a surface energy balance model. *Remote Sens.* **2019**, *11*, 508. [CrossRef]

36. Lu, Z.; Zhao, Y.; Wei, Y.; Feng, Q.; Xie, J. Differences among Evapotranspiration Products Affect Water Resources and Ecosystem Management in an Australian Catchment. *Remote Sens.* **2019**, *11*, 958. [CrossRef]

37. Yi, Z.; Zhao, H.; Jiang, Y. Continuous Daily Evapotranspiration Estimation at the Field-Scale over Heterogeneous Agricultural Areas by Fusing ASTER and MODIS Data. *Remote Sens.* **2018**, *11*, 1694. [CrossRef]

38. Wang, T.; Tang, R.; Li, Z.L.; Jiang, Y.; Liu, M.; Niu, L. An Improved Spatio-Temporal Adaptive Data Fusion Algorithm for Evapotranspiration Mapping. *Remote Sens.* **2019**, *11*, 761. [CrossRef]

39. Dhungel, S.; Barber, M. Estimating Calibration Variability in Evapotranspiration Derived from a Satellite-Based Energy Balance Model. *Remote Sens.* **2018**, *10*, 1695. [CrossRef]
40. Jung, H.C.; Getirana, A.; Arsenault, K.R.; Holmes, T.R.; McNally, A. Uncertainties in Evapotranspiration Estimates over West Africa. *Remote Sens.* **2019**, *11*, 892. [CrossRef]

Article

Intercomparison of Three Two-Source Energy Balance Models for Partitioning Evaporation and Transpiration in Semiarid Climates

Yongmin Yang [1,2], Jianxiu Qiu [3,*], Renhua Zhang [4], Shifeng Huang [2], Sheng Chen [2], Hui Wang [2], Jiashun Luo [3] and Yue Fan [3]

[1] State Key Laboratory of Simulation and Regulation of Water Cycle in River Basin, Beijing 100038, China; yangym@iwhr.com

[2] Research Center on Flood and Drought Disaster Reduction of the Ministry of Water Resources, China Institute of Water Resources and Hydropower Research, Beijing 100038, China; huangsf@iwhr.com (S.H.); chenshenge@hotmail.com (S.C.); wanghui8169@foxmail.com (H.W.)

[3] Guangdong Provincial Key Laboratory of Urbanization and Geo-simulation, School of Geography and Planning, Sun Yat-sen University, Guangzhou 510275, China; luojsh3@mail2.sysu.edu.cn (J.L.); fany26@mail2.sysu.edu.cn (Y.F.)

[4] Key Laboratory of Water Cycle & Related Land Surface Processes, Institute of Geographic Sciences and Natural Resources Research, Beijing 100101, China; zhangrh@igsnrr.ac.cn

[*] Correspondence: qiujianxiu@mail.sysu.edu.cn

Received: 10 May 2018; Accepted: 18 July 2018; Published: 20 July 2018

Abstract: Evaporation (E) and transpiration (T) information is crucial for precise water resources planning and management in arid and semiarid areas. Two-source energy balance (TSEB) methods based on remotely-sensed land surface temperature provide an important modeling approach for estimating evapotranspiration (ET) and its components of E and T. Approaches for accurate decomposition of the component temperature and E/T partitioning from ET based on TSEB requires careful investigation. In this study, three TSEB models are used: (i) the TSEB model with the Priestley-Taylor equation, i.e., TSEB-PT; (ii) the TSEB model using the Penman-Monteith equation, i.e., TSEB-PM, and (iii) the TSEB using component temperatures derived from vegetation fractional cover and land surface temperature (VFC/LST) space, i.e., TSEB-T_C-T_S. These models are employed to investigate the impact of component temperature decomposition on E/T partitioning accuracy. Validation was conducted in the large-scale campaign of Heihe Watershed Allied Telemetry Experimental Research-Multi-Scale Observation Experiment on Evapotranspiration (HiWATER-MUSOEXE) in the northwest of China, and results showed that root mean square errors (RMSEs) of latent and sensible heat fluxes were respectively lower than 76 W/m^2 and 50 W/m^2 for all three approaches. Based on the measurements from the stable oxygen and hydrogen isotopes system at the Daman superstation, it was found that all three models slightly overestimated the ratio of E/ET. In addition, discrepancies in E/T partitioning among the three models were observed in the kernel experimental area of MUSOEXE. Further intercomparison indicated that different temperature decomposition methods were responsible for the observed discrepancies in E/T partitioning. The iterative procedure adopted by TSEB-PT and TSEB-PM produced higher LE_C and lower T_C when compared to TSEB-T_C-T_S. Overall, this work provides valuable insights into understanding the performances of TSEB models with different temperature decomposition mechanisms over semiarid regions.

Keywords: component temperature decomposition; evapotranspiration partitioning; two-source energy balance model

---

1. Introduction

Evapotranspiration (ET) observations and modeling are crucial in water cycle studies [1–4]. Water scarcity is one of the main factors constraining agricultural development in arid or semiarid areas. The knowledge of ET, as well as the mechanism of ET partitioning into evaporation (E) and transpiration (T), is very important for precise quantification of the water balance in water resources planning and management, optimizing crop production, identifying crop stress and drought impacts, and evaluating the effects of climate change on water yields [5–9]. As satellite remote sensing is widely used to obtain information on the regional water and heat balance, it has been used to derive several global multi-decadal ET datasets that arouse extensive concern [10–15].

Over the last few decades, several remote sensing-based ET models have been proposed to estimate regional surface heat fluxes via satellite-derived land surface temperature (LST) [5,6,16–20]. Specifically, a one-source model, such as the Surface Energy Balance Algorithm for Land (SEBAL) [16,17], Simplified Surface Energy Balances Index (S-SEBI) [21], Surface Energy Balance System (SEBS) [20,22], and Mapping EvapoTranspiration at high Resolution with Internalized Calibration (METRIC) [5,6], often use LST and empirical resistance corrections to estimate surface heat fluxes. Two-source models, such as the two-source energy balance model (TSEB) [19], the two-source time-integrated model (TSTIM) [23], Pixel Component Arranging and Comparing Algorithm (PCACA) [24], and the enhanced two-source evapotranspiration model for land (ETEML) [25] were developed to make use of LST to estimate a sensible heat flux (H) and latent heat flux (LE) for the soil and canopy separately. Other alternative practical approaches have been proposed based on vegetation fractional cover and LST (VFC/LST) space [26,27]. Related research includes the work by Moran et al. [28], Jiang and Islam [29,30], Stisen et al. [31], and Shu et al. [32]. Extensive reviews of remote sensing-based methodologies on surface heat flux estimation can be found in the works of Courault et al. [33], Kalma et al. [34] and Li et al. [35].

As an appealing modeling scheme, two-source energy balance models can estimate evapotranspiration (ET), evaporation (E), and transpiration (T) of vegetated surfaces, which has important applications in water resources management for irrigated crops in arid and semiarid areas [36]. To estimate E and T separately, the soil temperature (T_S) and canopy temperature (T_C) need to be accurately measured or derived, making temperature decomposition the core of the two-source modeling approach. Generally, four categories of methods have been developed to decompose remotely-sensed temperature into soil and canopy temperatures. (1) The first category calculates soil and vegetation component temperatures using dual- or multi-angular thermal infrared measurements [37]. For instance, a dual-source model using bi-angular thermal infrared measurements was developed by Jia [38] and Jia et al. [39]. However, currently limited sensors have bi- or multi-angular thermal infrared channels and thus constrained the availability of remote sensing data source in this category of method. (2) The second category assumed that LST is the sum of vegetation and soil temperatures weighted by the vegetation fractional cover (f_c), i.e., LST $\approx f_c T_C + T_S(1 - f_c)$. This kind of method assumed that the LST of a highly vegetated pixel in the neighborhood would be a reasonable approximation of T_C, and a selected neighborhood area of the target pixel based on certain thresholds in the corresponding NDVI image [40,41]. As T_C is approximately determined, T_S can be derived accordingly using $T_S = (\text{LST} - f_c T_C)/(1 - f_c)$. (3) The third category retrieves component temperatures using an iterative procedure based on an energy balance and resistance network. Typically, the TSEB model proposed by Norman et al. [19] uses a system of temperature gradient-resistance equations that are solved by an iterative procedure, in which an initial estimate of plant transpiration is determined by the Priestley-Taylor equation (TSEB-PT). The procedure terminates when soil evaporation (LEs) exceeds zero, and Tc and Ts are recalculated until energy balance closure is reached. Recently, TSEB was revised by Colaizzi et al. [42] and Colaizzi et al. [43] using the Penman-Monteith equation (TSEB-PM) instead of the Priestley-Taylor equation in order to characterize vegetation transpiration, and this revised version was found to be more applicable for advective semiarid climates. (4) The fourth category derives component temperatures based on VFC/LST space [24,28,44,45]. Moran et al. [28] introduced a water deficit index (WDI) based on the VFC/LST trapezoid space and extended the application of a crop water stress index (CWSI) over fully- to partially-vegetated surface areas. Zhang et al. [24] proposed a method to retrieve land surface component temperatures based on the interpretation of soil wetness isolines within a VFC/LST trapezoidal space. The main assumption of this method is that the isolines in the VFC/LST trapezoid space is mainly controlled by soil water availability and the isolines within a VI/LST trapezoidal space are used to decompose the composite temperature into component temperatures. This method was recently revised and adopted by Sun et al. [46], Long and Singh [47], Yang and Shang [48], Yang et al. [25], and Sun [49] to develop two-source models.

Combined with satellite remote sensing, the TSEB models have been extensively used to estimate regional ET. However, how the accuracy of ET estimations from TSEB models is affected by its component temperature estimation and E/T partitioning are rarely reported, and requires careful study. The objective of this study is to evaluate the capabilities of three TSEB models in predicting surface fluxes under various ranges of soil moisture contents, and then analyze how the different performances of three models are attributed to the different adopted schemes in component temperature decomposition and E/T partitioning. Section 2 presents the description of the three two-source models: TSEB-PT, TSEB-PM, and TSEB-T_C-T_S. Section 3 introduces the pertinent experiment campaign in the study area, i.e., Heihe Watershed Allied Telemetry Experimental Research-The Multi-Scale Observation Experiment on Evapotranspiration (HiWATER-MUSOEXE) Campaign, remotely-sensed data used for driving the TSEB models, and the ground measurements for the models' assessment. Section 4 first reports the fluxes estimation accuracies from three TSEB models with retrievals from ASTER imagery. In Section 4.2, E/T partitioning is intercompared within three models and evaluated against the benchmark obtained using the stable oxygen and hydrogen isotopes technique. In the remainder of Section 4, intercomparison using component temperatures generated from three models was conducted to further investigate the uncertainty in E/T partitioning. Section 5 discusses the advantages and limitations in this work, and the final section provides a conclusion.

2. Theory and Methodology

2.1. TSEB-PT Model

The TSEB model was originally developed by Norman et al. [19] to make use of remotely-sensed radiometric surface temperatures to estimate soil evaporation and canopy transpiration. This model has been modified by Kustas and Norman [50,51] through improving the soil surface resistance formulation and net radiation partitioning between the soil and canopy components. The net radiation is partitioned between the vegetated canopy and soil, and can be expressed as:

$$R_n = R_{ns} + R_{nc} = H + LE + G \tag{1}$$

where R_n is net radiation (W/m^2), R_{ns} and R_{nc} are the net radiation for soil and vegetation canopy (W/m^2), respectively; H and LE are the sensible and latent heat fluxes (W/m^2), respectively, and G is the soil heat flux (W/m^2). The energy balance for the soil and vegetated canopy can be expressed as:

$$R_{ns} = H_s + LE_s + G \tag{2}$$

$$R_{nc} = H_c + LE_c \tag{3}$$

where H_s and H_c are the sensible heat fluxes for the soil and canopy respectively (W/m^2), LE_s and LE_c are the latent heat fluxes for the soil and canopy, respectively (W/m^2). G is parameterized with the phase-difference approach proposed by Santanello and Friedl [52]:

$$G = R_{ns}\left\{ a\cdot \cos\left[\frac{2\pi}{b}(t+c)\right]\right\} \tag{4}$$

where t is the solar time angle (s), a is the amplitude parameter (dimensionless), b is the period length (s), and c is the shift (s). In this study, parameters a, b, and c take the values of 0.3, 86,400, and 10,800 following Colaizzi et al. [43] and Song et al. [53].

In this study, the series resistance network form was applied, in which H_c, H_s, and the sum of both terms are calculated as:

$$H_c = \rho C_P \frac{T_C - T_{AC}}{r_x} \tag{5}$$

$$H_s = \rho C_P \frac{T_S - T_{AC}}{r_s} \tag{6}$$

$$H = \rho C_P \frac{T_{AC} - T_A}{r_A} \tag{7}$$

where ρ is the air density (kg/cm^3), C_P is the specific heat of air (J/kg/K), T_S is the soil temperature (K), T_C is the canopy temperature (K), T_{AC} and T_A are the air temperature within the canopy boundary layer and air temperature (K), respectively, r_A is the aerodynamic resistance (s/m), r_x is the resistance in the boundary layer near the canopy (s/m), and r_s is the resistance to heat flux in the boundary layer above the soil surface (s/m). r_A, r_x, and r_s are calculated according to Norman et al. [19] and Kustas and Norman [50].

The TSEB-PT model uses a modified Priestley-Taylor formulation to parameterize the canopy transpiration:

$$LE_c = \alpha_{PT} f_G \frac{\Delta}{\Delta + \gamma} R_{nc} \tag{8}$$

where α_{PT} is the Priestley-Taylor parameter (dimensionless), f_G is the fraction of green vegetation (dimensionless), Δ is the slope of the saturation vapor pressure versus temperature curve (kPa/$^\circ$C), and γ is the psychrometric constant (kPa/$^\circ$C). An initial estimate of T_C can be derived as follows:

$$T_c = T_A + \frac{R_{nc} r_A}{\rho C_P} \left[1.0 - \alpha_{PT} f_G \frac{\Delta}{\Delta + \gamma} \right] \tag{9}$$

Accordingly, T_S is calculated with an initial estimate of T_S, and then r_s can be estimated with the temperature gradient between the soil and canopy described in Kustas and Norman [51]. From Equations (5) to (8), the component sensible heat flux H_S can be calculated and latent heat fluxes from canopy LE_C and soil surface LE_S are solved as residual terms. In order to obtain a realistic estimation of surface heat fluxes under water stressed conditions, the α_{PT} is iteratively decreased until LE_S exceeds zero. The detailed description of the TSEB model and the parameterization of the resistance network can be found in Norman et al. [19] and Kustas and Norman [51].

2.2. TSEB-PM Model

The TSEB model was revised by Colaizzi et al. [42] and Colaizzi et al. [43] using the Penman-Monteith equation instead of the Priestley-Taylor formulation to account for the impact of advection over semiarid climates. This revised version of the TSEB model is termed TSEB-PM. The effects of varying the vapor pressure deficit can thus be taken into account in the TSEB-PM model. The canopy transpiration is characterized using the Penman-Monteith equation:

$$LE_c = f_G \left(\frac{\Delta R_{nc}}{\Delta + \gamma^*} + \frac{\rho C_P (e_s - e_a)}{r_A (\Delta + \gamma^*)} \right) \tag{10}$$

and T_c is initialized as follows:

$$T_C = T_A + \frac{R_{nc} r_A \gamma^*}{\rho C_P (\Delta + \gamma^*)} - \frac{e_s - e_a}{\Delta + \gamma^*} \tag{11}$$

where $\gamma^* = \gamma(1 + r_c/r_A)$, r_c is the bulk canopy resistance (s/m), r_A is the aerodynamic resistance between the canopy and the air above the canopy (s/m), and e_a and e_s are the actual and saturation vapor pressure of the air (kPa), respectively. Similar to TSEB-PT, the TSEB-PM model was iteratively implemented as described in Section 2.1. During the iterative procedure, r_c increases from 10 s/m with an increment of 20 s/m and terminates at 5000 s/m, or until LE_S exceeds zero. The comprehensive introduction of the TSEB-PM can be found in Colaizzi et al. [42] and Colaizzi et al. [43].

2.3. TSEB-T_C-T_S Model

TSEB was modified by Kustas and Norman [54] to calculate turbulent heat fluxes using canopy and soil component temperatures that were measured or derived from other methods. This modified TSEB model is called TSEB-T_C-T_S in this study. The Priestley-Taylor iteration procedure is not applied in TSEB-T_C-T_S, but the remainder of the physical framework of TSEB-T_C-T_S is identical to that of TSEB-PT. The within-canopy temperature (T_{AC}) is estimated from derived component temperatures as follows:

$$T_{AC} = \frac{\frac{T_A}{r_A} + \frac{T_s}{r_s} + \frac{T_C}{r_x}}{\frac{1}{r_A} + \frac{1}{r_s} + \frac{1}{r_x}} \tag{12}$$

Consequently, the component sensible heat fluxes H_S and H_C are directly calculated from Equations (5) and (6), and the component latent heat fluxes LE_C and LE_S are calculated as residual terms from Equations (2) and (3).

The VFC/LST space is adopted to retrieve component temperatures by Zhang et al. [24], Zhang et al. [55], Sun et al. [40], Merlin et al. [44], and Merlin et al. [45]. Based on the theoretical determination of the dry and wet edges of the VFC/LST space, this method was further revised and adopted by Long and Singh [47] and Yang and Shang [48] to develop two-source models. In the traditional method, the VFC/LST trapezoid is mainly determined manually based on the 2-D VFC/LST scatter plot, which would cause great uncertainty. Recently, Yang et al. [25] improved this method and proposed to make use of the VFC/LST trapezoid space for each pixel. Different from traditional method, the four theoretical points for each pixel are determined based on an energy balance model and the Penman-Monteith equations. In this study, this pixel-wise surface temperature decomposition method is adopted. The employed VFC/LST space is determined with four theoretical points using the following equations from Moran et al. [28].

For the vertex of dry bare soil, the difference between LST and Ta can be derived as follows:

$$(T_S - T_a)_{max} = \left[r_a(R_n - G)/\rho_a C_p \right] \tag{13}$$

For the vertex of saturated bare soil,

$$(T_S - T_a)_{min} = \left[(r_a(R_n - G))/(\rho_a C_p) \right] \left[\gamma/(\Delta + \gamma) \right] - \left[VPD/(\Delta + \gamma) \right] \tag{14}$$

For the vertex of well-watered fully-cover vegetation,

$$(T_C - T_a)_{min} = \left[(r_a(R_n - G))/(\rho_a C_p) \right] \left[\gamma(1 + r_{cp}/r_a)/(\Delta + \gamma(1 + r_{cp}/r_a)) \right] - \left[VPD/(\Delta + \gamma(1 + r_{cp}/r_a)) \right] \tag{15}$$

For the vertex of water-stressed fully-cover vegetation,

$$(T_C - T_a)_{max} = \left[(r_a(R_n - G))/(\rho_a C_p) \right] \left[\gamma(1 + r_{cx}/r_a)/(\Delta + \gamma(1 + r_{cx}/r_a)) \right] - \left[VPD/(\Delta + \gamma(1 + r_{cx}/r_a)) \right] \tag{16}$$

where VPD (kPa) is the vapor pressure deficit of the air at temperature T_a, γ is the psychrometric constant (kPa/°C), and r_a is aerodynamic resistance (s/m) estimated with the equations proposed by Thom [56]. r_{cp} is the canopy resistance at potential evapotranspiration (s/m), and r_{cx} is the maximum canopy resistance (s/m). As such, a VFC/LST trapezoid space is determined for each pixel and the measurements of (LST-Ta) and the fractional vegetation cover would be located within the trapezoid space.

Based on the VFC/LST trapezoid space determined above, the composite radiometric surface temperature of each pixel can be decomposed to soil and canopy component temperatures [24,25,55]:

$$T_S - T_a = (LST - T_a) - K_s * f_c \tag{17}$$

$$T_C - T_a = K_s * (1 - f_c) + (LST - T_a) \tag{18}$$

where K_s is the slope of the isoline that passes through the point located in VFC/LST space, and can be derived by interpolating the slope of the warm edge and that of the cold edge. The detailed description of this method can be found in Yang et al. [25]. Once T_S and T_C are determined, the component fluxes in TSEB-T_C-T_S can be estimated.

In addition to T_S and T_C, TSEB models can estimate evaporation (E) and transpiration (T) of vegetated surface, such information can be used to calculate component water stress of vegetation and soil. Following the equations from Yang et al. [25], the crop water stress index for the canopy component ($CWSI$c) and soil water deficit index for soil component ($SWDI_S$) are calculated as follows:

$$CWSI_C = 1 - LE_c / EP_c \tag{19}$$

$$SWDI_s = 1 - LE_s / EP_s \tag{20}$$

EP_c and EP_s are the potential transpiration rate and potential soil evaporation rate, respectively.

3. Study Area and Data Processing

3.1. HiWATER-MUSOEXE Campaign and Ground-Based Measurements

Heihe Watershed Allied Telemetry Experimental Research (HiWATER) was performed at the Heihe River Basin of northwestern China with airborne, satellite-borne, and ground-based remote sensing experiments at various scales during 2012–2015 [57]. As one of most important thematic experiments of the HiWATER, the Multi-Scale Observation Experiment on Evapotranspiration (MUSOEXE) over the Zhangye oasis provided multiscale data sets of meteorological elements and land surface parameters that facilitate the development and validation of ET models over heterogeneous surfaces [58,59]. Figure 1 shows the distribution of flux towers and the land use classifications in the MUSOEXE.

MUSOEXE involved a multi-scale observation campaign over heterogeneous surfaces by using an observation matrix composed of 21 stations. Each station was equipped with a set of eddy covariance (EC) system and automatic weather system (AWS). Turbulent heat fluxes are measured with EC system and the raw data were processed using EdiRe software and averaged over 30 min. Wind speed, wind direction, air temperature, vapor pressure, net radiation, and atmospheric pressure were measured in AWS with 10-min intervals. The soil heat fluxes were measured using three heat flux plates located 6 cm below the ground's surface at each site. In order to better represent the surface soil heat flux, the Thermal Diffusion Equation and Correction (TDEC) method proposed by Yang and Wang [60] was applied to correct the observed soil heat flux with the observed soil moisture and temperature profile. The residual method [61] was adopted to adjust sensible and latent heat fluxes by forcing the energy balance closure.

The isotopic composition of atmospheric water vapor provided rich information on the hydrological cycle and gaseous exchange processes between the terrestrial vegetation and atmosphere. Due to technical and instrumental limitations, the measurements of the isotopic composition of water vapor were limited to discrete campaigns and discrete samples. During the HiWATER program, the isotopic composition of water vapor from the surface air was measured in a corn cropland (100.3722°E, 38.8555°N) at Daman superstation using a flux gradient method, using a cavity ring-down spectroscopy (CRDS) water vapor isotope analyzer. Considering water vapor as a mixture of ET from an ecosystem that carried unique isotopic signals from plant transpiration and soil evaporation separately, the measured isotopic composition of water was used to partition ET into evaporation and transpiration. Details of the isotope experiment and its calibration procedure are given by Huang and Wen [62] and Wen et al. [63]. In this study, the ratio of T over ET was collected to evaluate the reliability of three TSEB models on evaporation and transpiration partitioning.

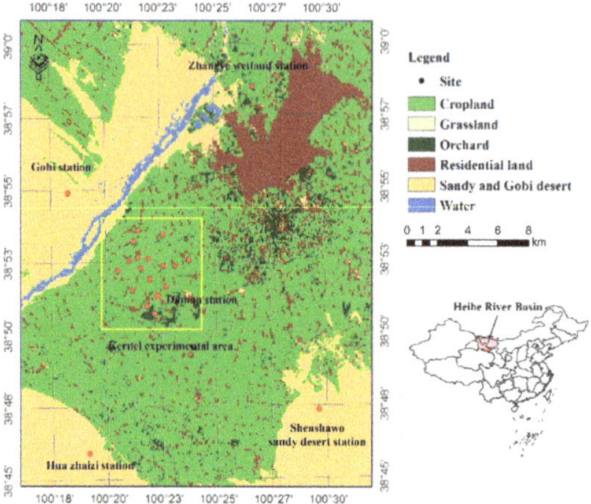

Figure 1. The distribution of flux towers and the land use classifications in MUSOEXE over the Zhangye oasis. The yellow rectangular in the left shows the kernel experimental area in MUSOEXE, and the subset figure in the lower right shows the location of MUSOEXE (marked in red triangle) in the Heihe River Basin (marked by pink polygon) and in China.

3.2. Remote Sensing Data and Derivation of Related Variables

In this study, nine scenes from the Advance Spaceborne Thermal Emission and Reflection Radiometer (ASTER) on board Terra during the experiment period were collected. These nine scenes were acquired on 15 June, 24 June, 10 July, 2 August, 11 August, 18 August, 27 August, 3 September, and 12 September of 2012 (DOYs: 167, 176, 192, 215, 224, 231, 240, 247, 256, respectively). The LST and the land surface albedo were provided by "Heihe Plan Science Data Center, National Natural Science Foundation of China" (http://www.heihedata.org). The LST data were retrieved by Li et al. [64] using a temperature/emissivity separation (TES) algorithm proposed by Gillespie et al. [65], combined with the water vapor scaling atmospheric correction method [66]. Land surface albedo was retrieved from a Charge Coupled Device (CCD) camera on HJ-1 satellite by Sun et al. [67].

Three data sets including the visible, near-infrared (NIR) bands of ASTER and albedo from HJ-1, were all re-sampled to 90 m to be consistent with the thermal infrared band in spatial resolution. In addition, the leaf area index (LAI), crop height, and fractional vegetation cover for HiWATER-MUSOEXE were derived based on the empirical relationships proposed by Yang et al. [68].

4. Results

4.1. Validation of Three TSEB Models over MUSOEXE

TSEB-PT, TSEB-PM, and TSEB-T_C-T_S were applied to the Zhangye oasis using ground-based and satellite-derived observations introduced in Section 3, and the model performances were evaluated using flux measurements from the MUSOEXE observation matrix. As a first step of model validation, the flux estimates were averaged over the upwind source area for each flux tower [69], and flux measurements were linearly interpolated to temporally match the time of satellite overpass. Figure 2 shows the validation scatter plot for each energy balance component (R_n, G, LE, and H) from TSEB-PT, TSEB-PM, and TSEB-T_C-T_S estimations. Validation statistics comparing the three models' performances are summarized in Table 1. Results show that estimated R_n from all three models were in good agreement with tower observations, and the absolute mean biases in estimated R_n for three models are all below 10 W/m^2. Besides, all models slightly overestimated G, with the RMSEs all slightly exceeding 37 W/m^2. Despite the different parameterization schemes used in three models (for instance, both TSEB-PT and TSEB-PM use the iterative procedure to calculate component surface temperatures, whereas TSEB-T_C-T_S employs the theoretical VFC/LST trapezoid for component temperature decomposition), they all exhibited comparable skills in estimation of H and LE, which can be indicated by the similar RMSEs ($\approx$44.9–47.9 W/m^2 for H, and $\approx$61.8–75.3 W/m^2 for LE) from Table 1.

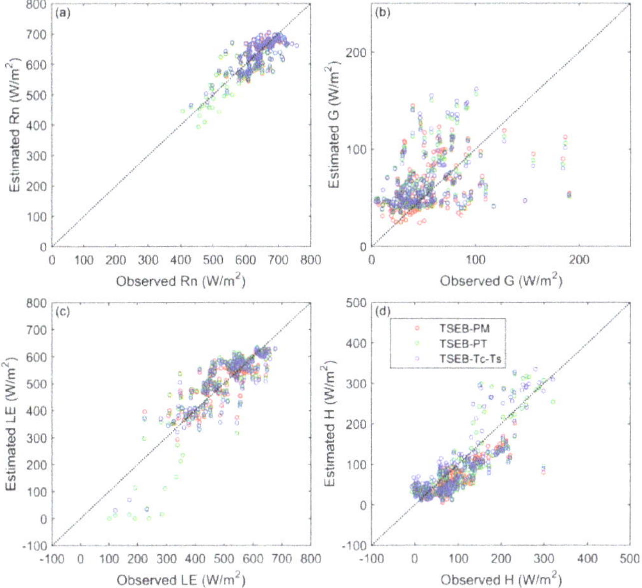

Figure 2. Validation of energy balance components of TSEB-PT, TSEB-PM, and TSEB-T_C-T_S during the HiWATER experiment at times of ASTER overpass. Energy balance components are (**a**) R_n, (**b**) G, (**c**) LE and (**d**) H.

Table 1. Validation statistics of TSEB-PM, TSEB-PT and TSEB-T_C-T_S estimations.

Flux Component	TSEB-PM			TSEB-PT			TSEB-T_C-T_S		
	RMSE (W/m^2)	Bias (W/m^2)	R	RMSE (W/m^2)	Bias (W/m^2)	R	RMSE (W/m^2)	Bias (W/m^2)	R
R_n	37.6	−8.5	0.84	37.4	−7.2	0.84	35.5	−5.7	0.76
G	37.9	11.5	0.37	37.5	12.8	0.34	37.7	12.7	0.33
LE	70.6	−2.1	0.86	75.3	−0.2	0.85	61.8	3.2	0.82
H	44.9	−14.3	0.84	47.5	−16.2	0.83	47.9	−8.6	0.81

The spatial distributions of H and LE over the Zhangye oasis based on TSEB-PT, TSEB-PM, and TSEB-T_C-T_S for the satellite overpass on July 10 of 2012 are shown in Figure 3. Generally, the spatial patterns of surface fluxes were similar between the three models. In addition, the contrasting features over the oasis and the surrounding sandy and Gobi desert were clearly observed from all three models. Specifically, the Zhangye oasis, which mainly comprises of irrigated farmland, exhibited an average *LE* over 400 W/m^2. On the contrary, across the sandy and Gobi desert, *LE* was typically below 300 W/m^2 and *H* was over 100 W/m^2.

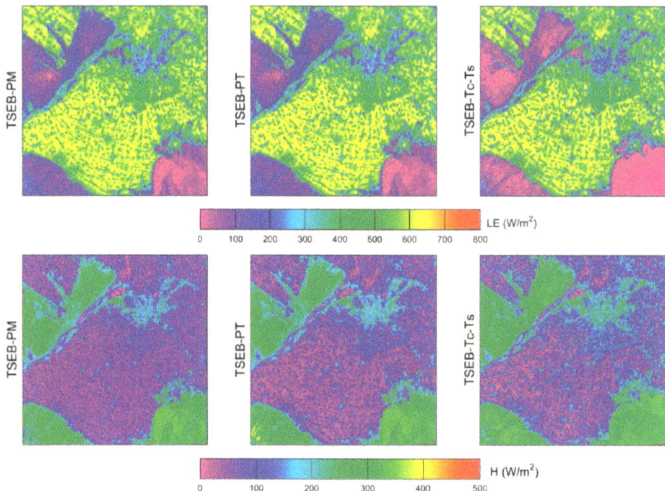

Figure 3. The spatial distributions of *LE* (first row) and *H* (second row) over the Zhangye oasis based on TSEB-PT, TSEB-PM, and TSEB-T_C-T_S for the satellite overpass time on 10 July 2012.

In summary, all three employed models performed similarly in estimating *H* and *LE* despite using different schemes for modeling canopy transpiration and deriving component surface temperatures. With substantial ground observations from the tower-based network, the performances of the three models over MUSOEXE were proven reliable. The difference in derived component temperatures and partitioned evaporation and transpiration among the three models are analyzed in the following two sections.

4.2. Intercomparison of E/T Partitioning from Three TSEB Models

The spatial distributions of canopy transpiration (*LE*c) and soil evaporation (*LE*s) based on TSEB-PT, TSEB-PM, and TSEB-T_C-T_S on 10 July of 2012 are shown in Figure 4. The spatial patterns of *LE*c and *LE*s derived from the three models are similar, and the range of *LE*c for irrigated farmland was about ≈350–500 W/m^2. With respect to *LE*s, the difference derived from the three models can be visually discerned from Figure 4. In comparison with TSEB-PT and TSEB-PM, TSEB-T_C-T_S tended to produce higher *LE*s, especially over the sparsely vegetated area around the residential area and the sandy Gobi desert.

In this study, both CWSIc and SWDIs were further derived to compare the performances of the three TSEB models on detecting vegetation and soil water stresses. The spatial distributions of CWSIc and SWDIs on 10 July 2012 based on three models are shown in Figure 5. The sandy and Gobi desert pixels are masked in CWSIc images in order to more clearly reveal the difference of the three TSEB models in detecting vegetation water stress over the oasis. TSEB-PM and TSEB-PT show similar performances regarding the detection of vegetation stress with both CWSIc values close to zero. On the contrary, TSEB-T_C-T_S seemed to detect a higher level of vegetation water stress, with a CWSIc peak at 0.45. In addition, the spatial heterogeneity was more prominent in TSEB-T_C-T_S, since the CWSIc gradient in the oasis and urban areas are clearly seen from the upper right subplot of Figure 5.

With respect to SWDIs, the spatial distributions were quite similar between the three TSEB models. The SWDIs values for the oasis was smaller than the surrounding sandy and Gobi desert, and the contrasting features over the oasis and the surrounding sandy and Gobi desert were clearly observed from SWDIs subplots from all three models.

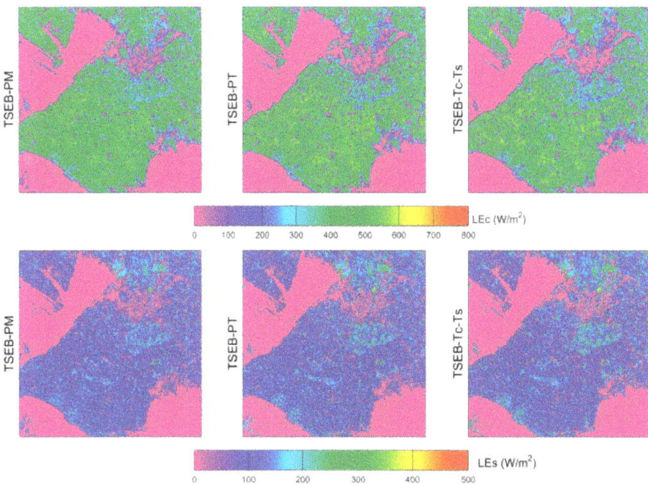

Figure 4. The spatial distribution of *LEc* (first row) and *LEs* (second row) over the Zhangye oasis based on TSEB-PT, TSEB-PM, and TSEB-T_C-T_S for the satellite overpass time on 10 July 2012.

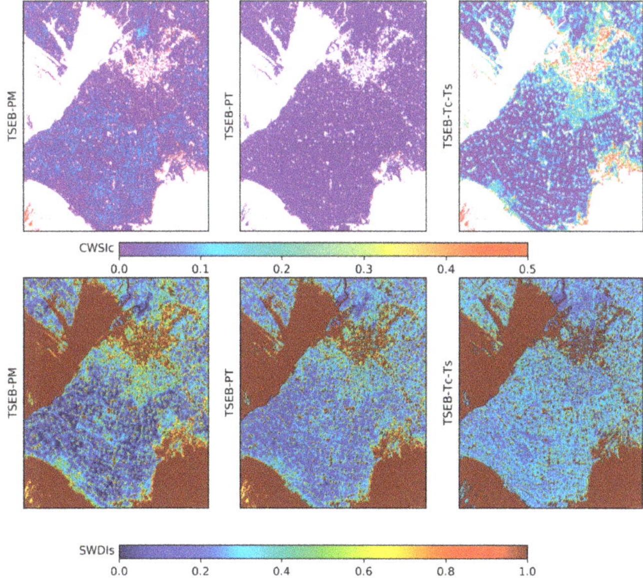

Figure 5. The spatial distribution of *CWSIc* (first row) and *SWDIs* (second row) over the Zhangye oasis based on TSEB-PM, TSEB-PT, and TSEB-T_C-T_S for the satellite overpass time on 10 July 2012. The white areas correspond to the sandy and Gobi desert, and these pixels are masked.

The ratios of LE_C/LE (i.e., T/ET) measured using the stable oxygen and hydrogen isotopes technique during HiWATER-MUSOEXE are utilized to evaluate the performances of three TSEB models on partitioning E and T. Figure 6 shows the intercomparison of LE_C/LE between the three models and ground measurements. The ratios of LE_C/LE were underestimated on DOYs of 176, 192, 215, 224, 231, and 240, while slight overestimation of LE_C/LE occurred on DOYs of 247 and 256 in all three models. The LE_C/LE ratios derived from three models were very close in most cases at this site. The mean observed LE_C/LE ratio was 84.7%, while the mean estimated LE_C/LE ratios were 76.7%, 76.9% and 77.0% for TSEB-PT, TSEB-PM, TSEB-T_C-T_S. All three models seemed to slightly underestimate the LE_C/LE ratio. However, the observed LE_C/LE ratios exhibited a decline during September (DOYs of 247 and 256), mainly due to leaf senescence, which was not characterized by all three models.

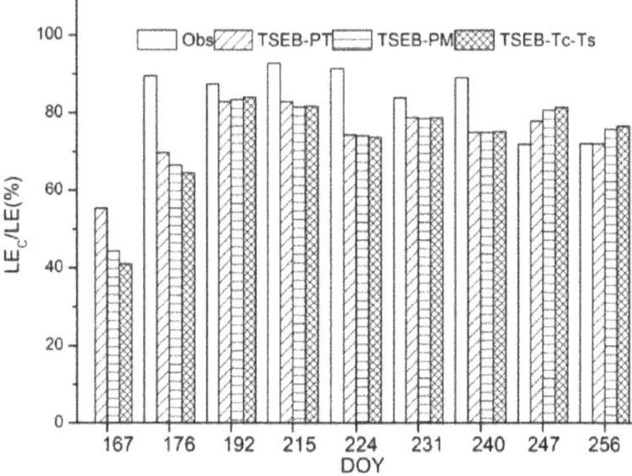

Figure 6. Comparison of LE_C/LE (%) between the three TSEB models and ground measurements by stable oxygen and hydrogen isotopes technique at Daman superstation.

In order to further investigate the difference regarding E and T partitioning between the three models, a pixel-based comparison of LE_C and LE_S in the kernel experimental area of MUSOEXE was conducted and shown in Figure 7, with color shading indicating pixel density. The statistics for a pixel-based comparison of LE_C and LE_S in the kernel experimental area are summarized in Table 2. Most points are under the 1:1 line in Figure 7a,b, which suggests that TSEB-PM tended to produce higher LE_C than TSEB-PT and TSEB-T_C-T_S. The mean differences (MDs) for the pairs of TSEB-PT/TSEB-PM and TSEB-T_C-T_S/TSEB-PM were 2.9 W/m^2 and 18.1 W/m^2, respectively (mean differences were calculated by subtracting the former model of the pair from the latter model). LE_C estimated from TSEB-PT was comparatively higher than that from TSEB-T_C-T_S, with MD being 15.2 W/m^2 for the TSEB-T_C-T_S/TSEB-PT pair. Correspondingly, comparing to TSEB-PT and TSEB-T_C-T_S, TSEB-PM tended to underestimate LE_S as shown in Figure 7e,f, with MDs of -4.2 W/m^2 and -2.8 W/m^2 for the pairs of TSEB-PT/TSEB-PM and TSEB-T_C-T_S/TSEB-PM, respectively. The LE_S estimations from TSEB-PT and TSEB-T_C-T_S for the kernel experimental area were similar, with a correlation coefficient R being approximately 1.0, MD being 1.4 W/m^2, and the mean absolute difference (MAD) being 6.2 W/m^2. Since the component temperature decomposition has a major impact on E/T partitioning, to explore the mechanism underlying the observed difference on LE_C (LE_S) from three models, a further intercomparison on the derived component temperature is conducted in the next section.

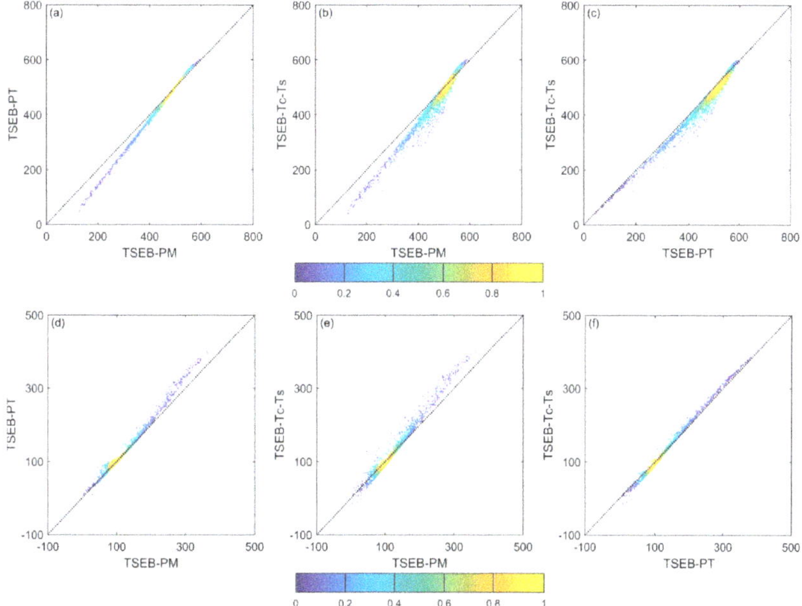

Figure 7. The intercomparison of LE_C (first row) and LE_S (second row) derived from TSEB-PT, TSEB-PM, and TSEB-T_C-T_S in the kernel experimental area on 10 July 2012. The intercompared LE_C pairs are (**a**) TSEB-PM vs. TSEB-PT; (**b**) TSEB-PM vs. TSEB-T_C-T_S; (**c**) TSEB-PT vs. TSEB-T_C-T_S. The intercompared LEs pairs are (**d**) TSEB-PM vs. TSEB-PT; (**e**)TSEB-PM vs. TSEB-T_C-T_S; (**f**) TSEB-PT vs. TSEB-T_C-T_S.

Table 2. Statistics summarizing the intercomparison of LE_C and LE_S derived from TSEB-PT, TSEB-PM, and TSEB-T_C-T_S in the kernel experimental area.

		TSEB-PT vs. TSEB-PM	TSEB-*Tc*-*Ts* vs. TSEB-PM	TSEB-*Tc*-*Ts* vs. TSEB-PT
LEc	RMSE (W/m^2)	18.8	33.9	23.2
	MD * (W/m^2)	2.9	18.1	15.2
	MAD (W/m^2)	13.2	24.6	16.5
	R	1.00	0.98	0.99
LEs	RMSE (W/m^2)	10.8	16.2	7.6
	MD (W/m^2)	−4.2	−2.8	1.4
	MAD (W/m^2)	6.1	11.1	6.2
	R	0.99	0.99	1.00

* MD is calculated by subtracting the former model of the pair from the latter model.

4.3. Intercomparison of Tc and Ts Derived from Three TSEB Models

Intercomparison of the component temperatures derived from the three models is shown in Figure 8. The T_C derived from TSEB-PM showed a relatively homogenous pattern over the Zhangye oasis and the average of T_C approximated 301 K. On the contrary, T_C derived from TSEB-PT and TSEB-T_C-T_S showed a much larger spatial variability, and the contrast between farmland and residential area is more discernable in both models. However, the spatial contrast between farmland and residential area was less significant for the TSEB-T_C-T_S derived T_S (Figure 8).

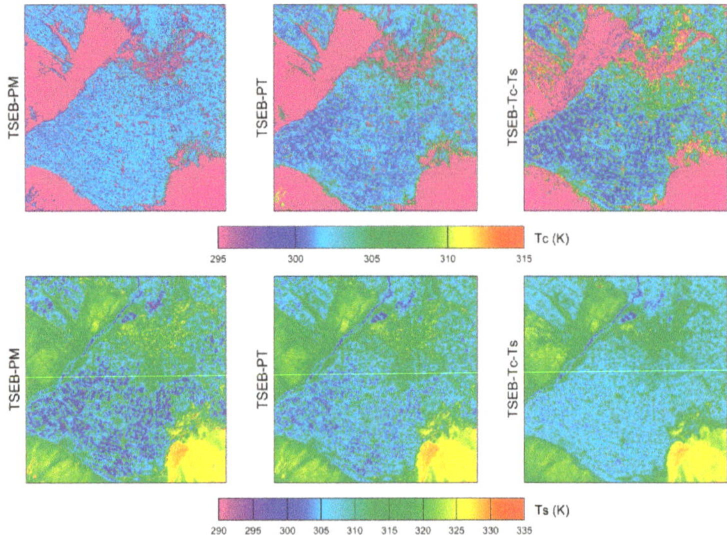

Figure 8. The spatial distribution of T_C (first row) and Ts (second row) over HiWATER-MUSOEXE derived from TSEB-PT, TSEB-PM, and TSEB-T_C-T_S on 10 July 2012.

A pixel-based comparison of decomposed T_C and T_S in the kernel experimental area was conducted and is shown in Figure 9. It is noticed that the component temperatures were much more scattered compared to the component fluxes in the scatter plot of Figure 7. The statistics for the pixel-based comparison of T_C and T_S in the kernel experimental area are listed in Table 3. Generally, TSEB-T_C-T_S estimates higher T_C compared to TSEB-PT and TSEB-PM, and TSEB-PT tended to overestimate T_C in relation to TSEB-PM, with the MD of TSEB-PT/TSEB-PM pair being -0.2 K.

Table 3. Statistics summarizing the intercomparison of T_C and T_S derived from TSEB-PT, TSEB-PM, and TSEB-T_C-T_S in the kernel experimental area.

		TSEB-PT vs. TSEB-PM	TSEB-Tc-Ts vs. TSEB-PM	TSEB-Tc-Ts vs. TSEB-PT
	RMSE (K)	1.4	2.4	1.0
Tc	MD (K)	-0.2	-0.48	-0.3
	MAD (K)	0.8	1.5	0.6
	R	0.13	-0.09	0.97
	RMSE (K)	2.0	3.4	1.6
Ts	MD (K)	-0.6	-0.3	0.3
	MAD (K)	1.5	2.7	1.3
	R	0.95	0.80	0.94

Different strategies for deriving T_C and T_S were responsible for the observed differences in surface fluxes between the three modeling approaches. As previously stated, both TSEB-PT and TSEB-PM applied an iterative approach to derive component temperatures. Although the identical temperature decomposition method was applied to TSEB-PT and TSEB-PM, the component temperatures derived from both models showed a significant difference, with correlation coefficient R being only 0.13 for T_C. Different from TSEB-PT and TSEB-PM, TSEB-T_C-T_S adopted the VFC/LST trapezoid space to estimate the component temperatures, which would explain the distinct characteristics shown in Figure 8. Specifically, for the irrigated farmland, the component temperatures derived from TSEB-T_C-T_S were comparable to those derived from the other two models. Due to the lack of constraint on adjusting the canopy transpiration, the vegetation component temperature from TSEB-PT and TSEB-PM would be very close to pixels in well-watered and fully-covered vegetation areas.

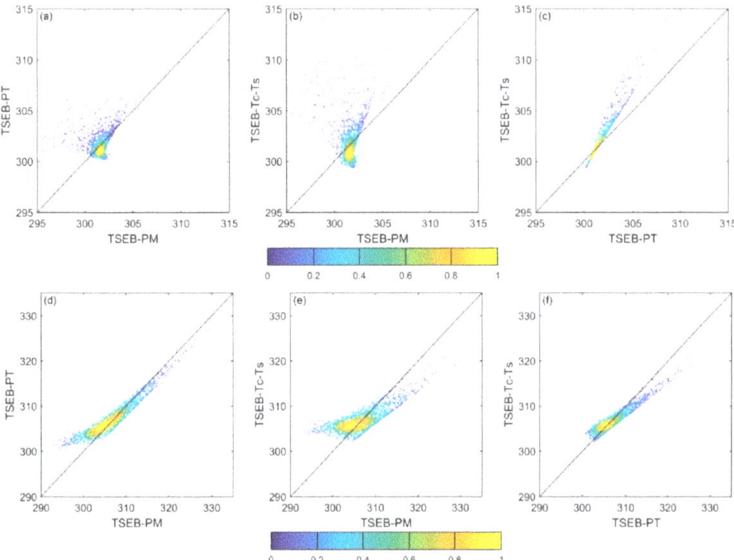

Figure 9. Intercomparison of T_c (first row) and T_s (second row) derived from TSEB-PT, TSEB-PM, and TSEB-T_c-T_s in the kernel experimental area on 10 July 2012. The intercompared T_c pairs are (**a**) TSEB-PM vs. TSEB-PT; (**b**) TSEB-PM vs. TSEB-T_C-T_S; (**c**) TSEB-PT vs. TSEB-T_C-T_S. The intercompared T_s pairs are (**d**) TSEB-PM vs. TSEB-PT; (**e**)TSEB-PM vs. TSEB-T_C-T_S; (**f**) TSEB-PT vs. TSEB-T_C-T_S.

The scientific rationale can be explained as follows. Based on the VFC/LST space, Figure 10 illustrates the temperature decomposition methods adopted in TSEB-PT, TSEB-PM, and TSEB-T_C-T_S. Both TSEB-PT and TSEB-PM applied an iterative approach to derive the component temperatures and this approach assumed vegetation transpiration at the potential rate as an initial value. Due to the lack of constraint on adjusting the canopy transpiration, the vegetation component temperature from TSEB-PT and TSEB-PM would be very close to Point C (Figure 10) in well-watered and fully vegetated areas. Among TSEB-PM and TSEB-PT, the former employs the Penman-Monteith equation to consider the varying VPD under advective semiarid climates, and this would lead to higher LE_C compared to TSEB-PT. Consequently, the soil component temperature from TSEB-PM was higher than that from TSEB-PT. Different from TSEB-PM and TSEB-PT, TSEB-T_C-T_S adopted the VFC/LST trapezoid space to estimate component temperatures. TSEB-T_C-T_S assumed that the vegetation and soil share the same water pool, and the slope of each isoline in the VFC/LST space could be derived by interpolating the slopes of both dry and cold edges. The temperature decomposition methods adopted in the three TSEB models is illustrated in Figure 10, the soil surface temperatures derived from TSEB-PM, TSEB-PT, and TSEB-T_C-T_S are denoted by T_{s1}, T_{s2}, and T_{s3}, and the vegetation canopy temperatures derived from the same three models are denoted by T_{v1}, T_{v2}, and T_{v3}. It is clear that T_{v1} and T_{v2} were underestimated compare to T_{v3}, and this led to a higher T/ET partition for TSEB-PM and TSEB-PT compared to TSEB-T_C-T_S, which consequently overestimated LE_C when compared to TSEB-T_C-T_S. A component temperatures decomposition method based on isolines in the VFC/LST space provided a further constraint for vegetation transpiration and would be a good substitution for an iterative approach.

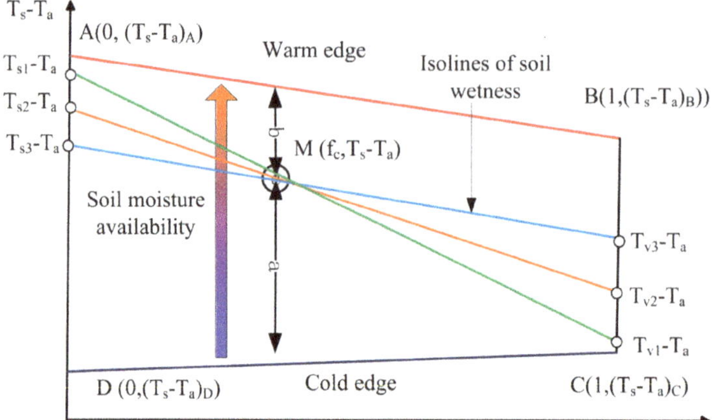

Figure 10. Illustration of the temperature decomposition methods adopted in the three TSEB models. T_{s1}, T_{s2}, and T_{s3} denote the soil surface temperatures derived from TSEB-PM, TSEB-PT, and TSEB-T_C-T_S, respectively, and T_{v1}, T_{v2}, and T_{v3} denote the vegetation canopy temperatures derived from the same three models.

5. Discussion

5.1. Reliability of the Employed TSEB Models in Estimating Surface Fluxes

The HiWATER–MUSOEXE data have been used extensively to validate the land surface flux models, including one-source and two-source models [70–72]. Using the same set of ground-based observations, Ma et al. [70] applied a revised SEBS model to estimate regional heat fluxes in the Heihe River Basin, and their assessment indicates that the RMSEs of the modeled H and LE are 56.9 W/m^2 and 74.8 W/m^2, respectively. Huang et al. [71] integrated a Normalized Difference Water Index (NDWI) as a water stress index into SEBS through the modification of the parameter kB^{-1}, and showed RMSEs of 79.8 W/m^2 in H and 84.1 W/m^2 in LE for the revised SEBS. In this study, three two-source models were applied to the middle reach of the Heihe River Basin. The RMSEs of H and LE from TSEB-PT are 47.5 and 75.3 W/m^2, respectively, the RMSEs of H and LE from TSEB-PM are 44.9 and 70.6 W/m^2, respectively, and the values from TSEB-T_C-T_S are 47.9 and 61.8 W/m^2, respectively. Overall, the performances of two-source modeling approaches were reliable in relation to previously published studies.

5.2. Discrepancies in E/T Partitioning between the Three TSEB Models

Despite the comparable skills of ET estimations within the three TSEB models, different assumptions and formulations were adopted by the three different models, and discrepancies in E/T partitioning among the three models were observed in the kernel experimental area of MUSOEXE. In this study, both CWSI$_C$ and SWDI$_S$ were derived to compare the performances of three models regarding the detection of vegetation and soil water stresses. The results indicated that different from TSEB-PT and TSEB-PM, TSEB-T_C-T_S had the potential to detect vegetation water stress. In addition, the E/T partitioning efficacies of the three TSEB modeling approaches were evaluated using the measurements from the stable oxygen and hydrogen isotopes system. It was found that all three models tended to slightly underestimate the ratio of T/ET at the Daman site. Aside from the intercomparison between the three two-source models, it was found that TSEB-PM tended to generate a higher *LE*c estimation than the other two models, especially for the partially vegetated areas. *LE*$_S$ derived from

TSEB-PT and TSEB-T_C-T_S were quite similar. The differences in component temperatures derived from the three models, as well as the aerodynamic resistance, were responsible for this divergence. Besides, the declined LE_C/LE ratios observed due to leaf senescence by the end of the growing season was not well characterized by all three models. This indicated the necessity of adding a green vegetation fraction in the future work, as suggested by Kustas et al. [73].

5.3. Impact of Temperature Decomposition Accuracies on ET Estimations

To separately estimate E and T, component temperatures are indispensable in the TSEB modeling approach and the temperature decomposition method is the core in E/T partitioning process. In this study, two categories of temperature decomposition methods were intercompared: (1) an iterative procedure based on an energy balance resistance network; and (2) a VFC/LST space method with the assumption that the isolines could be used to decompose composite temperatures. The iterative procedure is commonly adopted in the TSEB model and this approach is verified by different researchers [69,74,75]. The VFC/LST-based approach and related contextual-based method are applied by different researchers, such as Zhang et al. [24], Zhang et al. [55], Merlin et al. [44], Merlin et al. [45], Long and Singh [47], Yang et al. [48], Song et al. [53], and Sun [49].

Further intercomparison indicated that the differences in component temperatures derived from two categories of models were responsible for the discrepancy in ET estimates. As there was no constraint for canopy transpiration to terminate the procedure when LE_S exceeded zero in the iterative procedure, this procedure adopted by TSEB-PT and TSEB-PM may have derived a higher LEc and lower Tc compared to TSEB-T_C-T_S. This is consistent with the findings of Anderson et al. [76], who pointed out the same case under conditions of moderate stomatal closure for TSEB model. In this study, we found the temperature decomposition based on VFC/LST added a further constraint on vegetation transpiration, and this category of method could be a substitute for the iterative method. Besides, it was found that TSEB-T_C-T_S performed better at detecting the vegetation stress than the other two models. Although a similar temperature decomposition approach was applied to TSEB-PT and TSEB-PM, a significant difference was observed between them. For instance, the R between the two models was only 0.13 for the derived T_C. Two reasons may be responsible for this: (1) TSEB-PM applied the Penman-Monteith equation to characterize the canopy transpiration and thus took into account the effect of varying VPD over different underlying surfaces, which was a different case than in TSEB-PT. (2) Resistances were involved in the iterative procedure and the resistance would impact the temperature decomposition. In this study, the theoretically defined VFC/LST trapezoid space was adopted to estimate the component temperatures in TSEB-T_C-T_S. A small amount of points may have been located outside the trapezoid, and this was mainly caused by ill-parameterized aerodynamic resistance. However, this portion was limited to 1% of the total pixels. Future studies are required to further investigate the influence of resistance on the component temperature decomposition.

6. Conclusions

In this study, three two-source modeling approaches were evaluated using the ground-based observations from the HiWATER-MUSOEXE campaign. Validated with tower observations, the RMSEs of H and LE were lower than 50 W/m^2 and 76 W/m^2, respectively, and the results demonstrated that the three TSEB models were capable of predicting reliable surface heat fluxes. The measurements from the stable oxygen and hydrogen isotopes system were used to evaluate the capabilities of three models in E and T partitioning, and it was found that all three models appeared to slightly underestimate the ratio of T/ET.

A further intercomparison of the component temperature decomposition among the three models was conducted to explore the underlying mechanism for the observed differences. Results indicated that the interactive methods applied in TSEB and TSEB-PM may have produced higher LE_C and lower T_C compared to TSEB-T_C-T_S due to lack of constraint on vegetation transpiration. Based on the soil moisture isoline in the VFC/LST space, the VFC/LST-based temperature decomposition method added a further constraint on vegetation transpiration, and could be used as a substitution for the interactive procedure adopted in the original TSEB model.

Author Contributions: Conceptualization, Y.Y. and J.Q.; Data curation, Y.Y. and J.Q.; Formal analysis, Y.Y. and J.Q.; Funding acquisition, Y.Y. and J.Q.; Investigation, Y.Y. and J.Q.; Methodology, Y.Y. and J.Q.; Project administration, R.Z., S.H., S.C., H.W., J.L., and Y.F.; Validation, Y.Y.; Visualization, J.Q.; Writing–original draft, Y.Y.; Writing–review & editing, J.Q.

Funding: This research was funded by the National Natural Science Foundation of China (grant numbers 41501415, 41501450, 51420105014), the IWHR Research & Development Support Program (grant number. JZ0145B032017), the 13th five-year plan of National Scientific Research and Development (2017YFC0405803, 2017YFC1502406), the Natural Science Foundation of Guangdong Province, China (grant number 2016A030310154), the Fundamental Research Funds for the Central Universities (grant number 16lgpy06).

Acknowledgments: We are very thankful to the researchers in HiWATER-MUSOEXE for their efforts on data acquisition and sharing. The data set used in this study is provided by Heihe Plan Science Data Center, National Natural Science Foundation of China (http://www.heihedata.org).

Conflicts of Interest: The authors declare no conflict of interest.

References

1. Brutsaert, W. *Evaporation into the Atmosphere: Theory, History and Applications*; Springer: New York, NY, USA, 1982.
2. Brutsaert, W. *Hydrology: An Introduction*; Cambridge University Press: Cambridge, UK, 2005.
3. Eagleson, P.S. *Ecohydrology: Darwinian Expression of Vegetation Form and Function*; Cambridge University Press: Cambridge, UK, 2002.
4. Kabat, P. *Vegetation, Water, Humans and the Climate: A New Perspective on an Internactive System*; Springer Science & Business Media: Berlin, Germany, 2004.
5. Allen, R.G.; Tasumi, M.; Morse, A.; Trezza, R.; Wright, J.L.; Bastiaanssen, W.; Kramber, W.; Lorite, I.; Robison, C.W. Satellite-based energy balance for mapping evapotranspiration with internalized calibration (metric)—applications. *J. Irrig. Drain. Eng.* **2007**, *133*, 395–406. [CrossRef]
6. Allen, R.G.; Tasumi, M.; Trezza, R. Satellite-based energy balance for mapping evapotranspiration with internalized calibration (metric)—model. *J. Irrig. Drain. Eng.* **2007**, *133*, 380–394. [CrossRef]
7. Glenn, E.P.; Huete, A.R.; Nagler, P.L.; Hirschboeck, K.K.; Brown, P. Integrating remote sensing and ground methods to estimate evapotranspiration. *Crit. Rev. Plant Sci.* **2007**, *26*, 139–168. [CrossRef]
8. Kustas, W.; Anderson, M. Advances in thermal infrared remote sensing for land surface modeling. *Agric. For. Meteorol.* **2009**, *149*, 2071–2081. [CrossRef]
9. Wang, K.; Dickinson, R.E. A review of global terrestrial evapotranspiration: Observation, modeling, climatology, and climatic variability. *Rev. Geophys.* **2012**, *50*. [CrossRef]
10. Jimenez, C.; Prigent, C.; Mueller, B.; Seneviratne, S.I.; McCabe, M.F.; Wood, E.F.; Rossow, W.B.; Balsamo, G.; Betts, A.K.; Dirmeyer, P.A. Global intercomparison of 12 land surface heat flux estimates. *J. Geophys. Res. Atmos.* **2011**, *116*. [CrossRef]
11. McCabe, M.; Kustas, W.; Anderson, M.; Kongoli, C.; Ershadi, A.; Hain, C.R. Global-scale estimation of land surface heat fluxes from space: Current status, opportunities, and future directions. In *Remote Sensing of Energy Fluxes and Soil Moisture Content*; CRC Press: Boca Raton, FL, USA, 2013; pp. 447–462.
12. Mu, Q.; Heinsch, F.A.; Zhao, M.; Running, S.W. Development of a global evapotranspiration algorithm based on modis and global meteorology data. *Remote Sens. Environ.* **2007**, *111*, 519–536. [CrossRef]
13. Mu, Q.; Zhao, M.; Running, S.W. Improvements to a modis global terrestrial evapotranspiration algorithm. *Remote Sens. Environ.* **2011**, *115*, 1781–1800. [CrossRef]
14. Mueller, B.; Hirschi, M.; Jimenez, C.; Ciais, P.; Dirmeyer, P.A.; Dolman, A.J.; Fisher, J.B.; Jung, M.; Ludwig, F.; Maignan, F. Benchmark products for land evapotranspiration: Landflux-eval multi-data set synthesis. *Hydrol. Earth Syst. Sci.* **2013**, *17*, 3707–3720. [CrossRef]
15. Zhang, Y.; Leuning, R.; Chiew, F.H.S.; Wang, E.; Zhang, L.; Liu, C.; Sun, F.; Peel, M.C.; Shen, Y.; Jung, M. Decadal trends in evaporation from global energy and water balances. *J. Hydrometeorol.* **2012**, *13*, 379–391. [CrossRef]

16. Bastiaanssen, W.G.M.; Menenti, M.; Feddes, R.A.; Holtslag, A.A.M. A remote sensing surface energy balance algorithm for land (SEBAL). 1. Formulation. *J. Hydrol.* **1998**, *212*, 198–212. [CrossRef]

17. Bastiaanssen, W.G.M.; Pelgrum, H.; Wang, J.; Ma, Y.; Moreno, J.F.; Roerink, G.J.; Van der Wal, T. A remote sensing surface energy balance algorithm for land (SEBAL): Part 2: Validation. *J. Hydrol.* **1998**, *212*, 213–229. [CrossRef]

18. Boegh, E.; Soegaard, H.; Thomsen, A. Evaluating evapotranspiration rates and surface conditions using landsat tm to estimate atmospheric resistance and surface resistance. *Remote Sens. Environ.* **2002**, *79*, 329–343. [CrossRef]

19. Norman, J.M.; Kustas, W.P.; Humes, K.S. Source approach for estimating soil and vegetation energy fluxes in observations of directional radiometric surface temperature. *Agric. For. Meteorol.* **1995**, *77*, 263–293. [CrossRef]

20. Su, Z. The surface energy balance system (sebs) for estimation of turbulent heat fluxes. *Hydrol. Earth Syst. Sci.* **2002**, *6*, 85–100. [CrossRef]

21. Roerink, G.J.; Su, Z.; Menenti, M. S-sebi: A simple remote sensing algorithm to estimate the surface energy balance. *Phys. Chem. Earth Part B Hydrol. Oceans Atmos.* **2000**, *25*, 147–157. [CrossRef]

22. Su, Z. Hydrological applications of remote sensing. Surface fluxes and other derived variables–surface energy balance. In *Encyclopedia of Hydrological Sciences*; John Wiley and Sons: Hoboken, NJ, USA, 2005.

23. Anderson, M.C.; Norman, J.M.; Diak, G.R.; Kustas, W.P.; Mecikalski, J.R. A two-source time-integrated model for estimating surface fluxes using thermal infrared remote sensing. *Remote Sens. Environ.* **1997**, *60*, 195–216. [CrossRef]

24. Zhang, R.H.; Sun, X.M.; Wang, W.M.; Xu, J.P.; Zhu, Z.L.; Tian, J. An operational two-layer remote sensing model to estimate surface flux in regional scale: Physical background. *Sci. China Ser. D* **2005**, *48*, 225–244.

25. Yang, Y.; Su, H.; Zhang, R.; Tian, J.; Li, L. An enhanced two-source evapotranspiration model for land (ETEML): Algorithm and evaluation. *Remote Sens. Environ.* **2015**, *168*, 54–65. [CrossRef]

26. Carlson, T. An overview of the "triangle method" for estimating surface evapotranspiration and soil moisture from satellite imagery. *Sensors* **2007**, *7*, 1612–1629. [CrossRef]

27. Petropoulos, G.; Carlson, T.N.; Wooster, M.J.; Islam, S. A review of Ts/Vi remote sensing based methods for the retrieval of land surface energy fluxes and soil surface moisture. *Prog. Phys. Geogr.* **2009**, *33*, 224–250. [CrossRef]

28. Moran, M.S.; Clarke, T.R.; Inoue, Y.; Vidal, A. Estimating crop water deficit using the relation between surface-air temperature and spectral vegetation index. *Remote Sens. Environ.* **1994**, *49*, 246–263. [CrossRef]

29. Jiang, L.; Islam, S. Estimation of surface evaporation map over southern great plains using remote sensing data. *Water Resour. Res.* **2001**, *37*, 329–340. [CrossRef]

30. Jiang, L.; Islam, S. An intercomparison of regional latent heat flux estimation using remote sensing data. *Int. J. Remote Sens.* **2003**, *24*, 2221–2236. [CrossRef]

31. Stisen, S.; Sandholt, I.; Nørgaard, A.; Fensholt, R.; Jensen, K.H. Combining the triangle method with thermal inertia to estimate regional evapotranspiration—Applied to msg-seviri data in the senegal river basin. *Remote Sens. Environ.* **2008**, *112*, 1242–1255. [CrossRef]

32. Shu, Y.; Stisen, S.; Jensen, K.H.; Sandholt, I. Estimation of regional evapotranspiration over the north china plain using geostationary satellite data. *Int. J. Appl. Earth Obs. Geoinf.* **2011**, *13*, 192–206. [CrossRef]

33. Courault, D.; Seguin, B.; Olioso, A. Review on estimation of evapotranspiration from remote sensing data: From empirical to numerical modeling approaches. *Irrig. Drain. Syst.* **2005**, *19*, 223–249. [CrossRef]

34. Kalma, J.D.; McVicar, T.R.; McCabe, M.F. Estimating land surface evaporation: A review of methods using remotely sensed surface temperature data. *Surv. Geophys.* **2008**, *29*, 421–469. [CrossRef]

35. Li, Z.-L.; Tang, R.; Wan, Z.; Bi, Y.; Zhou, C.; Tang, B.; Yan, G.; Zhang, X. A review of current methodologies for regional evapotranspiration estimation from remotely sensed data. *Sensors* **2009**, *9*, 3801–3853. [CrossRef] [PubMed]

36. Anderson, M.C.; Allen, R.G.; Morse, A.; Kustas, W.P. Use of landsat thermal imagery in monitoring evapotranspiration and managing water resources. *Remote Sens. Environ.* **2012**, *122*, 50–65. [CrossRef]

37. Rauwerda, J.; Roerink, G.J.; Su, Z. *Estimation of Evaporative Fractions by the Use of Vegetation and Soil Component Temperature Determined by Means of Dual-Looking Remote Sensing*; Alterra: Osborne Park, WA, USA, 2002.

38. Jia, L. *Modeling Heat Exchanges at the Land-Atmosphere Interface Using Multi-Angular Thermal Infrared Measurements*; Wageningen University: Wageningen, The Netherlands, 2004.

39. Jia, L.; Li, Z.L.; Menenti, M.; Su, Z.; Verhoef, W.; Wan, Z. A practical algorithm to infer soil and foliage component temperatures from bi-angular atsr-2 data. *Int. J. Remote Sens.* **2003**, *24*, 4739–4760. [CrossRef]
40. Sun, Z.; Wang, Q.; Matsushita, B.; Fukushima, T.; Ouyang, Z.; Watanabe, M. A new method to define the vi-ts diagram using subpixel vegetation and soil information: A case study over a semiarid agricultural region in the north China plain. *Sensors* **2008**, *8*, 6260–6279. [CrossRef] [PubMed]
41. Bisquert, M.; Sánchez, J.M.; López-Urrea, R.; Caselles, V. Estimating high resolution evapotranspiration from disaggregated thermal images. *Remote Sens. Environ.* **2016**, *187*, 423–433. [CrossRef]
42. Colaizzi, P.D.; Agam, N.; Tolk, J.A.; Evett, S.R.; Howell, T.A.; Gowda, P.H.; O'Shaughnessy, S.A.; Kustas, W.P.; Anderson, M.C. Two-source energy balance model to calculate e, t, and et: Comparison of priestley-taylor and penman-monteith formulations and two time scaling methods. *Trans. ASABE* **2014**, *57*, 479–498.
43. Colaizzi, P.D.; Kustas, W.P.; Anderson, M.C.; Agam, N.; Tolk, J.A.; Evett, S.R.; Howell, T.A.; Gowda, P.H.; O'Shaughnessy, S.A. Two-source energy balance model estimates of evapotranspiration using component and composite surface temperatures. *Adv. Water Resour.* **2012**, *50*, 134–151. [CrossRef]
44. Merlin, O.; Escorihuela, M.J.; Mayoral, M.A.; Hagolle, O.; Al Bitar, A.; Kerr, Y. Self-calibrated evaporation-based disaggregation of smos soil moisture: An evaluation study at 3 km and 100 m resolution in Catalunya, Spain. *Remote Sens. Environ.* **2013**, *130*, 25–38. [CrossRef]
45. Merlin, O.; Rudiger, C.; Al Bitar, A.; Richaume, P.; Walker, J.P.; Kerr, Y.H. Disaggregation of SMOS soil moisture in southeastern Australia. *IEEE Trans. Geosci. Remote Sens.* **2012**, *50*, 1556–1571. [CrossRef]
46. Sun, Z.; Wang, Q.; Matsushita, B.; Fukushima, T.; Ouyang, Z.; Watanabe, M. Development of a simple remote sensing evapotranspiration model (sim-reset): Algorithm and model test. *J. Hydrol.* **2009**, *376*, 476–485. [CrossRef]
47. Long, D.; Singh, V.P. A two-source trapezoid model for evapotranspiration (TTME) from satellite imagery. *Remote Sens. Environ.* **2012**, *121*, 370–388. [CrossRef]
48. Yang, Y.; Shang, S. A hybrid dual-source scheme and trapezoid framework–based evapotranspiration model (htem) using satellite images: Algorithm and model test. *J. Geophys. Res. Atmos.* **2013**, *118*, 2284–2300. [CrossRef]
49. Sun, H. A two-source model for estimating evaporative fraction (TMEF) coupling priestley-taylor formula and two-stage trapezoid. *Remote Sens.* **2016**, *8*, 248. [CrossRef]
50. Kustas, W.P.; Norman, J.M. Evaluation of soil and vegetation heat flux predictions using a simple two-source model with radiometric temperatures for partial canopy cover. *Agric. For. Meteorol.* **1999**, *94*, 13–29. [CrossRef]
51. Kustas, W.P.; Norman, J.M. A two-source energy balance approach using directional radiometric temperature observations for sparse canopy covered surfaces. *Agron. J.* **2000**, *92*, 847–854. [CrossRef]
52. Santanello, J.A., Jr.; Friedl, M.A. Diurnal covariation in soil heat flux and net radiation. *J. Appl. Meteorol.* **2003**, *42*, 851–862. [CrossRef]
53. Song, L.; Liu, S.; Kustas, W.P.; Zhou, J.; Xu, Z.; Xia, T.; Li, M. Application of remote sensing-based two-source energy balance model for mapping field surface fluxes with composite and component surface temperatures. *Agric. For. Meteorol.* **2016**, *230*, 8–19. [CrossRef]
54. Kustas, W.P.; Norman, J.M. A two-source approach for estimating turbulent fluxes using multiple angle thermal infrared observations. *Water Resour. Res.* **1997**, *33*, 1495–1508. [CrossRef]
55. Zhang, R.; Tian, J.; Su, H.; Sun, X.; Chen, S.; Xia, J. Two improvements of an operational two-layer model for terrestrial surface heat flux retrieval. *Sensors* **2008**, *8*, 6165–6187. [CrossRef] [PubMed]
56. Thom, A.S. Momentum, mass and heat exchange of plant communities. *Veg. Atmos.* **1975**, *1*, 57–109.
57. Li, X.; Cheng, G.; Liu, S.; Xiao, Q.; Ma, M.; Jin, R.; Che, T.; Liu, Q.; Wang, W.; Qi, Y. Heihe watershed allied telemetry experimental research (hiwater): Scientific objectives and experimental design. *Bull. Am. Meteorol. Soc.* **2013**, *94*, 1145–1160. [CrossRef]
58. Liu, S.; Xu, Z.; Song, L.; Zhao, Q.; Ge, Y.; Xu, T.; Ma, Y.; Zhu, Z.; Jia, Z.; Zhang, F. Upscaling evapotranspiration measurements from multi-site to the satellite pixel scale over heterogeneous land surfaces. *Agric. For. Meteorol.* **2016**, *230*, 97–113. [CrossRef]
59. Xu, Z.; Liu, S.; Li, X.; Shi, S.; Wang, J.; Zhu, Z.; Xu, T.; Wang, W.; Ma, M. Intercomparison of surface energy flux measurement systems used during the hiwater-musoexe. *J. Geophys. Res. Atmos.* **2013**, *118*, 13–140. [CrossRef]

60. Yang, K.; Wang, J. A temperature prediction-correction method for estimating surface soil heat flux from soil temperature and moisture data. *Sci. China Ser. D Earth Sci.* **2008**, *51*, 721–729. [CrossRef]

61. Twine, T.E.; Kustas, W.P.; Norman, J.M.; Cook, D.R.; Houser, P.; Meyers, T.P.; Prueger, J.H.; Starks, P.J.; Wesely, M.L. Correcting eddy-covariance flux underestimates over a grassland. *Agric. For. Meteorol.* **2000**, *103*, 279–300. [CrossRef]

62. Huang, L.; Wen, X. Temporal variations of atmospheric water vapor δd and δ18o above an arid artificial oasis cropland in the Heihe river basin. *J. Geophys. Res. Atmos.* **2014**, *119*, 11456–11476. [CrossRef]

63. Wen, X.; Yang, B.; Sun, X.; Lee, X. Evapotranspiration partitioning through in-situ oxygen isotope measurements in an oasis cropland. *Agric. For. Meteorol.* **2016**, *230*, 89–96. [CrossRef]

64. Li, H.; Sun, D.; Yu, Y.; Wang, H.; Liu, Y.; Liu, Q.; Du, Y.; Wang, H.; Cao, B. Evaluation of the viirs and modis lst products in an arid area of northwest china. *Remote Sens. Environ.* **2014**, *142*, 111–121. [CrossRef]

65. Gillespie, A.; Rokugawa, S.; Matsunaga, T.; Cothern, J.S.; Hook, S.; Kahle, A.B. A temperature and emissivity separation algorithm for advanced spaceborne thermal emission and reflection radiometer (aster) images. *IEEE Trans. Geosci. Remote Sens.* **1998**, *36*, 1113–1126. [CrossRef]

66. Tonooka, H. Accurate atmospheric correction of aster thermal infrared imagery using the WVS method. *IEEE Trans. Geosci. Remote Sens.* **2005**, *43*, 2778–2792. [CrossRef]

67. Sun, C.; Liu, Q.; Wen, J. An algorithm for retrieving land surface albedo from hj-1 CCD data. *Remote Sens. Land Resour.* **2013**, *25*, 58–63.

68. Yang, Y.; Long, D.; Guan, H.; Liang, W.; Simmons, C.; Batelaan, O. Comparison of three dual-source remote sensing evapotranspiration models during the musoexe-12 campaign: Revisit of model physics. *Water Resour. Res.* **2015**, *51*, 3145–3165. [CrossRef]

69. Gonzalez-Dugo, M.P.; Neale, C.M.U.; Mateos, L.; Kustas, W.P.; Prueger, J.H.; Anderson, M.C.; Li, F. A comparison of operational remote sensing-based models for estimating crop evapotranspiration. *Agric. For. Meteorol.* **2009**, *149*, 1843–1853. [CrossRef]

70. Ma, Y.; Liu, S.; Zhang, F.; Zhou, J.; Jia, Z.; Song, L. Estimations of regional surface energy fluxes over heterogeneous oasis–desert surfaces in the middle reaches of the Heihe river during hiwater-musoexe. *IEEE Geosci. Remote Sens. Lett.* **2015**, *12*, 671–675.

71. Huang, C.; Li, Y.; Gu, J.; Lu, L.; Li, X. Improving estimation of evapotranspiration under water-limited conditions based on sebs and modis data in arid regions. *Remote Sens.* **2015**, *7*, 16795–16814. [CrossRef]

72. Yang, Y.; Qiu, J.; Su, H.; Bai, Q.; Liu, S.; Li, L.; Yu, Y.; Huang, Y. A one-source approach for estimating land surface heat fluxes using remotely sensed land surface temperature. *Remote Sens.* **2017**, *9*, 43. [CrossRef]

73. Kustas, W.P.; Nieto, H.; Morillas, L.; Anderson, M.C.; Alfieri, J.G.; Hipps, L.E.; Villagarcía, L.; Domingo, F.; Garcia, M. Revisiting the paper "using radiometric surface temperature for surface energy flux estimation in mediterranean drylands from a two-source perspective". *Remote Sens. Environ.* **2016**, *184*, 645–653. [CrossRef]

74. Choi, M.; Kustas, W.P.; Anderson, M.C.; Allen, R.G.; Li, F.; Kjaersgaard, J.H. An intercomparison of three remote sensing-based surface energy balance algorithms over a corn and soybean production region (Iowa, US) during smacex. *Agric. For. Meteorol.* **2009**, *149*, 2082–2097. [CrossRef]

75. Tang, R.; Li, Z.-L.; Jia, Y.; Li, C.; Chen, K.-S.; Sun, X.; Lou, J. Evaluating one-and two-source energy balance models in estimating surface evapotranspiration from landsat-derived surface temperature and field measurements. *Int. J. Remote Sens.* **2013**, *34*, 3299–3313. [CrossRef]

76. Anderson, M.C.; Norman, J.M.; Kustas, W.P.; Houborg, R.; Starks, P.J.; Agam, N. A thermal-based remote sensing technique for routine mapping of land-surface carbon, water and energy fluxes from field to regional scales. *Remote Sens. Environ.* **2008**, *112*, 4227–4241. [CrossRef]

 *remote sensing*

Article

Mapping Maize Evapotranspiration at Field Scale Using SEBAL: A Comparison with the FAO Method and Soil-Plant Model Simulations

Carla Grosso [1], Gabriele Manoli [2], Marco Martello [3], Yann H. Chemin [4], Diego H. Pons [5], Pietro Teatini [6], Ilaria Piccoli [3] and Francesco Morari [3,*

[1] Mario Gulich Institute, National Agency of Spatial Activities (CONAE), X5000 Córdoba, Argentina; cachigrosso@hotmail.com
[2] Institute of Environmental Engineering, ETH Zurich, 8093 Zurich, Switzerland; manoli@ifu.baug.ethz.ch
[3] Department of Agronomy, Food, Natural Resources, Animals and Environment, University of Padova, 35020 Legnaro PD, Italy; martello84@gmail.com (M.M.); ilaria.piccoli@unipd.it (I.P.)
[4] Harvesting, Silicon Valley, CA 94027, USA; dr.yann.chemin@gmail.com
[5] INTA EEA Manfredi, 5988 Córdoba, Argentina; pons.diego@inta.gob.ar
[6] Department of Civil, Environmental and Architectural Engineering, University of Padova, 35131 Padova, Italy; teatini@dmsa.unipd.it
* Correspondence: francesco.morari@unipd.it; Tel.: +39-049-827-2857

Received: 8 July 2018; Accepted: 7 September 2018; Published: 11 September 2018

Abstract: The surface energy balance algorithm for land (SEBAL) has been successfully applied to estimate evapotranspiration (ET) and yield at different spatial scales. However, ET and yield patterns have never been investigated under highly heterogeneous conditions. We applied SEBAL in a salt-affected and water-stressed maize field located at the margin of the Venice Lagoon, Italy, using Landsat images. SEBAL results were compared with estimates of evapotranspiration by the Food and Agriculture Organization (FAO) method (ET_c) and three-dimensional soil-plant simulations. The biomass production routine in SEBAL was then tested using spatially distributed crop yield measurements and the outcomes of a soil-plant numerical model. The results show good agreement between SEBAL evapotranspiration and ET_c. Instantaneous ET simulated by SEBAL is also consistent with the soil-plant model results (R^2 = 0.7047 for 2011 and R^2 = 0.6689 for 2012). Conversely, yield predictions (6.4 t/ha in 2011 and 3.47 t/ha in 2012) are in good agreement with observations (8.64 t/ha and 3.86 t/ha, respectively) only in 2012 and the comparison with soil-plant simulations (8.69 t/ha and 5.49 t/ha) is poor. In general, SEBAL underestimates land productivity in contrast to the soil-plant model that overestimates yield in dry years. SEBAL provides accurate predictions under stress conditions due to the fact that it does not require knowledge of the soil/root characteristics.

Keywords: surface energy balance algorithm for land (SEBAL); evapotranspiration; yield; remote sensing; heterogeneous conditions

---

1. Introduction

The growing world population needs more food, possibly with less water available for agriculture [1], making the wise management of water resources one of the great challenges of our times. This problematic situation can improve only if water is managed more effectively leading to increased crop yield per unit of water consumed (i.e., improving water use efficiency). In arid and semiarid regions cropland irrigation is the major consumer of water and efficient and reliable methods for determining water consumption by crops are crucial for sustainable management [2–5]. Evapotranspiration (ET) is the largest sink of irrigation water and, thanks to well-established crop-specific relations between ET and yield [6,7], it provides a measure of both water demand

and land productivity. Yield is thus the ultimate indicator to describe crop response to water resource management [8] and the quantification of field scale *ET* is fundamental for managers to maximize land productivity while minimizing water losses [9–11]. Crop yield is also a key element for rural development and national food security. For these reasons, forecasting crop yield a few months before harvest can be of paramount importance for timely initiation of the food trade, securing national demand, and organizing food transport within countries [6,7,12].

The yield of many agricultural crops is generally predicted from the amount of water used by the crop, i.e., *ET* [6,7]. Traditionally, *ET* from fields has been estimated according to the Food and Agriculture Organization (FAO) method [13], i.e., by multiplying a weather-based reference ET_0 by a crop coefficient (K_c) determined according to crop type and growth stage. However, the suitability of the idealized K_c coefficient to describe the actual vegetative and growing conditions, especially in water limited areas, was questioned by many authors [14]. In addition, it is difficult to predict the correct growth stage dates for large populations of crops and fields [15].

A viable alternative for mapping evaporation at field and regional scales is the use of satellite images that can provide an excellent tool to detect the spatial and temporal structure of *ET* [16]. Remote sensing (RS) is a reliable and cost-effective method to forecast crop *ET* and yield over large areas [17,18] and the integrated use of remote-sensing data and crop modeling for yield prediction has been applied for many years [18]. Applications can be found in the literature for different crop types and regions using various data assimilation schemes [19–21]. However, these applications are based on medium resolution satellite data (MODIS, MERIS) and valid for regional assessments only [18]. Common RS models are the surface energy balance algorithm for land (SEBAL; [22]), mapping evapotranspiration at high resolution with internalized calibration (METRIC; [15]), remote sensing of evapotranspiration (ReSET; [23]), analytical land atmosphere radiometer model (ALARM; [24]) and surface aerodynamic temperature ([25]). Most of these models use the land surface energy balance equation:

$$R_n = LE + G + H \tag{1}$$

where R_n is net radiation, LE is the latent heat flux, G is the soil heat flux and H is the sensible heat flux. When using satellite imagery, the sensed surface radiances are converted into surface properties such as albedo, vegetation indices, surface emissivity and surface temperature. These products are then used to estimate the various components of Equation (1) [26].

SEBAL is capable of estimating *ET* (from the latent heat flux) without prior knowledge of the soil, crop and management conditions [27], and it has been used to estimate the surface energy fluxes at different spatial and temporal resolutions in more than 30 countries [27–30] *ET* estimated by SEBAL has been utilized to quantify spatial variation of soil moisture [31], biomass production and crop yield [6] using high- and low-resolution satellite images within a field, across fields, and at regional scale [32]. The energy balance model in SEBAL uses a near-surface temperature gradient, *dT*, which eliminates the need for absolute surface temperature calibration, a "major stumbling block in operational satellite *ET*" [15]. Typical accuracies of the estimated *ET* by SEBAL are 85%, 95%, and 96% at daily, seasonal and annual timescales, respectively [27,30]. The advantages of using SEBAL have already been highlighted by [30]: (1) it uses minimal ground-based data; (2) calculation of the near-surface air temperature is not mandatory; and (3) a self-calibration process is automated in each region of interest. Despite a recent version of SEBAL model (SEBAL2008) having incorporated major refinements, such as correction of the advection effect or improved estimates of surface albedo and soil heat flux [33], the ability of SEBAL to describe *ET* and yield patterns at field scale has never been tested under highly heterogeneous conditions. The purpose of this study is to apply SEBAL for calculating *ET* in a salt-affected and water-stressed maize field located at the margin of the Venice Lagoon, Italy. SEBAL results are compared with *ET* estimates by the FAO method and the outcome of a three-dimensional soil-plant model. The biomass production routine in SEBAL is then tested using spatially distributed crop yield measurements and the soil-plant model results.

2. Materials and Methods

2.1. Study Area

The study site (Figure 1) is a ca. 21 ha field located in Ca' Bianca (45°10′57″N, 12°13′55″E), near the town of Chioggia (Italy) at the southern margin of the Venice Lagoon. The area is in proximity to the Brenta and Bacchiglione Rivers and approximately 5 km from the Adriatic Sea. With an elevation ranging between 1 and 3.3 m below mean sea level, the site has a silt-clay soil with peat and sandy drifts (i.e., paleochannels) [34,35] (Figure 2). The area is known to be affected by saltwater contamination down to about 20 m depth with the presence of a first confined fresh-water aquifer 45–50 m below mean sea level. The climate is continental with annual rainfall around 780 mm. Rainfall is more intense in spring and autumn and snow is not very frequent during winter. The average daily temperature is 13.4 °C. Rainfed maize (*Zea mais* L.) was cultivated in 2011 (seeding 4 April and harvest 2 September), and 2012 (seeding 21 March and harvest 11 September). Precipitation differed substantially over the two growing seasons and was also quite unusual with respect to the average April-to-September rainfall from 1993 to 2012, which amounts to 360 mm. Indeed, the two growing seasons were rather dry (i.e., in the 1st quartile, 199.8 mm in 2011, 150.6 mm in 2012). Rainfall in 2011 was evenly spread throughout the season. Contrarily, the low 2012 precipitation occurred almost exclusively during the maize vegetative phase. No precipitation occurred during the early tassel and kernel blister reproductive stages, which are known to be among the most critical stages of maize growth [36]. The daily average reference evapotranspiration (ET_0) was 4.42 and 4.08 mm d^{-1}, whereas the ET_0 over the entire season was 672 and 717 mm in 2011 and 2012, respectively. The study site was divided into five site-specific management units (SSMUs, Figure 1b) by [35] according to the spatial variability of soil characteristics and salinity (Figure 2) and has been extensively studied by soil sampling, hydro-geophysical monitoring and soil-plant modeling [34–38]. More precisely, the classification identified a peaty, acidic, moderately saline and sandy zone (SSMU 1); a very saline zone (SSMU 2); a non-saline zone comprising the coarser portions of the paleochannels (SSMU 3); a zone with the best conditions for maize growth (SSMU 4) with mid to low salinity, mid to low peat content and the highest clay content; and a peaty, acidic, moderately saline and silty zone (SSMU 5) (Figure 3).

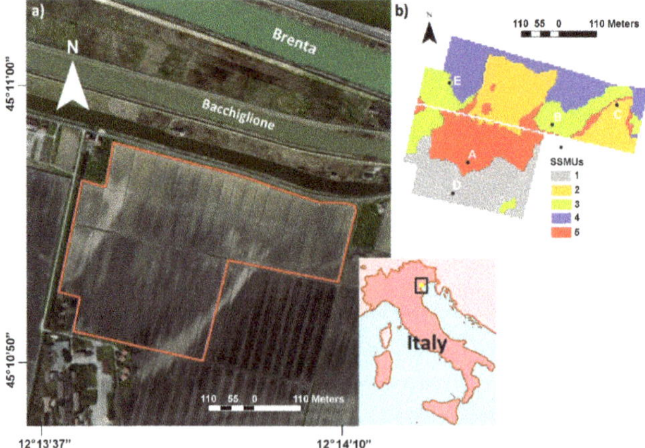

Figure 1. The study area: (**a**) aerial image of the study area at the southern edge of the Venice Lagoon, Italy; (**b**) delineation of site-specific management units (SSMUs) based on [35] and monitoring stations (A, B, C, D, E).

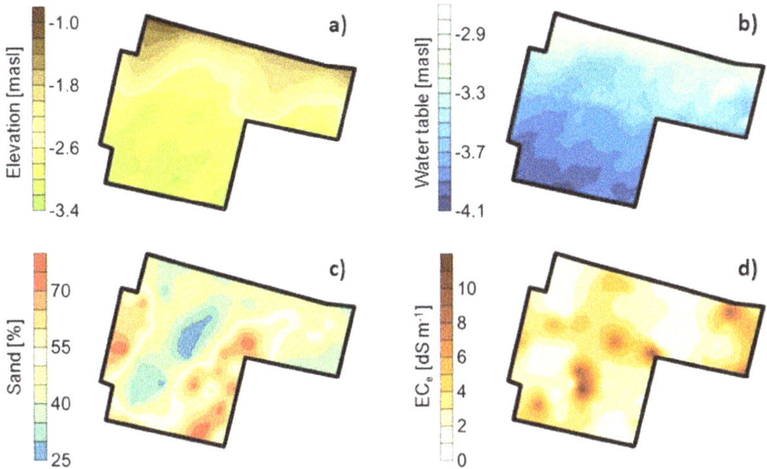

Figure 2. Site characterization: (**a**) land and (**b**) water table elevation [m above sea level, masl], average (**c**) soil texture (as sand percentage [%]), and (**d**) soil salinity (in terms of EC_e [dS m^{-1}]) in the 0–1.2 m soil profile.

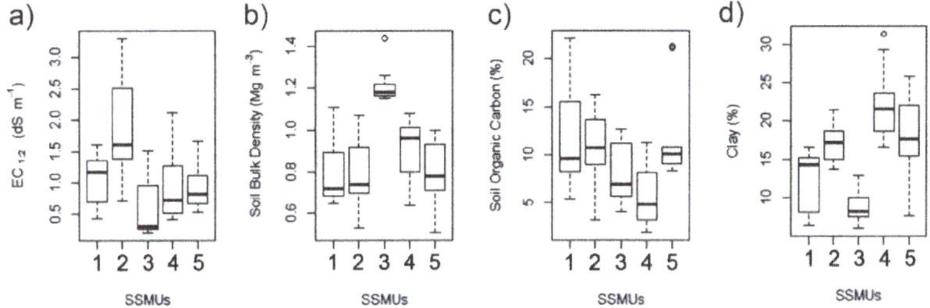

Figure 3. Boxplots for (**a**) EC1:2, electrical conductivity of a soil extract with a soil to water ratio of 1:2, (**b**) soil bulk density, (**c**) soil organic carbon content, and (**d**) clay content.

2.2. Data

2.2.1. Acquisition of Satellite Imagery

Satellite images of the study area were obtained from Landsat 7 ETM+ (Path = 192, Row = 029). These images are made publicly available by the U.S. Geological Survey (at http://glovis.usgs.gov/ USGS) as GeoTIFF with a level correction 1T (terrain corrected), providing a systematic radiometric and geometric accuracy through the use of point ground control (GCPs). For 2011, images were available for the dates 16 and 25 April, 2 and 18 May, 21 July, 6 and 31 August, while for 2012 the images were obtained on 18 and 27 April, 4 May, 5 June, 7 and 16 July, 1, 8, 17 August and 9 September. All dates had favorable clear-sky weather conditions. Visible bands (bands 1–5, 7) were used for albedo (α), and vegetation index calculations. Albedo was calculated by integrating surface reflectivity values, using different weighting coefficients for each band [2]. A "top of atmosphere" bidirectional reflectance was converted into at-surface reflectance, by using simple humidity and sun-angle based algorithms [39]. This simple approach for atmospheric correction could originate errors in albedo, which are anyway compensated by the internal calibration procedures of SEBAL [40]. Thermal band (band 6) was used for surface temperature (T_s), computed using a modified Plank equation following [41], and sensible heat (H). The spatial resolution is 30 × 30 m on the visible bands and 60 × 60 m on the thermal band.

Due to the presence of gaps in ETM+7 scan-line corrector (SLC)–off, only images covering 100% of the study area were considered. We decided not to fill the Landsat 7 gaps to avoid the risk of generating additional uncertainties. Seasonal *ET* was calculated from the Landsat images using the method for temporal integration implemented in the Geographic Resources Analysis Support System Geographic Information System (GRASS GIS 7) [42].

2.2.2. Hydrological Monitoring

To assess the impact of water stress on crop productivity, five monitoring stations were set up in the study site. Each station was equipped with capacitance-resistance probes (ECH2O-5TE, Decagon Devices, Pullman, WA, USA) to measure water content and pore-water salinity [38], and electronic tensiometers (T4e, UMS GmbH, Munich, Germany) to record soil-water potential at 10, 30, 50, and 70 cm depths. The sensors were connected to a data logger and recorded hourly. The water table level at each station was monitored every second week using phreatic wells. The five monitoring stations were named A, B, C, D and E (Figure 1b). Unfortunately, since the SSMUs were defined after station installation, SSMU 4 was not monitored by any of the stations, whereas two stations (B and E) were located in SSMU 3.

2.2.3. Meteorological Data and Estimation of Evapotranspiration by the Food and Agriculture Organization (FAO) Method (ET_c)

In addition to satellite images, the SEBAL model requires the following input data: wind speed, precipitation, air humidity (vapor pressure or dew point temperature), solar radiation, air temperature, ET_0, and meteorological station characteristics (height of wind measurement and vegetation height). Hourly meteorological data were acquired from a nearby automatic station (Regional Agency for Environmental Protection and Prevention of the Veneto) and used to calculate the reference evapotranspiration ET_0 with the FAO-56 method [13]. Crop evapotranspiration (ET_c), which differs distinctly from ET_0 as it accounts for crop-specific ground cover, canopy properties and aerodynamic resistance, is then estimated by means of the crop coefficient K_c, as $ET_c = K_c \times ET_0$ [13].

Where field conditions differ from the standard conditions, correction factors are required to adjust ET_c. This adjustment reflects the fact that the real crop evapotranspiration often deviates from ET_c due to non-optimal field conditions such as the presence of pests and diseases, soil salinity, low soil fertility, water shortage or waterlogging. The crop evapotranspiration under non-standard conditions ($ET_{c,adj}$) was calculated at the five monitoring stations by using a water stress coefficient K_s [13]. The estimation of K_s was done by daily water balance computation for the root zone [13] each SSMU using soil moisture data recorded at 10 to 70 cm depth. The field capacity and wilting point were calculated from the soil characteristics [35] and for each SSMU using the "Rosetta" model [43]. K_s was finally calculated on a daily basis using the FAO-56 method [13] when $Dr > TAW$:

$$K_{s=} \frac{TAW - Dr}{(1-p)TAW} \tag{2}$$

where Dr is the root zone depletion, TAW the total available soil water in the root zone, and p the fraction of TAW that a crop can extract from the root zone before reaching water stress. Crop evapotranspiration under non-optimal conditions is then calculated as:

$$ET_{c,adj} = K_s K_c ET_0 \tag{3}$$

2.2.4. Yield Data

Maize grain yield was measured with a combine harvester equipped with a yield monitor (Agrocom, Claas, Germany) and a differential global positioning system (DGPS). The harvester had an 8-m bar and took yield measurements every 5 m [36].

2.3. Surface Energy Balance Algorithm for Land (SEBAL) Method

The SEBAL model was developed using existing and newly created modules in the GRASS open-source GIS [42,44]. SEBAL utilizes spectral raster images from the visible, near infrared and thermal infrared energy spectrum to compute the energy balance on a pixel-by-pixel basis. In SEBAL, R_n is computed from satellite-measured broad-band reflectance and surface temperature, while G is estimated from R_n, surface temperature and vegetation indices. The sensible heat flux H is considered proportional to the ratio between the surface-air temperature difference (dT) and bulk aerodynamic resistance (r_{ah}) [22]. SEBAL uses the partitioning of H and LE as described in [22], where the evaporative fraction (Λ) is calculated as:

$$\Lambda = \frac{LE}{R_n - G} = \frac{R_n - G - H}{R_n - G} \tag{4}$$

The underlying assumption is that Λ is constant during the day or, in other words, the instantaneous partition of LE and H is equal to the average diurnal partitioning ratio. The difference between Λ at the moment of satellite overpass and Λ derived from the 24-h integrated energy balance is considered as non-significant [45].

The latent heat flux is then calculated multiplying Λ by the diurnal net radiation at the land surface ($R_{n,day}$) [42].

$$LE = \Lambda \times R_{n,day} \tag{5}$$

A fundamental advantage of SEBAL is the calibration of the result using pixels of extreme meteorological conditions: a very wet pixel with negligible H ($H_{wet} \sim 0$) and a very dry pixel with negligible LE ($LE_{dry} \sim 0$).

The calibration is done by manually selecting a dry and wet pixel to define the range of vertical temperature gradients (dT) above the ground surface [19]. The wet pixel was selected according to the following criteria: low temperature, low albedo and high Normalized Difference Vegetation Index (NDVI). The dry pixel was searched for in a dry and bare agricultural field, in poorly-vegetated areas presenting a low NDVI, high temperature value and high albedo. For each image specific wet/dry pixels were selected.

In the recent literature [42,46] a method was suggested for ET temporal integration. The fraction $ETrF_j = ET_{Sebal,j}/ET_{0,j}$ (where j refers to the satellite image acquisition date and $ET_{0,j}$ to the potential evapotranspiration for day j) is considered constant for the time period between two consecutive satellite images, and the seasonal ET (termed ET_s) is computed using the following equation:

$$ET_s = \sum_j \left(ETrF_j \sum_{i=t_j}^{t_{j+1}} ETo_i \right) \tag{6}$$

where t_j and t_{j+1} delimit a short time period around the acquisition date j.

The above equation integrates the time component (ET_0, estimated with the FAO-56 method [13]) and spatial component ($ETrF$) to describe the daily fluctuation of ET_{Sebal} across the study area during the cropping season.

2.4. Biomass Production and Maize Yield

Crop biomass production was calculated from photosynthetically active radiation (PAR) data using a GRASS GIS module [44,47]. Only a fraction of PAR is absorbed by the canopy (APAR) and used for carbon dioxide assimilation. Biomass growth is therefore a function of APAR, light use efficiency (e') and a water stress index (the evaporative fraction Λ, defined in Equation (4)). The APAR/PAR fraction (fPAR) can be directly estimated from the NDVI as:

$$f\text{PAR} = -0.161 + 1.257 \times \text{NDVI} \tag{7}$$

Aboveground biomass production on a single image acquisition day, *bio* (kg/ha/d), is then calculated as:

$$bio = e' \times \Lambda \times f\text{PAR} \times 0.84 \tag{8}$$

where the light use efficiency factor e' was set at 4 g/MJ [48]. Grain yield is obtained by multiplying accumulated seasonal aboveground biomass production with a specific Harvest index (HI) for maize, assumed constant at a value of 0.4 [36].

2.5. 3D Soil-Plant Model

ET and crop productivity simulated by a 3D soil-plant model [49] were also compared to the SEBAL results. Given the lack of spatially distributed measurements of *ET*, model simulations provide a viable tool to evaluate the spatial structure of water consumption estimated by SEBAL.

The soil-plant model simulates soil moisture dynamics, plant photosynthesis and transpiration and it was calibrated and validated at the study site [49]. The transpiration flux is modeled in terms of water potentials in the soil, root xylem and leaf, and is regulated by an optimal stomatal conductance that maximizes carbon assimilation while minimizing water losses [50]. A detailed description of the soil-plant model is given in [36] while the application at the study site is presented in [49]. Model simulations were run for 2011–2012 with the same dataset as that used here for the SEBAL simulations.

3. Results and Discussion

Figure 4 shows the seasonal cycle of daily *ET* obtained by the two methods used in this study. The comparison between evapotranspiration estimated by the FAO method (ET_c) and SEBAL (ET_{Sebal}) shows good agreement, with an $R^2 = 0.87$ for 2011 and $R^2 = 0.89$ for 2012. The cumulative evapotranspiration from 16 April to 31 August, 2011 and 18 April to 9 September, 2012, as obtained by spatially integrating the ET_{Sebal} seasonal values (Figure 5), was 515 mm and 472 mm, respectively. A comparison between Figures 2 and 5 shows that soil texture was generally one of the main factors controlling the ET_{Sebal} spatial distribution, with higher values found in fine soil areas. However, the spatial patterns of *ET* are the results of complex interactions between crops, soil texture, water table level, salinity and climate inter-annual variability. The average difference between ET_c and ET_{Sebal} over the whole season was 6.4% for 2011 and 21% for 2012. This comparison is satisfactory given that the K_c coefficient method does not account for water stress factors, meaning that ET_c represents an accurate estimation of *ET* only under well-watered conditions.

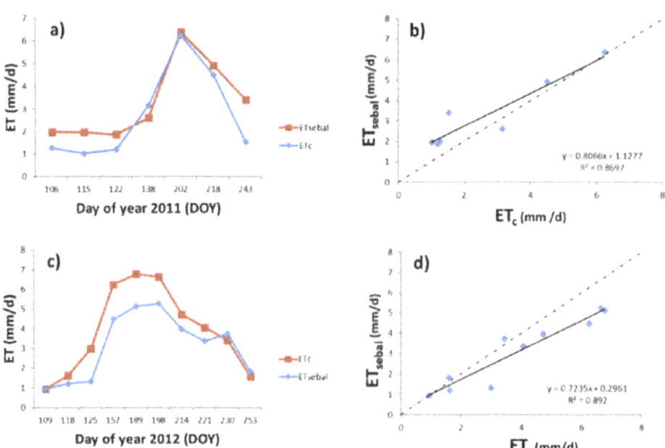

Figure 4. Temporal variation of evapotranspiration (*ET*) (mm/day) computed by surface energy balance algorithm for land (SEBAL) and Food and Agriculture Organization (FAO-56) approaches from April to September: (**a**) 2011, (**b**) 2012. ET_{Sebal} versus ET_c results: (**c**) 2011 and (**d**) 2012. Dotted lines represent the identity line (1:1).

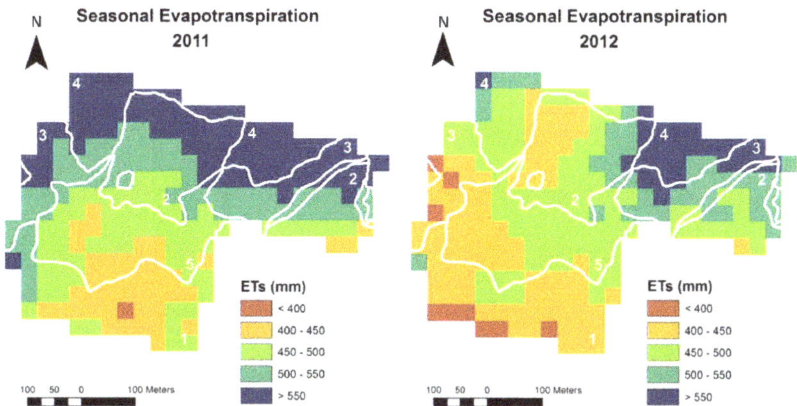

Figure 5. Map of seasonal evapotranspiration (ET_s) (mm) for the study site in 2011 (16 April–31 August) and 2012 (18 April–9 September) as obtained by Equation (6). SSMUs are overlaid (white lines and numbers).

The evapotranspiration reduction coefficient K_s was used to quantify water/salinity stresses at the five monitoring stations during the 2012 growing season. Figure 6 shows daily ET for the different images calculated by SEBAL and the FAO-56 method ($ET_{c,adj}$). The comparison shows a good agreement at stations B ($R^2 = 0.83$), D ($R^2 = 0.78$) and E ($R^2 = 0.85$). The coefficient of determination at stations A and C were smaller (0.33 and 0.56, respectively) due to a shift in time of the ET fluxes. In station E, despite a high R^2, SEBAL underestimated the ET_c peak by about 30%.

Plants were affected by soil salinity stress at Stations A, C, and D. Because Station C had a very shallow water table, the daily K_s was influenced only by salt stress, while both salinity and water stress affected crop development at Stations A and D. Stations B and E were not affected by salt stress. However, severe water stress was experienced at these stations because they were both located on paleochannels in SSMU3 [36]. Consistent with other authors [51,52], SEBAL underpredicted ET with respect to the FAO dual coefficient method, even if the difference was within 20% [53]. Note that FAO-56 has been demonstrated to overestimate the actual ET values. Many authors (e.g., [54,55]) in the last 15 years have shown that the FAO-56 K_c and K_{cb} tabulated coefficients, even if adjusted using the specific procedure based on local meteorological, irrigation and crop data suggested by FAO-56, tend to overestimate the observed crop coefficients and actual ET in humid and semi-humid regions. Differences of up to ±40% especially during the middle growth cycle are reported in the literature, mainly due to the complexity of the crop coefficient that actually integrates several physical and biological factors [54,55]. Furthermore, other authors [56] questioned FAO-56 accuracy in estimating soil salinity effect on ET in the different growth stages.

Spatial ET dynamics calculated by SEBAL was confirmed by the 3D soil-plant model. Because of the different spatial resolution of SEBAL (30 m by 30 m) and the soil-plant model (20 m by 20 m), ordinary kriging was used to resample the 3D model results on a common 30 m by 30 m grid and compare evapotranspiration simulated by the 3D model (ET_m) with ET_{Sebal}. Figure 7 shows instantaneous ET (mm/s) calculated with both models. In general, good agreement is observed between the two spatial patterns, with the lowest values in the southern sandy zone and the highest in the northern fine-texture zone. A quantitative comparison of average ET at the times of satellite overpass shows good agreement between the two methodologies with $R^2 = 0.70$ in 2011 and $R^2 = 0.67$ in 2012 (Figure 8). The cumulative ET_m values computed by the model were in good agreement with the values provided by SEBAL.

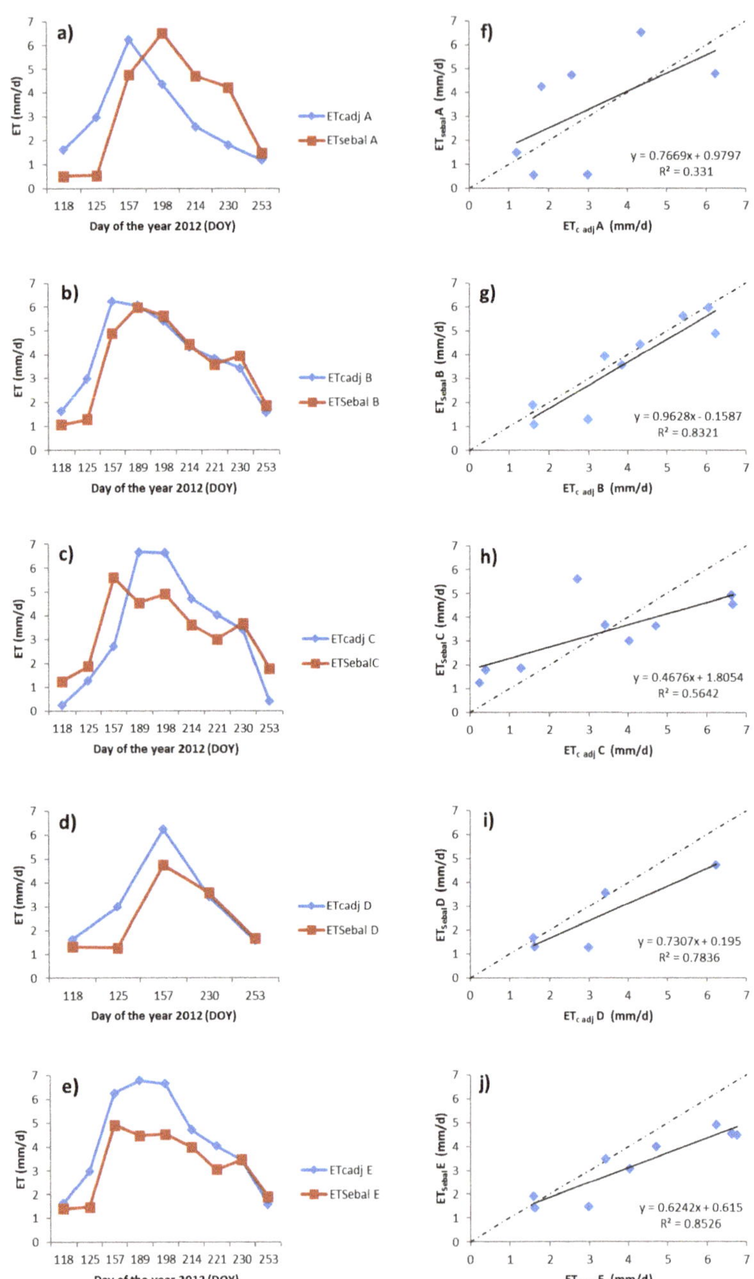

Figure 6. Temporal variation of ET_{Sebal} and $ET_{c,adj}$ (mm/day) at the five monitoring stations: (**a**) Station A, (**b**) Station B, (**c**) Station C, (**d**) Station D, (**e**) Station E. ET_{Sebal} versus $ET_{c,adj}$: (**f**) Station A, (**g**) Station B, (**h**) Station C, (**i**) Station D, (**j**) Station E.

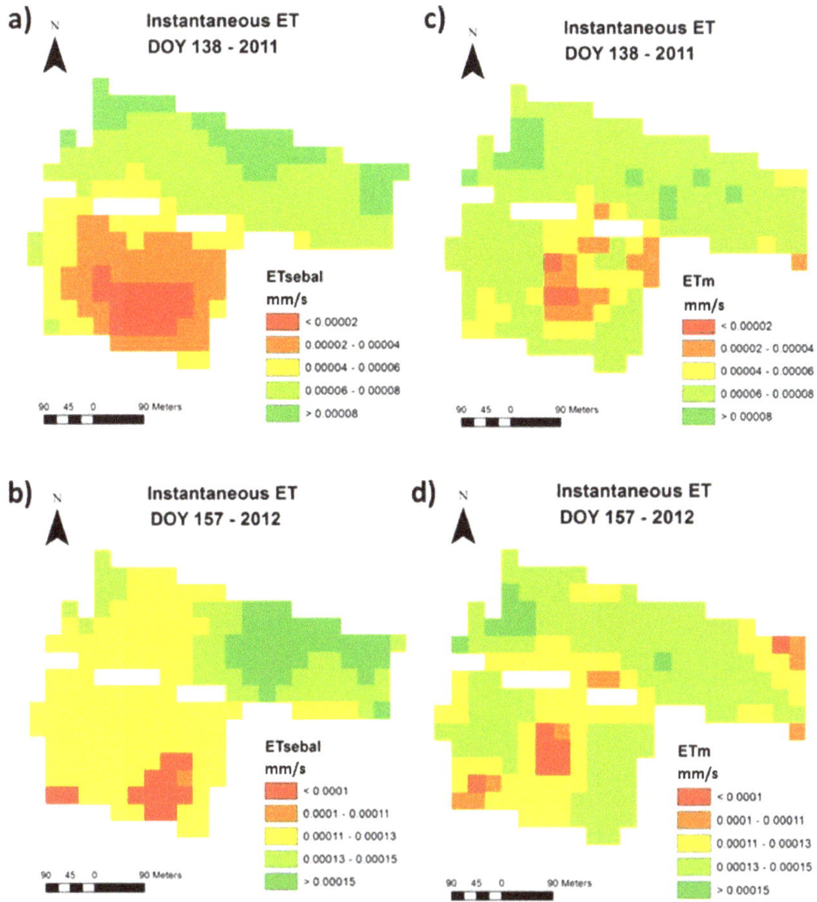

Figure 7. Detailed map of instantaneous ET (mm/s) distribution at the study site as predicted by SEBAL and the soil-plant model: (**a**) ET_{Sebal} Day of Year (DOY) 138 2011; (**b**) ET_{Sebal} DOY 157 2012; (**c**) ET_m DOY 138 2011; (**d**) ET_m DOY 157 2012.

A further model validation was performed based on the results of biomass production and maize yield. A comparison between the measured values and those simulated by the SEBAL biomass routine and the 3D soil-plant model is provided in Figure 9a for the two years. In general, the SEBAL method slightly underestimated land productivity, in contrast to the soil-plant model that overestimated yield, especially under stress conditions (2012). Measured yield data in 2011 had higher average (8.64 t/ha) and maximum (13.67 t/ha) values than in 2012 (average: 3.86 t/ha; maximum: 7.09 t/ha). SEBAL results exhibit a similar behavior, i.e., higher average (6.4 t/ha) and maximum (8.2 t/ha) values in 2011 than in the following dry year (average: 3.47 t/ha; maximum: 5.5 t/ha). Consistently, the 3D soil-plant model provided 8.69 t/ha and 11.11 t/ha for the average and maximum values in 2011, respectively, and 5.49 t/ha and 6.67 t/ha in 2012. Both methodologies are, therefore, sensitive to the observed inter-annual climate variability, but comparisons are not always satisfactory.

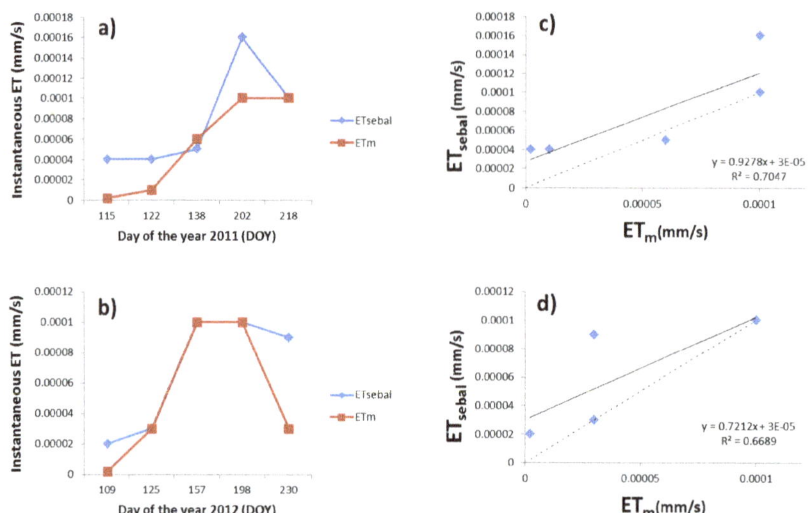

Figure 8. Temporal variation of instantaneous *ET* (mm/s) calculated with SEBAL and the 3D soil-plant model: (**a**) 2011, (**b**) 2012. Instantaneous *ET* from SEBAL versus that from the 3D soil-plant model: (**c**) 2011, (**d**) 2012.

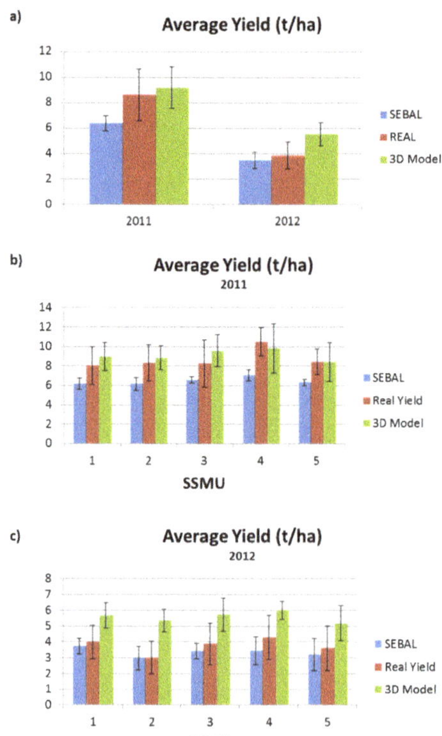

Figure 9. (**a**) Measured and computed average yield in 2011 and 2012. Measured and computed average yield in the SSMUs in (**b**) 2011 and (**c**) 2012.

The yield data in 2011 and 2012 were then classified at the scale of the management units (Figure 9b,c, respectively). The results suggest that climate variability has a stronger impact on productivity than soil characteristics, differences between years being higher than those among SSMs. This is usually expected, even in areas characterized by contrasting soils [57].

The SEBAL model responded better in 2012 (average absolute error: 0.39 t/ha; average error: 10%) than in 2011 (average absolute error: 2.24 t/ha, average error: 26%), a general performance that can be evaluated as good according to [58]. It should be recalled that rainfall was spread evenly throughout the season in 2011. On the contrary, no precipitation occurred in 2012 during the early tassel and kernel blister reproductive stages, which are known to be among the most critical maize growth stages [37].

Maize yield showed high variability across the study site in both 2011 and 2012. The most productive area in both years was SSMU 4, because of the mid to low salinity and the highest clay content (Figure 10). Salinity affected maize yield in SSMU 2, which was the least productive area, while yield was intermediate in the other SSMUs.

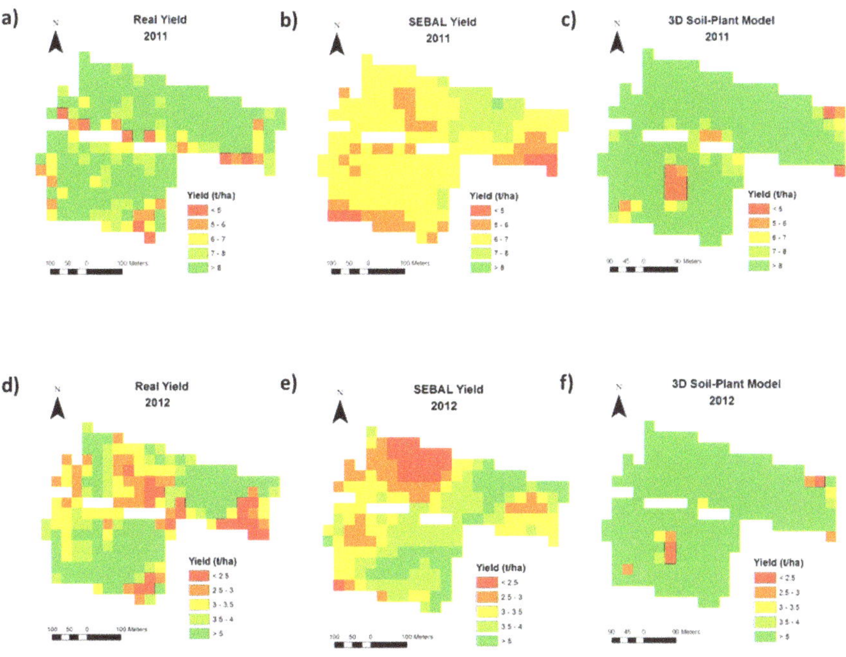

Figure 10. Maps of maize yield (t/ha) in the study area: (**a**–**c**): 2011 measured yield, SEBAL yield, and 3D soil-plant model yield, respectively; (**d**–**f**): 2012 measured yield, SEBAL yield, and 3D soil-plant model yield, respectively.

As noted by [49], model results (both SEBAL and the 3D model) provide a good estimate of field productivity at large scales (i.e., field or SSMU) but are less effective at capturing the small-scale heterogeneities in field crop yield (Figure 11). This can be explained by a resolution problem, because both models operate on a ~10 m grid size, thus neglecting the sub-grid variability of input data (e.g., soil characteristics). However, while the 3D model is outperformed during dry conditions, due to a coarse representation of soil/root characteristics that are not required by SEBAL, this latter underestimates productivity during the wet year, probably due to the uncertainties in the utilized harvest index. In particular, even if HI may vary temporally and spatially (i.e., across the field), a single value was used here for the whole field and for both 2011 and 2012. According to the results shown in

Figure 9 and published in [36], SSMU 4 was the most productive area over the 2 investigated years. This management unit may thus require a larger HI compared to the other SSMUs since it is the zone with the best conditions for maize growth. It should be noted that the HI is an important factor to estimate yield and a better knowledge on the relation between biomass production and HI will improve the accuracy of yield maps [32]. In particular, future research should clarify the impact of soil moisture conditions during flowering and crop nutrition on the harvest index [59].

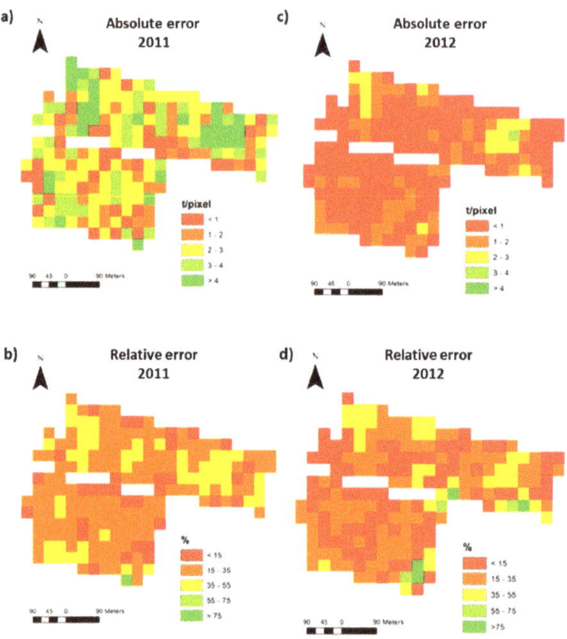

Figure 11. Yield maps: absolute and relative errors in 2011 (**a,b**) and 2012 (**c,d**) using the SEBAL model.

4. Conclusions

In this study we used the SEBAL model in a salt-affected, water-stressed maize field in the surroundings of the Venice Lagoon, Italy, to map the spatial structure of water fluxes and crop yield. Due to the lack of measured *ET* data, SEBAL modeled data were not validated but only compared with *ET* estimates by the FAO method and three-dimensional soil-plant simulations. Nevertheless, results confirm that SEBAL is a viable tool for calculating *ET* at field scale even under highly heterogeneous conditions. The comparison between daily *ET* estimated by the FAO method and SEBAL shows good agreement in 2011 and 2012. However, in terms of crop production SEBAL responds better in 2012 (dry year) than in 2011 (wet year). SEBAL is, therefore, outperformed by a 3D soil-plant model in the case of wet conditions, but it provides far more accurate predictions in dry periods due to the fact that it does not require knowledge of the soil/root characteristics (which are crucial for soil-plant simulations, especially during drought). *ET* fluxes calculated at five monitoring stations installed in the area also reveal that SEBAL provides better predictions under severe water stress rather than soil salinity stress conditions.

These results suggest that the integration of SEBAL with field observations and soil-plant simulations can be very beneficial for precision agriculture practices (e.g., precision irrigation), particularly in environments where water availability is scarce or the quality of irrigation water is poor. The method presented here can be used to design and optimize variable rate irrigation strategies, thus maximizing land productivity while minimizing water losses and management costs.

However, further research is needed to improve crop yield predictions at high spatial resolution in highly heterogeneous settings.

Author Contributions: F.M. and P.T. conceived and designed the experiments; C.G., G.M., M.M. performed the experiments; C.G., G.M., F.M., Y.H.C., analyzed the data; all the authors wrote the paper.

Funding: This research was supported by grants from National Commission for Space Activities (CONAE, Argentina), Córdoba National University (UNC, Argentina) and Italian Space Agency (ASI, Italy). G. Manoli acknowledges funding from the National Science Foundation (NSF-EAR-1344703).

Acknowledgments: We thank the Department of Agronomy Food Natural Resources Animals and Environment (DAFNAE), Italy and the Gulich Institute, Argentina. The authors would like to thank Elia Scudiero for conducting the field work.

Conflicts of Interest: The authors declare no conflict of interest. The funding sponsors had no role in the design of the study, in the collection, analyses, or interpretation of data, in the writing of the manuscript, nor in the decision to publish the results.

References

1. Serageldin, I. Looking ahead: Water, life and the environment in the twenty-first century. *Int. J. Water Resour. Dev.* **1999**, *15*, 17–28. [CrossRef]
2. Allen, R.; Tasumi, M.; Trezza, R. SEBAL (Surface Energy Balance Algorithms for Land)—Idaho Implementation—Advanced Training and Users Manual. Available online: http://www.posmet.ufv.br/wp-content/uploads/2016/09/MET-479-Waters-et-al-SEBAL.pdf (accessed on 11 September 2018).
3. Yang, J.; Mei, X.; Huo, Z.; Yan, C.; Ju, H.; Zhao, F.; Liu, Q. Water consumption in summer maize and winter wheat cropping system based on SEBAL model in Huang-Huai-Hai Plain, China. *J. Integr. Agric.* **2015**, *14*, 2065–2076. [CrossRef]
4. Perry, C. Accounting for water use: Terminology and implications for saving water and increasing production. *Agric. Water Manag.* **2011**, *98*, 1840–1846. [CrossRef]
5. Burke, M.; Lobell, D.B. Satellite-based assessment of yield variation and its determinants in smallholder African systems. *Proc. Natl. Acad. Sci. USA* **2017**, *114*, 2189–2194. [CrossRef] [PubMed]
6. Bastiaanssen, W.G.M.; Ali, S. A new crop yield forecasting model based on satellite measurements applied across the Indus Basin, Pakistan. *Agric. Ecosyst. Environ.* **2003**, *94*, 321–340. [CrossRef]
7. Abtew, W.; Melesse, A. Crop Yield Estimation Using Remote Sensing and Surface Energy Flux Model. In *Evaporation and Evapotranspiration: Measurements and Estimations*, 1st ed.; Abtew, W., Melesse, A., Eds.; Springer: Dordrecht, The Netherlands, 2013; pp. 161–175.
8. Molden, D.; Sakthivadivel, R. Water accounting to assess use and productivity of water. *Int. J. Water Resour. Dev.* **1999**, *15*, 55–71. [CrossRef]
9. Burnett, B. A Procedure for Estimating Total Evapotranspiration Using Satellite-Based Vegetation Indices with Separate Estimates from Bare Soil. Master's Thesis, University of Idaho, Moscow, ID, USA, 2007.
10. Rwasoka, D.T.; Gumindoga, W.; Gwenzi, J. Estimation of actual evapotranspiration using the Surface Energy Balance System (SEBS) algorithm in the Upper Manyame catchment in Zimbabwe. *Phys. Chem. Earth* **2011**, *36*, 736–746. [CrossRef]
11. Bansouleh, B.; Karimi, A.; Hesadi, H. Evaluation of SEBAL and SEBS algorithms in the estimation of maize evapotranspiration. *Int. J. Plant Soil Sci.* **2015**, *6*, 350–358. [CrossRef]
12. Li, Y.; Zhou, Q.; Zhou, J.; Zhang, G.; Chen, C.; Wang, J. Assimilating remote sensing information into a coupled hydrology-crop growth model to estimate regional maize yield in arid regions. *Ecol. Model.* **2014**, *291*, 15–27. [CrossRef]
13. Allen, R.G.; Pereira, L.S.; Raes, D.; Smith, M. *Crop Evapotranspiration. Guidelines for Computing Crop Water Requirements*; FAO Irrigation and Drainage Paper No. 56; FAO: Rome, Italy, 1998; p. 300.
14. Morton, C.G.; Huntington, J.L.; Pohll, G.M.; Allen, R.G.; Mcgwire, K.C.; Bassett, S.D. Assessing calibration uncertainty and automation for estimating evapotranspiration from agricultural areas using METRIC. *J. Am. Water Resour. Assoc.* **2013**, *49*, 549–562. [CrossRef]
15. Allen, R.G.; Tasumi, M.; Trezza, R. Satellite-based energy balance for mapping evapotranspiration with Internalized Calibration (METRIC)—Model. *J. Irrig. Drain. Eng.* **2007**, *133*, 380–394. [CrossRef]

16. Anderson, M.; Gao, F.; Knipper, K.; Hain, C.; Dulaney, W.; Baldocchi, D.; Eichelmann, E.; Hemes, K.; Yang, Y.; Medellin-Azuara, J.; et al. Field-scale assessment of land and water use change over the California Delta using remote sensing. *Remote Sens.* **2018**, *10*, 889. [CrossRef]

17. Hamada, Y.; Ssegane, H.; Negri, M.C. Mapping intra-field yield variation using high resolution satellite imagery to integrate bioenergy and environmental stewardship in an agricultural watershed. *Remote Sens.* **2015**, *7*, 9753–9768. [CrossRef]

18. Hank, T.B.; Bach, V.; Mauser, W. Using a remote sensing-supported hydro-agroecological model for field-scale simulation of heterogeneous crop growth and yield: Application for wheat in Central Europe. *Remote Sens.* **2015**, *7*, 3934–3965. [CrossRef]

19. Vazifedoust, M.; van Dam, J.C.; Bastiaanssen, W.G.M.; Feddes, R.A. Assimilation of satellite data into agrohydrological models to improve crop yield forecasts. *Int. J. Remote Sens.* **2009**, *30*, 2523–2545. [CrossRef]

20. De Wit, A.; Duveiller, G.; Defourny, P. Estimating regional winter wheat yield with WOFOST through the assimilation of green area index retrieved from MODIS observations. *Agric. For. Meteorol.* **2012**, *164*, 39–52. [CrossRef]

21. Dente, L.; Satalino, G.; Mattia, F.; Rinaldi, M. Assimilation of leaf area index derived from ASAR and MERIS data into CERES-Wheat model to map wheat yield. *Remote Sens. Environ.* **2008**, *112*, 1395–1407. [CrossRef]

22. Bastiaanssen, W.G.M.; Menenti, M.; Feddes, R.A.; Holtslag, A.A.M. A remote sensing surface energy balance algorithm for land (SEBAL): 1. Formulation. *J. Hydrol.* **1998**, *212–213*, 198–212. [CrossRef]

23. Elhaddad, A.; Garcia, L. Surface Energy Balance-Based Model for Estimating Evapotranspiration Taking into Account Spatial Variability in Weather. *J. Irrig. Drain. Eng.* **2008**, *134*, 681–689. [CrossRef]

24. Suleiman, A.A.; Bali, K.M.; Kleissl, J. Comparison of ALARM and SEBAL evapotranspiration of irrigated alfalfa. In Proceedings of the 2009 ASABE Annual International Meeting, Grand Sierra Resort and Casino, Reno, NV, USA, 21–24 June 2009; pp. 2318–2325. [CrossRef]

25. Chávez, J.; Neale, C.M.U.; Hipps, L.E.; Prueger, J.H.; Kustas, W.P. Comparing aircraft-based remotely sensed energy balance fluxes with eddy covariance tower data using heat flux source area functions. *J. Hydrometeorol.* **2005**, *6*, 923–940. [CrossRef]

26. Gowda, P.H.; Howell, T.A.; Paul, G.; Colaizzi, P.D.; Marek, T.H. *Sebal for Estimating Hourly ET Fluxes over Irrigated and Dryland Cotton during BEAREX08*; World Environmental and Water Resources Congress: Reston, VA, USA, 2011; pp. 2787–2795.

27. Bastiaanssen, W.G.M.; Noordman, E.M.; Davids, G.; Thoreson, B.P.; Allen, R.G. SEBAL model with remotely sensed data to improve water-resources management under actual field conditions. *J. Irrig. Drain. Eng.* **2005**, *131*, 85–93. [CrossRef]

28. Singh, R.K.; Irmak, A.; Irmak, S.; Martin, D.L. Application of SEBAL model for mapping evapotranspiration and estimating surface energy fluxes in South-Central Nebraska. *J. Irrig. Drain. Eng.* **2008**, *134*, 273–285. [CrossRef]

29. De Teixeira, A.H.C.; Bastiaanssen, W.G.M.; Ahmad, M.D.; Bos, M.G. Reviewing SEBAL input parameters for assessing evapotranspiration and water productivity for the Low-Middle Sao Francisco River basin, Brazil. Part A: Calibration and validation. *Agric. For. Meteorol.* **2009**, *149*, 462–476. [CrossRef]

30. Tang, R.; Li, Z.L.; Chen, K.; Jia, Y.; Li, C.; Sun, X. Spatial-scale effect on the SEBAL model for evapotranspiration estimation using remote sensing data. *Agric. For. Meteorol.* **2013**, *174–175*, 28–42. [CrossRef]

31. Scott, C.A.; Bastiaanssen, W.G.M.; Ahmad, M.D. Mapping root zone soil moisture using remotely sensed optical imagery. *J. Irrig. Drain. Eng.* **2003**, *129*, 326–335. [CrossRef]

32. Zwart, S.J.; Bastiaanssen, W.G.M. SEBAL for detecting spatial variation of water productivity and scope for improvement in eight irrigated wheat systems. *Agric. Water Manag.* **2007**, *89*, 287–296. [CrossRef]

33. Bastiaanssen, W.; Thoreson, B.; Byron, C.; Davids, G. Discussion of "Application of SEBAL Model for Mapping Evapotranspiration and Estimating Surface Energy Fluxes in South-Central Nebraska" by Ramesh K. Singh, Ayse Irmak, Suat Irmak, and Derrel L. Martin. *J. Irrig. Drain. Eng.* **2010**, *136*, 282–283. [CrossRef]

34. Scudiero, E.; Deiana, R.; Teatini, P.; Cassiani, G.; Morari, F. Constrained optimization of spatial sampling in salt contaminated coastal farmland using EMI and continuous simulated annealing. *Procedia Environ. Sci.* **2011**, *7*, 234–239. [CrossRef]

35. Scudiero, E.; Teatini, P.; Corwin, D.L.; Deiana, R.; Berti, A.; Morari, F. Delineation of site-specific management units in a saline region at the Venice Lagoon margin, Italy, using soil reflectance and apparent electrical conductivity. *Comput. Electron. Agric.* **2013**, *99*, 54–64. [CrossRef]

36. Scudiero, E.; Teatini, P.; Corwin, D.L.; Dal Ferro, N.; Simonetti, G.; Morari, F. Spatio-temporal response of maize yield to edaphic and meteorological conditions in a saline farmland. *Agron. J.* **2014**, *106*, 2163–2174. [CrossRef]

37. Manoli, G.; Bonetti, S.; Domec, J.C.; Putti, M.; Katul, G.; Marani, M. Tree root systems competing for soil moisture in a 3D soil-plant model. *Adv. Water Resour.* **2014**, *66*, 32–42. [CrossRef]

38. Scudiero, E.; Berti, A.; Teatini, P.; Morari, F. Simultaneous monitoring of soil water content and salinity with a low-cost capacitance-resistance probe. *Sensors* **2012**, *12*, 17588–17607. [CrossRef] [PubMed]

39. Tasumi, M.; Allen, R.G.; Trezza, R. At-surface albedo from Landsat and MODIS satellites for use in energy balance studies of evapotrans-piration. *J. Hydrol. Eng.* **2008**, *13*, 51–63. [CrossRef]

40. Tasumi, M.; Trezza, R.; Allen, R.G.; Wright, J.L. US Validation tests on the SEBAL model for evapotranspiration via satellite. In Proceedings of the 54th IEC Meeting of the International Commission on Irrigation and Drainage (ICID), Workshop Remote Sensing of ET for Large Regions, Montpellier, France, 14–19 September 2003.

41. Markham, B.L.; Barker, J.L. *Landsat MSS and TM Post-Calibration Dynamic Ranges, Exoatmospheric Reflectances and At-Satellite Temperatures*; EOSAT Landsat Technical Notes 1:3-8; Earth Observation Satellite Company: Lanham, MD, USA, 1986.

42. Alexandridis, T.K.; Cherif, I.; Chemin, Y.; Silleos, G.N.; Stavrinos, E.; Zalidis, G.C. Integrated methodology for estimating water use in mediterranean agricultural areas. *Remote Sens.* **2009**, *1*, 445–465. [CrossRef]

43. Schaap, M.G.; Leij, F.J.; Van Genuchten, M.T. Rosetta: A computer program for estimating soil hydraulic parameters with hierarchical pedotransfer functions. *J. Hydrol.* **2001**, *251*, 163–176. [CrossRef]

44. GRASS Development Team. GRASS GIS v7 Download. Available online: http://grass.itc.it/download/index.php (accessed on 9 July 2009).

45. Brutsaert, W.; Sugita, M. Application of self-preservation in the diurnal evolution of the surface energy budget to determine daily evaporation. *J. Geophys. Res.* **1992**, *97*, 377–382. [CrossRef]

46. Chemin, Y.; Alexandridis, T. Improving spatial resolution of et seasonal for irrigated rice in Zhanghe, China. In Proceedings of the 22nd Asian Conference on Remote Sensing, Singapore, 5–9 November 2011.

47. Chemin, Y.; Platonov, A.; Abdullaev, I.; Ul-Hassan, M. Supplementing farm-level water productivity assessment by remote sensing in transition economies. *Water Int.* **2005**, *30*, 513–521. [CrossRef]

48. Giardini, L.; Berti, A.; Morari, F. Simulation of two cropping systems with EPIC and CropSyst models. *Ital. J. Agron.* **1998**, *1–2*, 29–38.

49. Manoli, G.; Bonetti, S.; Scudiero, E.; Morari, F.; Putti, M.; Teatini, P. Modeling soil—Plant dynamics: Assessing simulation accuracy by comparison with spatially distributed crop yield measurements. *Vadose Zone J.* **2015**, *14*. [CrossRef]

50. Katul, G.; Manzoni, S.; Palmroth, S.; Oren, R. A stomatal optimization theory to describe the effects of atmospheric CO2 on leaf photosynthesis and transpiration. *Ann Bot.* **2010**, *105*, 431–442. [CrossRef] [PubMed]

51. Ramos, J.G.; Cratchley, C.R.; Kay, J.A.; Casterad, M.A.; Martínez-Cob, A.; Domínguez, R. Evaluation of satellite evapotranspiration estimates using ground-meteorological data available for the Flumen District into the Ebro Valley of N.E. Spain. *Agric. Water Manag.* **2009**, *96*, 638–652. [CrossRef]

52. Kite, G.W.; Droogers, P. Comparing evapotranspiration estimates from satellites, hydrological models and field data. *J. Hydrol.* **2000**, *229*, 3–18. [CrossRef]

53. Allen, R.G. Using the FAO-56 dual crop coefficient method over an irrigated region as part of an evapotranspiration intercomparison study. *J. Hydrol.* **2000**, *229*, 27–41. [CrossRef]

54. Katerji, N.; Rana, G. Modelling evapotranspiration of six irrigated crops under Mediterranean climate conditions. *Agric. Meteorol.* **2006**, *138*, 142–155. [CrossRef]

55. Facchi, A.; Gharsallah, O.; Gandolfi, C. Evapotranspiration models for a maize agro-ecosystem in irrigated and rainfed conditions. *J. Agric. Eng.* **2013**, *44*. [CrossRef]

56. Abedinpour, M.; Sarangi, A.; Rajput, T.B.S.; Singh, M.; Pathak, H.; Ahmad, T. Performance evaluation of AquaCrop model for maize crop in a semi-arid environment. *Agric. Water Manag.* **2012**, *110*, 55–66. [CrossRef]

57. McBratney, A.B.; Minasny, B.; Whelan, B.M. Obtaining 'useful' high-resolution soil data from proximally sensed electrical conductivity/resistivity (PSEC/R) surveys. *Precis. Agric.* **2005**, *5*, 503–510.
58. Cannavo, P.; Recous, S.; Parnaudeau, V.; Reau, R. Modeling N dynamics to assess environmental impacts of cropped soils. *Adv. Agron.* **2008**, *97*, 131–174. [CrossRef]
59. Bastiaanssen, W.G.M.; Steduto, P. The water productivity score (WPS) at global and regional level: Methodology and first results from remote sensing measurements of wheat, rice and maize. *Sci. Total Environ.* **2017**, *575*, 595–611. [CrossRef] [PubMed]

Article

Assessment of Multi-Source Evapotranspiration Products over China Using Eddy Covariance Observations

Shijie Li [1], Guojie Wang [1,*], Shanlei Sun [2], Haishan Chen [2], Peng Bai [3], Shujia Zhou [2], Yong Huang [4], Jie Wang [2] and Peng Deng [2]

[1] School of Geographical Sciences, Nanjing University of Information Science and Technology (NUIST), Nanjing 210044, China; lishijie@nuist.edu.cn
[2] Collaborative Innovation Center on Forecast and Evaluation of Meteorological Disasters/Key Laboratory of Meteorological Disaster, Ministry of Education/International Joint Research Laboratory on Climate and Environment Change, NUIST, Nanjing 210044, China; ppsunsanlei@126.com (S.S.); haishan@nuist.edu.cn (H.C.); zsjzmvivian@163.com (S.Z.); wangjie0775@163.com (J.W.); dp@nuist.edu.cn (P.D.)
[3] Key Laboratory of Water Cycle & Related Land Surface Process, Institute of Geographic Sciences and Natural Resources Research, Chinese Academy of Sciences, Beijing 100101, China; baip@igsnrr.ac.cn
[4] The Anhui Province Meteorological Science Research Institute, Anhui Meteorological Bureau, Hefei 230031, China; hy121_2000@126.com
* Correspondence: gwang@nuist.edu.cn; Tel.: +86-025-5873-1418

Received: 8 August 2018; Accepted: 23 October 2018; Published: 26 October 2018

Abstract: As an essential variable in linking water, carbon, and energy cycles, evapotranspiration (ET) is difficult to measure. Remote sensing, reanalysis, and land surface model-based ET products offer comprehensive alternatives at different spatio-temporal intervals, but their performance varies. In this study, we selected four popular ET global products: The Global Land Evaporation Amsterdam Model version 3.0a (GLEAM3.0a), the Modern Era Retrospective-Analysis for Research and Applications-Land (MERRA-Land) project, the Global Land Data Assimilation System version 2.0 with the Noah model (GLDAS2.0-Noah) and the EartH2Observe ensemble (EartH2Observe-En). Then, we comprehensively evaluated the performance of these products over China using a stratification method, six validation criteria, and high-quality eddy covariance (EC) measurements at 12 sites. The aim of this research was to provide important quantitative information to improve and apply the ET models and to inform choices about the appropriate ET product for specific applications. Results showed that, within one stratification, the performance of each ET product based on a certain criterion differed among classifications of this stratification. Furthermore, the optimal ET (OET) among these products was identified by comparing the magnitudes of each criterion. Results suggested that, given a criterion (a stratification classification), the OETs varied among stratification classifications (the selected six criteria). In short, no product consistently performed best, according to the selected validation criterion. Thus, multi-source ET datasets should be employed in future studies to enhance confidence in ET-related conclusions.

Keywords: evapotranspiration; eddy covariance observations; latent heat flux; a stratification method; multi-source; China

---

1. Introduction

As an essential component of water balance, evapotranspiration (ET) can directly impact both regional and global hydrological processes. Globally, ET has changed over recent decades, owing to climate and vegetation changes, human activities, and other factors [1–3]. Additionally, ET plays a crucial role in the land–atmosphere interface, which is closely associated with various climate variables (e.g., humidity, cloud information, temperature, and precipitation) given its link with water, energy, and carbon cycles, thus further influencing the climate system [4,5]. Accurate estimation of ET is crucial to comprehensively understand the changes in regional and global hydrological cycles (including extreme events, such as floods and droughts) and climate, and to reasonably and accurately estimate ecosystem productivity and agricultural irrigation needs [6–9]. More importantly, this information is of practical significance for food security and sustainable development of the global socio-economy [10,11].

Despite its importance, direct and continuous measurements of ET are challenging [4,12,13]. With the development of theories on boundary layer meteorology and observation technology, short-term ET measurements have become available based on porometry and lysimeters [14], energy balance and micrometeorological techniques, such as the Bowen ratio [15], eddy covariance (EC) techniques [16], and scintillometry [17]. Undoubtedly, these measurements provide necessary materials for investigating ET processes and relevant mechanisms, as well as ET-related issues at specific locations and periods; however, owing to the sparse distribution of the observation sites and the shorter time span, the conclusions based on the limited ET observations may lack universality, especially for long time periods and for a large spatial span [8,18]. To that end, numerous remote sensing [19–23], reanalysis [24–27], and land surface model (LSM)-based ET products [28–30], as well as estimates from empirical up-scaling of in situ observations [31] with different spatio-temporal resolutions and spans have recently been developed. While these datasets provide an opportunity for use in long-term and large spatial ET-related studies, validations and inter-comparisons of the data are necessary. Usually, these ET products have different levels of uncertainties, which are associated with their distinct purposes and applications [5,7,32–34]. It is reported that the accuracy of remote sensing-based ET varies over space and time, with uncertainties between 15% and 30% [32,33]. Thus, to reduce the impacts of ET product uncertainties on the degree of confidence for ET-related results (e.g., hydrological cycle, land-atmosphere interaction, agriculture, and ecosystem), we should assess the suitability of the ET products.

Eddy covariance (EC) ET has been used as the typical reference data for validating various ET estimates at the site and pixel level (e.g., for a remote sensing-based product) or at grid (e.g., reanalysis- and LSM-based products) scales [4,35,36]. Yet, EC measurements are commonly flawed, particularly with respect to a lack of energy balance closure at some EC sites, relatively short periods, and sparse spatial coverage [37]. Recently, many studies have quantified the performance of various ET products across the globe [5,8,34,38–42]. For example, Michel et al. [40] used EC ET at 24 towers across the world as benchmark data to assess four remote sensing-based ET products, and stated that all of the products performed better in wet and moderately wet climate regimes than in dry regimes. Majozi et al. [8] evaluated the accuracy and precision of four ET estimates over two eco-regions of South Africa, and indicated that none of the ET products always performed better in the two biomes. Kim et al. [38] found that the Moderate Resolution Imaging Spectroradiometer (MODIS) MOD16 ET for forested land cover of Asia was more accurate than for other biomes. Ershadi et al. [34] concluded that the ET models in Europe and North America performed differently for certain biomes, and models with relatively higher accuracy varied among biomes.

The climate in China has greatly changed in recent decades, with obvious variations in precipitation, temperature, wind speed, sunshine duration or radiation, and humidity [43–51]. It is worth quantifying how and by what magnitudes the ET processes responded to the climate change in order to formulate climate change countermeasures (e.g., maintaining ecosystem health, planning agricultural irrigation, and reducing natural disasters to the socio-economy). While a number of ET products provide the necessary tools to examine this issue, the potential risks of inaccurate and even incorrect conclusions are still large, owing to a lack of validations of these products. Recently, some assessments have been conducted for various ET products, as well as the robustness of different ET algorithms across China based on limited EC observations [39,52–57]. In the work of Yang et al. [42], the validation results for the GLEAM ET showed that, relative to EC ET at eight sites, this product performed well, particularly for the grassland sites. On the basis of routine measurements at one EC site in a semi-arid environment of north China, Schneider et al. [52] analyzed the capabilities of four ET algorithms in estimating ET and suggested that the Hargreaves and Makkink methods outperformed others. Yang et al. [56] evaluated the performance of three dual-source ET products in the Heihe River Basin in Northwest China, and indicated that the MOD16 and HTEM (hybrid dual-source scheme and trapezoid framework-based ET model) ET performed the worst and best, respectively.

Undoubtedly, ET processes and variations are of theoretical significance in the development of disciplines and inter-disciplines and have practical application value for social sectors, especially for China with exacerbating climate change. Therefore, evaluations of existing and newly released ET products (e.g., the EartH2Observe ensemble) from various perspectives (e.g., performance in various biomes and climate regimes and at various elevation levels) are essential for comprehensively documenting the suitability of these available products and further improving them. Such evaluations will provide more accurate ET estimates for ET-related studies, and thus, enhance the robustness of ET-related results. For this purpose, we collected EC observations from 12 sites in China, which generally cover common biomes, climate regimes and elevation levels, and four popular or new ET global products (one remote sensing-based product, one LSM ensemble, and two reanalyzes-based ET products). A stratification method using the whole of all of the EC sites, biomes, climate regimes, and elevation levels was employed to comprehensively validate these products using EC ET as a benchmark reference. Then, the corresponding optimal ET product (OET) was identified by comparing the magnitude of each validation criterion. We will discuss the potential causes for the performance outcomes, as well as various aspects of the product uncertainties.

2. Data and Methods

2.1. Global Land Evaporation Amsterdam Model ET

A remote sensing-based product, the Global Land Evaporation Amsterdam Model (GLEAM) ET was among the products selected for this study. This model comprises a set of algorithms with inputs of various satellite observations and reanalysis forcings (Table 1), whose rationale is to maximize the recovery of information on ET contained in current satellite observations of climate and environmental elements [58]. It separately estimates three sources of ET (transpiration, soil evaporation, and interception) for bare soil, short vegetation, and vegetation with a tall canopy within each grid cell. First, potential ET (PET) was calculated based on the Priestley-Taylor formula and measurements of surface net radiation and near-surface temperature. For each fraction of bare soil, tall canopy, and short canopy, the estimated PET was then converted into actual ET by applying a multiplicative stress factor, which is a function of microwave vegetation optical depth (VOD; [59]) observations and soil moisture (SM) estimates from a multi-layer running water balance. Specifically, to minimize uncertainties from random forcing, satellite-based SM was assimilated into the soil profile. Regarding interception loss, a Gash analytical model was employed by GLEAM. In contrast, ET for water bodies and regions covered by ice and/or snow was obtained by a variant of the Priestley-Taylor equation. Three new datasets of ET with different forcings and spatio-temporal coverage were produced by GLEAM version 3.0 (v3.0). The GLEAM v3.0a (GLEAM3.0a) ET product was chosen because of the valuable potential of this data in climate change studies, given that the datasets have the longest temporal and the largest spatial spans of 1980–2014. This daily datasets have a spatial resolution of $0.25° \times 0.25°$ and are based on satellite-observed SM, VOD, and snow-water equivalent (SWE), reanalysis air temperature (T) and radiation, and a multi-source precipitation product.

Table 1. Overview of ET products, including their PET schemes, along with the number of soil layers, precipitation and radiation datasets and other forcings.

ET Products	PET Schemes	Major Forcing Datasets			References
		Precipitation	Radiation	Others	
GLEAM3.0a	Priestley-Taylor	MSWEP	ERA-Interim	ESA GLOBSNOW and NSIDC SWE, CCI-LPRM VOD, CCI SM and LIS/OTD LF	*Martens* et al. [58]
MERRA-Land	Penman-Monteith	CPC-U	MERRA version 1.0 outputs	T, W, Q and SP	*Reichle* et al. [60]
GLDAS2.0-Noah	Penman-Monteith	PUMFD	PUMFD	T, W, Q and SP	*Rodell* et al. [28]
EartH2Observe-En	Variable	WFDEI	WFDEI	T, W, Q and SP	*Schellekens* et al. [30]

Note: MSWEP: Multi-Source Weighted-Ensemble Precipitation; CPC-U: Climate Prediction Center Unified; ESA: European Space Agency; NSIDC: National Snow and Ice Data Center; CCI-LPRM: Climate Change Initiative-Land Parameter Retrieval Model; LIS/OTD LF: Lightning Imaging Sensor/Optical Transient Detector lighting frequency; W: wind speed; Q: relative or specific humidity; SP: surface pressure. Among the EartH2Oberve models, the PET schemes are different, including Penman-Monteith, Bulk ETP, Hamon (tier 1), modified Penman, Priestley-Taylor and net radiation-based algorithms. A detailed description about these models can be found in Dutra et al. [61] and their respective model papers (Table S1).

2.2. Modern Era Retrospective-Analysis for Research and Applications-Land ET

The Modern Era Retrospective-Analysis for Research and Applications (MERRA)-Land ET is a reanalysis-based product. MERRA is an addition to the suite of global, long-term reanalysis products generated by the National Aeronautics and Space Administration (NASA) Global Modeling, and Assimilation Office (GMAO) with the Goddard Earth Observing System (GEOS-5; [62]). This system combines the NASA Atmospheric General Circulation Model (AGCM) with a set of state-of-the-art physics packages and the National Centers for Environmental Prediction (NCEP) Gridpoint Statistical Interpolation (GSI) assimilation package, and incorporates information from ground and satellite-based observations of the atmosphere, including many modern satellite derivations (e.g., Atmospheric Infrared Sounder (AIRS) radiances and scatterometer-based wind retrievals). In particular, MERRA focuses on historical analyses of the hydrological cycle on a broad range of weather and climate time scales, and thus introduces the innovative GEOS-5 Catchment LSM [63], which can explicitly address the subgrid-scale SM variability and its impact on runoff and ET. Unlike common LSMs, this model is run at the basic computational unit of the topographically determined hydrological catchment or watershed. For the original MERRA, the precipitation is simulated from the system's AGCM following the assimilation of the atmospheric observations; however, significant errors exist in the amounts and timing of the model-generated precipitation and negatively influence the land surface hydrological variables [26]. To overcome this issue, offline, land-only reanalysis data (i.e., MERRA-Land) were produced based on merging gauge-based data from the NOAA Climate Prediction Center with MERRA precipitation and revised parameters in the original canopy precipitation interception model. This supplemental land surface data of the original MERRA, as noted by Reichle et al. [60], stated that the capability of MERRA-Land in the land hydrology estimates has been significantly improved. The monthly MERRA-Land ET, with a horizontal grid of 0.67° longitude × 0.5° latitude, is used here and covers the period from 1980–2016.

2.3. Global Land Data Assimilation System ET

The Global Land Data Assimilation System (GLDAS) is based on the North American Land Data Assimilation System (NLDAS), and is a global, high-resolution, offline (uncoupled to the atmosphere) terrestrial modeling system together with data assimilation techniques for producing fields of land surface states and fluxes (e.g., ET, SM, and latent, sensible, and ground heat flux) in near-real time. Importantly, for more optimal land surface products from different LSMs (i.e., Mosaic, Noah, the Community Land Model, and the Variable Infiltration Capacity model), the satellite and ground-based observations are used as constraints in both model forcing (to avoid biases in atmospheric model-based forcing) and parameterization (to curb unrealistic model states; [28]). To date, two versions of the GLDAS product (i.e., GLDAS1.0 and GLDAS2.0) have been released. Recently, increasing evidence has reported that GLDAS1.0 products have serious discontinuity issues owing to their forcing data (e.g., with large precipitation and temperature errors in 1996 and 2000–2005, respectively) [64]. Therefore, we use the monthly ET from the GLDAS2.0 coupled with the Noah LSM (GLDAS2.0-NOAH), which has a spatial resolution of $0.25° \times 0.25°$. This product is simulated using the Princeton University meteorological forcing dataset (PUMFD), which has been bias corrected via observation-based products for the period 1948–2010 [65].

2.4. EartH2Observe ET

Aiming to develop a global water resources reanalysis for multi-scale water resource assessments and research projects, the EartH2Observe project uses state-of-the-art meteorological reanalysis and five global hydrological models (GHMs), a simple water balance model, and four LSMs with extended hydrological schemes. These models run offline and are driven by the same reanalysis-based forcing (i.e., WATCH (Water and Global Change FP7 project) Forcing Dataset ERA-Interim (WFDEI)) [66]. This dataset is based on the European Centre for Medium-Range Weather Forecasts (ECMWF) ERA-Interim reanalysis and has been adjusted with the Climatic Research Unit (CRU) dataset by a sequential elevation correction of surface meteorological elements plus monthly bias correction from gridded measurements. The simulations were performed from 1979–2012 in a continuous run. It should be noted that because of the different nature of the models, the spin-up procedures differed and were performed respectively to match their requirements and reach the climatic equilibrium states [30]; detailed information about these models can be found in Dutra et al. [61] and in their respective model papers. For an individual model, the daily and monthly simulations of the state of the surface water storage and fluxes are provided at a spatial resolution of $0.5° \times 0.5°$, as well as the 10-model arithmetic mean (i.e., ensemble). The monthly multi-model ensemble (named EartH2Observe-EN) ET is used in this study, which can mitigate the potential errors and uncertainties from a single model [67].

Notably, in order to make inter-comparison possible, all selected ET products were aggregated to the same spatial resolution ($0.25° \times 0.25°$) with a widely used bilinear interpolation method and temporal (monthly) resolutions. More information about these products is listed in Table 1.

2.5. Eddy Covariance ET

The observed ET (generally reflected by latent heat flux) at 12 EC sites (Table 2 and Figure 1), commonly used to monitor CO_2, water vapor, and energy exchanges between the biosphere and atmosphere, were collected to examine the performance of the four ET products. Of these sites, one, eight, and three are from National Climatological Observatory of China Meteorological Administration (NCO-CMA), FLUXNET (http://fluxnet.fluxdata.org/), and ChinaFlux (http://www.chinaflux.org/), respectively. While half-hourly observations were obtained, the time spans of the EC site observations differed, ranging from 2 (24) to 4 years (48 months). Standardized procedures [68] and the gap-filled method [69] were used for quality control of the EC measurements. To obtain consistent temporal resolutions for the four ET products, we also aggregated the EC half-hourly measurements to monthly and annual values at each site for the following analyses. These sites are distributed across different International Geosphere-Biosphere Programme (IGBP)-based biomes (i.e., mixed forest (MF), evergreen needleleaf forest (ENF), evergreen broadleaf forest (EBF), crop-land (CRO), grassland (GRA), and wetland (WET)), climate regimes (arid and wet regions), and elevation levels (>500 m, 500–1500 m, and <1500 m). Notably, the aridity index, which has been widely used to create climate divisions over the globe (e.g., Reference [70]), is employed to define climate regimes here. Arid and wet regions correspond to climatological aridity indices (CAI; climatological value of PET divided by that of precipitation) above and below 1.0, respectively. In this study, CAI is computed based on the gridded monthly PET and observational precipitation with a spatial resolution of $0.25° \times 0.25°$. PET is calculated from the Food and Agriculture Organization (FAO)-56 Penman-Monteith equation [71] with the gridded monthly meteorological observations (i.e., sunshine duration, wind speed at 2 m height, and maximum and minimum temperatures, and relative humidity). The gridded datasets are produced based on routine meteorological observations at 1211 weather sites of CMA using an inverse distance weighted interpolation method.

Table 2. Overview of EC stations selected to validate ET products.

Full (Abbreviated) Name	Lon (°N)	Lat (°E)	Altitude (m)	Time Span	IGBP Biomes	Precipitation (mm)	PET (mm)	CAI
Changbaishan (Cbs) [a]	128.10	42.40	738	2003–2005	MF	682.80	667.33	0.99
Qianyanzhou (Qyz) [a]	115.06	26.74	110.8	2003–2005	ENF	1517.2	995.29	0.65
Dinghushan (Dhs) [a]	112.54	23.17	300	2003–2005	EBF	1730	1064.2	0.63
Xishuangbanna (Xsbn) [b]	101.27	21.95	750	2003–2005	EBF	1446.9	1130.1	0.83
Yucheng (Yc) [b]	116.57	36.83	28	2003–2005	CRO	531.61	822.85	1.49
Haibei Alpine Tibet (Haa) [a]	101.18	37.37	3250	2002–2004	GRA	428.15	760.93	1.99
Haibei Shrub-land (Has) [a]	101.33	37.61	3160	2003–2005	WET	433.08	755.62	1.85
Neimenggu (Nmg) [b]	116.67	44.53	1189	2004–2005	GRA	304.82	703.01	2.39
Dangxiong (Dx) [a]	91.07	30.50	4333	2004–2005	GRA	405.52	871.01	2.56
Changling (Cl) [a]	123.51	44.59	171	2007–2010	GRA	404.66	716.59	1.76
Duolun (Dl) [a]	116.28	42.05	1350	2006–2008	GRA	389.51	730.72	1.91
Shouxian (Sx) [c]	116.79	32.44	24	2007–2010	CRO	1021.1	918.35	0.92

Note: [a], [b] and [c] denote that this site is from FLUXNET, ChinaFlux and NCO-CMA, respectively.

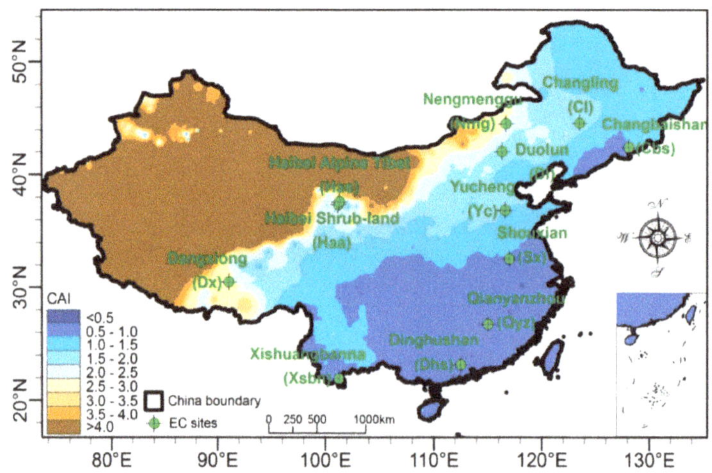

Figure 1. Locations of the twelve EC sites across China with the climatological aridity index (CAI).

To obtain the observed ET (mm/day), the daily EC latent heat flux (LE, W/m^2) from the twelve sites can be converted using the following equation [23,71,72]:

$$ET = \frac{LE}{\lambda} \tag{1}$$

where λ is the LE of vaporization with a fixed value of 2.45 MJ/kg. In fact, this parameter changes with temperature [22,73] and potentially influences the accuracy of the estimated EC ET with Equation (1). To measure the impacts of λ, comparisons of the estimated EC ET, with the constant of 2.45 MJ/kg and the variable λ (reflected by a function of temperature [73]), were conducted; detailed information is presented in Table S2. Briefly, the differences between the two estimations for each site were much smaller, implying that the impacts of the λ changes on the estimated EC ET are minimal. Thus, in this study, we do not consider the impacts of the λ changes due to temperature differences among sites. This study focuses on monthly and annual comparisons, and thus the daily EC ET estimates are integrated into monthly and annual values before conducting validations.

2.6. Validation Criteria

Several validation criteria are employed to comprehensively evaluate the performances of the four ET products. Mean Error (*ME*) provides a way to quantify the biases of the estimates relative to measurements, while Root-Mean-Square-Error (*RMSE*) can describe the accuracy of estimations. Due to spatio-temporal differences in ET magnitudes, it is difficult to directly compare ET products' performances among regions and during study periods using ME and RMSE, and therefore their relative values (i.e., *RME* and *RRMSE*) are also given. Alongside the criteria above, correlation coefficient (*R*) and Taylor Score (*TS*, between 0 and 1.0 [74]) are computed to measure the capability of capturing spatio-temporal ET variability, and the overall performance of each product, respectively. In general, the higher the *TS*, the better the ET product performs [74]. These validation metrics are expressed as:

$$ME = \frac{1}{n}\sum_{i=1}^{n}(S_i - O_i) \tag{2}$$

$$RME = \frac{ME}{O} \tag{3}$$

$$RMSE = \sqrt{\frac{\sum_{i=1}^{n}(S_i - O_i)^2}{n}} \tag{4}$$

$$RRMSE = \frac{RMSE}{O} \tag{5}$$

$$R = \frac{\sum_{i=1}^{n}[(S_i - S)(O_i - O)]}{\sqrt{\sum_{i=1}^{n}(S_i - S)^2}\sqrt{\sum_{i=1}^{n}(O_i - O)^2}} \tag{6}$$

$$TS = \frac{4\cdot(1 + R)}{\left(\sigma + \frac{1}{\sigma}\right)^2 \cdot(1 + R_0)} \tag{7}$$

where n represents the sample number; S is the mean of each ET product averaged among n samples, while O is for the observed ET; i denotes the ith sample; R_0 (=1.0 here) is the maximum theoretical R; and σ indicates the standard deviation of a certain ET product normalized by the standard deviation of the observed ET.

Furthermore, the mechanisms of energy and water exchanges between land and atmosphere are complex, and are often accompanied with strong variability in both space and time. Considering the relationships of ET with physical characteristics of land surface [75,76], it is necessary to conduct comprehensive evaluations from various perspectives, e.g., biome, elevation level and climate regime, which will enhance our knowledge on model performances, explaining possible causes and finally improving models. Therefore, we will employ a stratification method using the whole of all of the EC sites, biome, elevation level, and climate regime to conduct analyses of the four ET products in the coming sections. For each stratification, it has different classifications, i.e., 14 (1 for all monthly and annual data, and 12 for monthly data of 12 months) for the whole of all of the EC sites, 6 (MF, ENF, EBF, CRO, GRA, and WET) for biome, 3 (<500 m, 500–1500 m and >1500 m) for elevation level and 2 (wet and dry corresponding to CAI <1.0 and >1.0, respectively) for climate regime. Then, the validation criteria are calculated for each stratification classification.

3. Results

3.1. Validation by the Whole of All of the EC Sites

Figure 2a shows the intra-annual fluctuations of ET products and EC observations averaged over all of the sites. Considering the site-averaged monthly EC ET, there exists an evident seasonality, which is characterized by higher values (>40 mm) during April–September, with a peak in July of 83.69 mm. Intuitively, all four products can effectively capture the intra-annual changes, with the maximum in July ranging from 70.88 mm (EartH2Observe-En) to 97.76 mm (MERRA-Land). In Figure 3a–d, scatter-plots of monthly EC ET against the products are shown based on all of the samples (n = 420 site months) from the twelve sites. Except for EartH2Observe-En, the Rs of the other three products are all larger than 0.80, indicating that their monthly ET estimates can effectively reproduce the spatio-temporal variability of ET when taking all of the monthly data points as a whole. The fitted linear regression equations suggest that, except for MERRA-Land, which always overestimates ET, the other products underestimate ET. However, it should be noted that each product (excluding MERRA-Land) performs differently in estimating lower and higher ET values, i.e., lower ranges are overestimated, but higher ranges are underestimated. Moreover, MERRA-Land ET is overestimated for both lower and higher values, implying that there are potential systemic problems within this product. To further quantify product performance, various validation metrics were calculated against the EC data for the 420 site months; results are presented in the top left corner of each panel of Figure 3a–d. Evidently, MEs ($RMEs$) differ among these products, ranging between -5.48 mm (-12.55%) for EartH2Observe-En and 9.93 mm (22.71%) for MERRA-Land, which are closely related to their different performances in lower and higher ETs. For example, the negative ME of EartH2Observe-En is mainly because of its underestimates in higher ET (Figure 3d), while the highest and the moderate ME (RME) for MERRA-Land and GLEAM3.0a are closely associated with systemic biases (i.e., overestimates in both lower and higher ETs) and overestimates in lower ET, respectively (Figure 3a,b). Regarding the lowest ME (RME) of GLDAS2.0-Noah, it may be attributed to the bias offset (i.e., overestimates and underestimates in lower and higher ETs, respectively; Figure 3c). Relative to ME (RME) for each product, the $RMSE$ ($RRMSE$) is much larger; this may be due to both random errors plus different signs in biases, which can introduce additional randomness by aggregating EC sites from various ecosystems. Interestingly, despite the smallest ME (RME), GLDAS2.0-Noah $RMSE$ ($RRMSE$) is the largest (40.74 mm; 93.24%). Based on TS, the worst, the moderate, and the best overall performances in estimating monthly ET were found to correspond to the MERRA-Land, GLDAS2.0-Noah and EartH2Observe-En, and GLEAM3.0a products, respectively. On an annual scale (Figure 3e–h), lower (higher) values are underestimated (overestimated) in GLEAM3.0a, MERRA-Land, and GLDAS2.0-Noah, whereas EartH2Observe-En always tends to underestimate ET. The bias and error metrics indicate that the rankings of annual performances of the ET products (Figure 3e–h) are consistent with those on the monthly scale (Figure 3a–d). The lowest absolute value of ME (RME) exists in GLDAS2.0-Noah, but GLEAM3.0a has the minimum value of $RMSE$ ($RRMSE$). In addition, MERRA-Land outperforms the other datasets in terms of annual R, which is in contrast to the largest R in GLEAM3.0a on the monthly scale. This may result from the aggregation of monthly ET into annual values. Regarding the overall performance on the annual scale, EartH2Observe-En and MERRA-Land, respectively, correspond to the maximum and the minimum TS values.

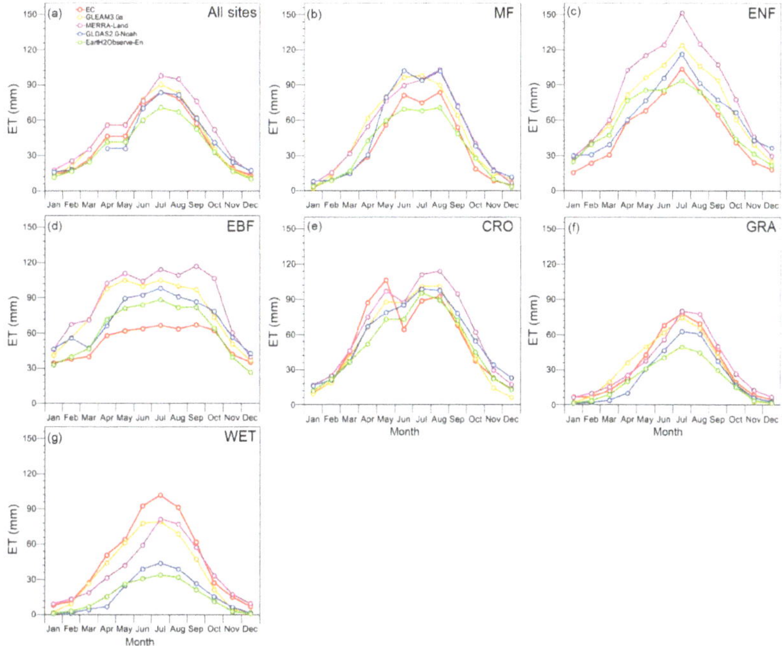

Figure 2. Intra-annual fluctuations of ET averaged across all of the sites, and MF, ENF, EBF, CRO, GRA, and WET sites.

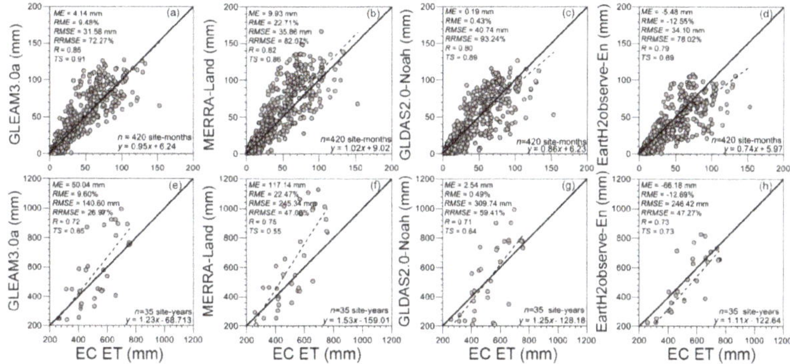

Figure 3. Scatter-plots of monthly and annual ET products against EC ET aggregated for all of the selected EC sites, accompanied by various validation criteria in the upper left corner of each panel.

As noted from the scatter-plots of ET products versus EC observation (not shown) and quantitative validation indicators at each month (Figure 4), intra-annual differences in the ET estimation performances are obvious among the four products. Within one year, *MEs* (in Figure 4a; *RMEs* in Figure 4b) for MERRA-Land are always positive, corresponding to larger biases during July–October (January–March and September–November). By contrast, EartH2Observe-En shows negative *MEs* (*RMEs*) for each month, and larger biases occur during May–August (May–August and November–January). Signs of *ME* or *RME* for the other two products vary among months, e.g., a negative *ME* or *RME* of GLEAM3.0a (GLDAS2.0-Noah) in January, November, and December (March–July) suggests underestimated ET in these months, while overestimated ET is found in the remaining months. Additionally, based on the magnitudes of *ME* (*RME*), GLEAM3.0a has larger values during March–May (February–May and December), but larger values for GLDAS2.0-Noah occur in March–May and September–December (March, April and October–January). Comparing the magnitude of monthly *ME* (*RME*) for each product, the maximum bias always occurs in MERRA-Land, excluding March–June and December; however, the ET product with the minimum bias changes among months. As shown in Figure 4c, the monthly *RMSE* for each ET product is above 10 mm, particularly in April–October, with a value larger than 20 mm. Except for January and December (June–September), MERRA-Land (EartH2Observe-En) always corresponds to the largest (lowest) *RMSE*. Due to the differences in ET magnitudes among months, intra-annual variation of *RRMSE* for each product differs from that of *RMSE*, mainly characterized by larger values in January–May and October–December (Figure 4d); the largest and the lowest *RRMSEs* in most months occur in MERRA-Land and EartH2Observe-En, respectively. Regarding the *R* for each product, it sharply declines from January and reaches the minimum (<0.25) in June, but increases rapidly from August (Figure 4e). Overall, all of the products have a higher *R* during January–April and September–December, and particularly in February and October with the largest value (>0.80). In January–July, GLDAS2.0-Noah (excluding January and May) and EartH2Observe-En (excluding March), respectively, correspond to the maximum and minimum *R*. By contrast, the smallest (largest) *R* during August–October exists in GLDAS2.0-Noah (MERRA-Land), and *R* in November and December is the largest in EartH2Observe-En, but the smallest in MERRA-Land. In Figure 4f, the monthly *TS* is above 0.50 for all of the products, particularly for January–May and September–December, which are generally higher than 0.70. In January–May and September–December, there are larger differences in *TS* among the products, and the maximum (~0.90) and the minimum (<0.80) are found in EartH2Observe-En (excluding April in GLDAS2.0-Noah) and MERRA-Land (excluding January in GLDAS2.0-Noah), respectively.

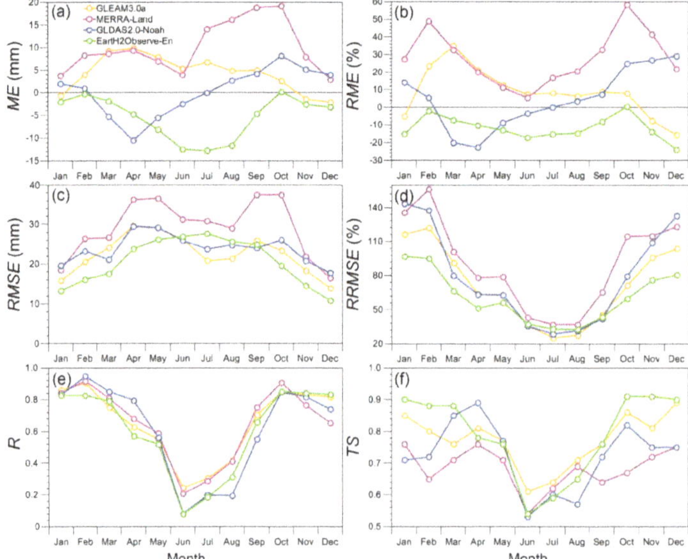

Figure 4. Validation criteria of ET products against EC ET for each month at all of the twelve sites. (**a–f**) is for Mean Error (*ME*), Relative Mean Error (*RME*), Root-Mean-Square-Error (*RMSE*), Relative Root-Mean-Square-Error (*RRMSE*), Correlation coefficient (*R*), and Taylor Score (*TS*), respectively.

3.2. Validation by Biome

The intra-annual ET variations of all biome types are illustrated in Figure 2b–g. Based on EC ET, characteristics of intra-annual fluctuations suggest apparent differences for the six biome types, i.e., two peaks for MF in June and August (Figure 2b), EBF in July and September (Figure 2d), and CRO in May and August (Figure 2e), with one for the other three biomes in July (Figure 2c,f,g). In spite of some differences in ET magnitude, intra-annual ET fluctuation can be well captured by each product for ENF (Figure 2c), CRO (Figure 2e), GRA (Figure 2f), and WET (Figure 2g); however, for MF (Figure 2b; EBF, Figure 2d), GLEAM3.0a and MERRA-Land (products excluding EartH2Observe-En) cannot reproduce the two ET peaks. With the exception of GLDAS2.0-Noah (EartH2Observe-En), which overestimates lower but underestimates higher ETs for ENF [Figure 5(b3); MF (Figure 5(a4)) and ENF (Figure 5(b4))], all of the ET products were generally overestimated for MF (Figure 5(a1–3)), ENF (Figure 5(b1–2)), and EBF (Figure 5(c1–4)). In CRO, GLEAM3.0a near-perfectly estimated ET (Figure 5(d1)), while MERRA-Land ET was generally overestimated (Figure 5(d2)); relative to the EC measurement, both estimates from the other two products were larger and smaller in lower and higher ETs, respectively (Figure 5(d3–4)). As depicted in Figure 5(e1–4) for GRA and Figure 5(f1–4) for WET, ETs were underestimated by all of the products, especially for WET ET estimates from GLDAS2.0-NOAH and EartH2Observe.

Quantitative validation results for different biome types are shown in the top left corner of each panel of Figure 5. With the exception of Earth2Observe-En with smaller negative *ME* (*RME*) in MF, the bias indicators for the other three products are close to or above 10 mm (30%; Figure 5(a1–4)). Correspondingly, Earth2Observe-En *RMSE* (*RRMSE*) is the smallest, but the remaining products present a comparable error. Based on *R* (*TS*), the ET products show no evident differences in performance, with a value of 0.97 (0.95). As for ENF (Figure 5(b1–4)) and EBF (Figure 5(c1–4)), larger differences in *ME* (*RME*) and *RMSE* (*RRMSE*) were observed among these products, respectively, corresponding to a range of 7.39–34.76 mm (14.35–65.96%) and 17.72–41.38 mm (34.40–78.52%). Moreover, the maximum for these four metrics always appeared in MERRA-Land, followed by GLEAM3.0a. Despite that, *R* (*TS*) in ENF is nearly equal and above 0.85 among these products, and this indicator in EBF is larger than 0.70, except for EartH2Observe-En (MERRA-Land). For CRO (Figure 5(d1–4)), *ME* (*RME*) for the ET estimates is different in sign and magnitude (i.e., underestimation for GLEAM3.0a and EartH2Observe-En versus overestimation for MERRA-Land and GLDAS2.0-Noah, and a larger magnitude in MERRA-Land and EartH2Observe-En versus a smaller magnitude in GLEAM3.0a and GLDAS2.0-Noah). In contrast, excluding GLEAM3.0a with a lower *RMSE* (*RRMSE*), the performances of the other three products are comparable based on these error indicators (~19 mm; ~34%). Regarding *R*, the largest value is in GLEAM3.0a, and the next largest is in MERRA-Land, but the other two products have the smallest *R*. For *TS*, each product corresponds to a value of approximately 0.90. In GRA (Figure 5(e1–4)) and WET (Figure 5(f1–4)), *MEs* (*RMEs*) for all of the products are below zero, accompanied by larger magnitudes for GRA and WET in GLDAS2.0-Noah and EartH2Observe-En. Consistently, GLDAS2.0-Noah and EartH2Observe-En, and GLEAM3.0a and MERRA-Land show larger and smaller errors for both GRA and WET, respectively. *R* for each product is above 0.83 in GRA (0.92 in WET), of which the minimum is found in GLDAS2.0-Noah. While all products perform differently with respect to the aforementioned five metrics for GRA, they have a comparable *TS* value of around 0.90. In WET, there is a larger *TS* range between 0.42 in EartH2Observe-En, and 0.96 in GLEAM3.0a.

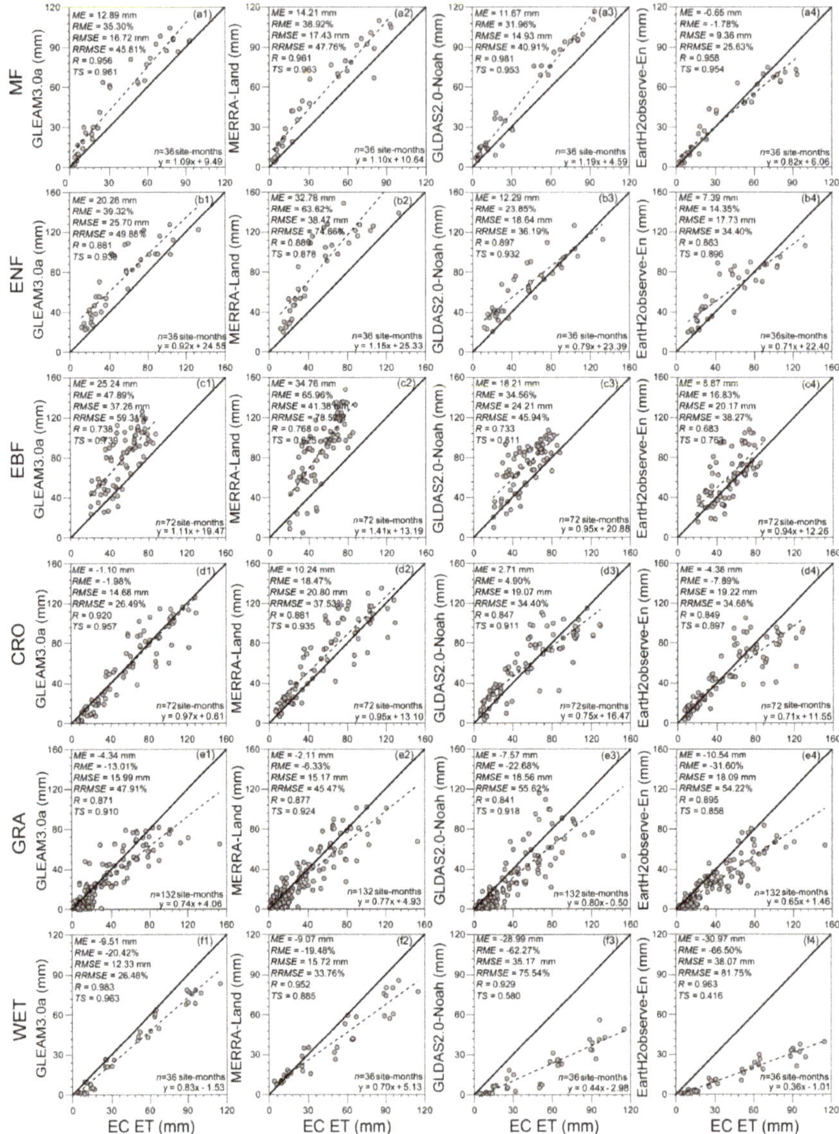

Figure 5. Scatter-plots of monthly ET products against EC ET aggregated for different biomes, accompanied by various validation criteria in the upper left corner of each panel.

3.3. Validation by Elevation Level

Validation results by elevation level (Figure 6) indicate that elevation has an influence on the performance of each ET product. For GLEAM3.0a and MERRA-Land, ET over sites below 1500 m is consistently overestimated (Figure 6(a1–2,b1–2)), while there are different overestimations and underestimations for lower and higher ETs at sites above 1500 m, respectively (Figure 6(c1–2)). For GLDAS2.0-Noah (Figure 6(a3,b3,c3)), overestimated ET is found in the two elevation levels below 500 m (except for some data points with higher ET) and between 500–1500 m; however, evident and systematic underestimations appear at elevations higher than 1500 m. Lower and higher ETs are, respectively, overestimated and underestimated by earth2Observe-En at a low elevation level (Figure 6(a4)), while the other two zones show underestimated ET (Figure 6(b4,c4)), especially for elevation levels above 1500 m.

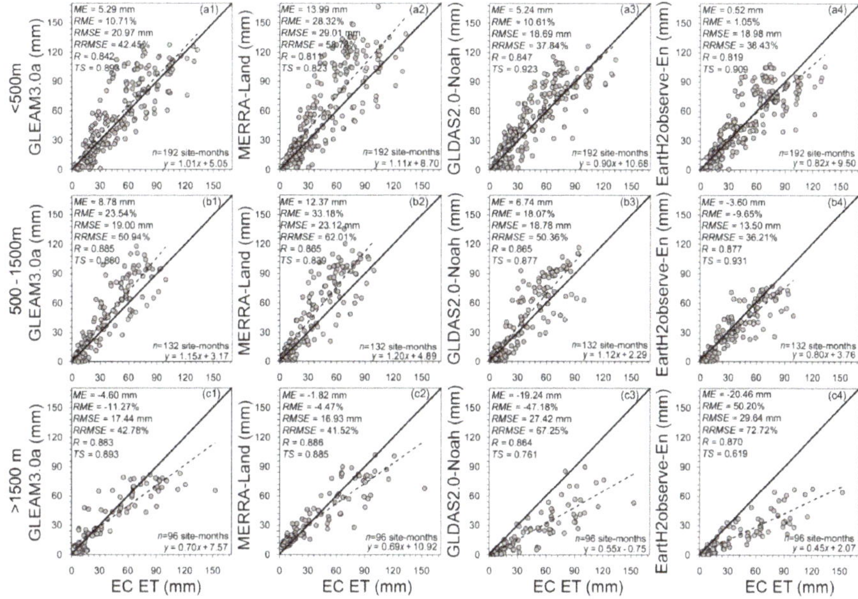

Figure 6. Scatter-plots of monthly ET products against EC ET aggregated for different elevation levels, accompanied by various validation criteria in the top left corner of each panel.

At elevation levels below 500 m (Figure 6(a1–4)), the *ME*s (*RME*s) of all of the ET datasets are positive, with a range between 0.52 mm (1.05%) in EartH2Observe-En and 13.99 mm (28.32%) in MERRA-Land. Comparing *RMSE*s (*RRMSE*s) of the four ET products, MERRA-Land corresponds to the largest value, while the other datasets have more similar values. Regarding *R* (*TS*), each product has a value above 0.80, in particular for GLEAM3.0a and GLDAS2.0-Noah (GLDAS2.0-Noah and EartH2Observe-En), which have values higher than 0.84 (0.90). Across the sites with an elevation of 500–1500 m (Figure 6(b1–4)), except for EartH2Observe-En with a slight negative *ME* (*RME*), the ET biases of the other datasets are positive and maximized in MERRA-Land. Correspondingly, the largest *RMSE* (*RRMSE*) is found in MERRA-Land, followed by the minimum value in EartH2Observe-En. Regarding *R* (*TS*), ET products with values near 0.86 (higher than 0.83) perform similarly; moreover, the maximum *TS* (0.93) occurs in EartH2Observe-En. Unlike the performance based on *ME* (*RME*) at the sites below 1500 m, all of the products have a negative bias for high elevation levels (Figure 6(c1–4)); in addition, both GLDAS2.0-Noah and EartH2Observe-En show larger magnitudes, corresponding to the larger *RMSE* (Figure 6(b3–4)). In spite of some differences in the bias and error metrics, these datasets have an approximate *R* of 0.87. Comparing *TS* values, the maximum values (~0.90) are in GLEAM3.0a and MERRA-Land, followed by the moderate (0.76) and the minimum (0.62) values in GLDAS2.0-Noah and EartH2Observe-En, respectively.

3.4. Validation by Climate Regime

The performance of each ET product varies in different climate regimes, i.e., systematic overestimations and underestimations in the wet (except for EartH2Observe-En, with overestimations and underestimations in lower and higher ETs, respectively) and the dry climate regimes, respectively (Figure 7). In the wet climate regime (Figure 7(a1–4)), the maximum *ME* (*RME*) of 25.19 mm (48.87%) occurs in MERRA-Land, while the smallest value of 3.11 mm (6.02%) exists in EartH2Observe-En. Of all four datasets, GLEAM3.0a and MERRA-Land exhibit a larger *RMSE* (*RRMSE*), while GLDAS2.0-Noah and EartH2Observe correspond to a smaller value. The *R* values indicate comparable performance among the ET products. Except for MERRA-Land, with the minimum *TS* of 0.86, the remaining products show a comparable *TS* of higher than 0.90. For the dry climate regime (Figure 7(b1–4)), MERRA-Land and EartH2Observe-En correspond to the largest and the smallest magnitudes of *ME*s (*RME*s), respectively. For GLDAS2.0-Noah and EartH2Observe-En, *RMSE*s (*RRMSE*s) are larger and close to each other, however, the other two datasets have smaller and approximate errors. The performance in *R* (*TS*) obviously differs among these ET products (i.e., *R*s (*TS* values) for GLEAM3.0a and MERRA-Land larger than 0.85 (0.90), but those for the other products are near 0.78 (smaller than 0.90)).

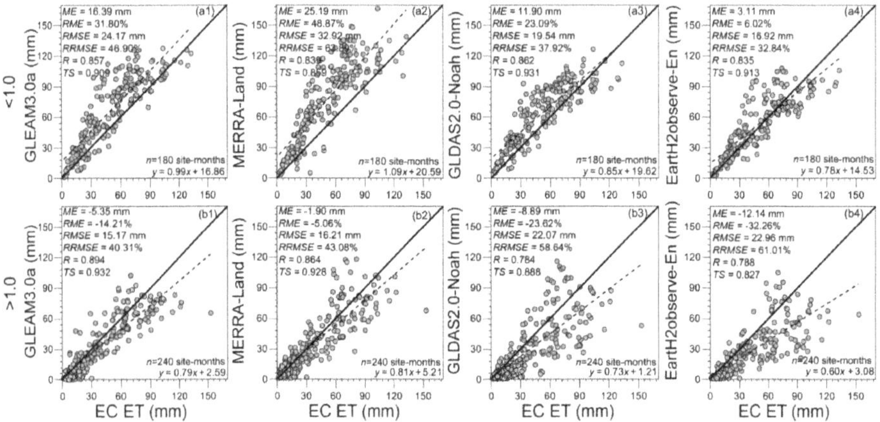

Figure 7. Scatter-plots of monthly ET products against EC ET aggregated for wet (climatological aridity index (CAI) < 1.0) and dry (CAI > 1.0) climate regimes.

3.5. Optimal ET Products

By comparing the magnitudes of each validation criterion among the four ET products, OET was identified for all 12 EC sites, biomes, elevation levels, and climate regimes (Figure 8). Taking all of the EC sites as a whole (Figure 8a), monthly OETs were GLDAS2.0-Noah (GLEAM3.0a) in view of *ME/RME* (other four criteria); however, annual OETs vary among these criteria (i.e., GLDAS2.0-Noah, GLEAM3.0a, MERRA-Land, and EartH2Observe-En based on *ME/RME*, *RMSE/RRMSE*, *R*, and *TS*, respectively). For all 12 months, the *ME/RME*-based OETs were GLEAM3.0a during January, November, and December; EartH2Observe-En in February–April and October; and GLDAS2.0-Noah from May to September, while EartH2Observe-En as the *RMSE/RRMSE*-based OET occurred in most months (January–May and October–December). In addition, most months show the *R*-based OETs of MERRA-Land (February, March, May, September, and October) and GLEAM3.0a (January and June and August), and the *TS*-based OETs of EartH2Observe-En (January–March and October–December) and GLEAM3.0a (June–September). As illustrated in Figure 8b, EartH2Observe-En were the *ME/RMSE*- and *RMSE/RRMSE*-based OETs for the forest biomes (i.e., MF, ENF, and EBF), while the *R*-based (*TS*-based) OETs were found to be GLADAS2.0-Noah (GLEAM2.0a) for MF and ENF and MERRA-Land (GLADAS2.0-Noah) for EBF. CRO and WET OETs (excluding the *ME/RME*-based OET of MERRA-Land) were found in GLEAM3.0a, based on all the validation criteria. Except for EartH2Observe-En (the *R*-based OET), GRA always had the OET for MERRA-Land for each validation criterion. Figure 8b shows that, given the six criteria, the performances of GLDAS2.0-Noah, EartH2Observe-En, and MERRA-Land were identified as OETs at low, moderate, and high elevation levels, respectively. Over the wet climate regime (Figure 8b), EartH2Observe-En (GLDAS2.0-Noah) was the OET with the smallest ME/RME and $RMSE/RRMSE$ (the highest R and TS), but the dry climate regime had the ME/RME-based OET of MERRA-Land and the $RMSE/RRMSE$-, R-, and TS-based OET of GLEAM3.0a.

The results noted above show that the performances of each ET product and the corresponding OET differ among classifications of each stratification and among criteria for a certain stratification classification. The differences may be caused by uncertainties of ET products due to simplifications, incomplete hypotheses of model structures and parameterizations, inaccurate models inputs, and uncertainties from the reference ET (i.e., EC ET). We will, therefore, discuss the potential causes of this in the next section.

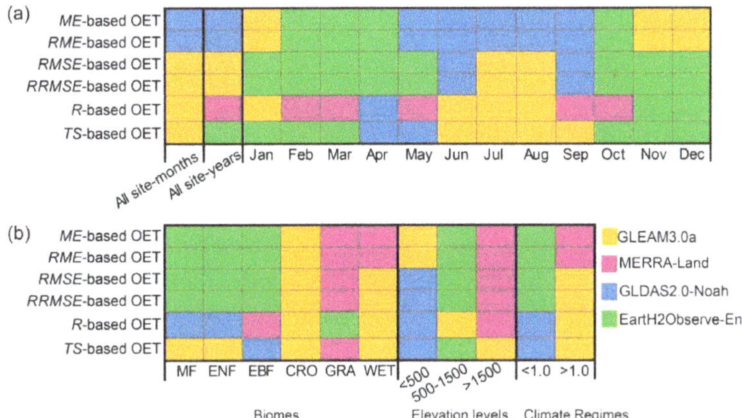

Figure 8. Validation criteria-based optimal ET products (OETs) for stratifications of (**a**) all of the 12 EC sites, and (**b**) biomes, elevation levels and climate regimes. Among the four ET datasets for one classification of each stratification, the OET of a given validation criteria can be specified as one product with the smallest (*ME, RME, RMSE,* and *RRMSE*) or the largest magnitude (*R* and *TS*) of this criteria.

4. Discussion

4.1. Sources of Uncertainties in ET Products

In the present study, we comprehensively compared and evaluated GLEAM3.0a, MERRA-Land, GLDAS2.0-Noah, and EartH2Observe-En ET products over China based on the EC measurements at twelve sites. From the perspective of all the EC sites, biome, elevation level, and climate regime, the performance of these products varies. Various hypotheses and simplifications of the ET processes, which control the land-atmosphere flux exchanges (e.g., water and energy), have been conducted for each model. Diversities in the complexity of both model structures and parameterizations among models are closely associated with specific applications and/or purposes. Moreover, a variety of inputs are required to run ET models; however, owing to specific requirements for each model and the availability of inputs, the number, types, and/or sources of inputs differ among models. Therefore, we would like to present possible explanations of uncertainties of the ET products from the perspectives of model structures and parameterizations and inputs [4,5,77–80].

4.1.1. Model Structures and Parameterizations

As shown in Table 1, different PET schemes for estimating ET are employed among the selected models. Thus, the behaviors of the ET products are likely to be directly related to differences in these schemes, which commonly have different levels of capability for capturing PET magnitude and variability given various structural complexities and parameterizations. Regarding the Penman-Monteith scheme, which has been widely regarded as a physically-based expression [71,81], a critical assumption and simplification is that the surface is a "big leaf", and thus, r_v (aerodynamic resistance to water transfer from the surface to the atmosphere) can be separated into r_c (canopy resistance) and r_h (aerodynamic resistance to heat transfer from the surface to the atmosphere). Even so, to directly run this equation is difficult because of a lack of observed relevant parameters (vegetation–specific parameters, e.g., r_c [82]). Therefore, many diagnostic and physiological equations were proposed based on environmental and biological controls (e.g., vapor pressure deficit, T, solar radiation incident on canopy, and SM) and then was used to estimate these parameters among different biomes [83–85]. As for the Priestley-Taylor scheme, it is a simplified variant of the Penman-Monteith equation, in which PET is linearly expressed as a so-called Priestley-Taylor parameter (i.e., α) multiplied by energy available to evaporate water [32]. Generally, the α parameter is between 1.2–1.3 under water unstressed conditions, but it can vary from 1.0 to 1.5; this value is mainly dependent on the degree of coupling between ET processes and the atmosphere, which can be reflected by W, vapor pressure deficit, and SM [4,32]. Komatsu [86] stated that to obtain this parameter, detailed information on canopy and micrometeorological conditions was required, but this knowledge could not be directly supplied, particularly for a larger spatial extent. For this reason alone, α is often set as 1.26 for some widely-used models, while its values of 1.26 in both short vegetation and bare soil fractions and 0.8 for the tall fraction are given by GLEAM3.0a [21,22]. In brief, both the Penman-Monteith and Priestley-Taylor PET schemes differ in their simplifications of some critical parameters, thus resulting in uncertainties and different performances for various ET datasets.

After employing the specified PET scheme for a model, it is vital to calculate the ET fractions from soil and interception evaporation and transpiration, which are summed to estimate ET. Generally speaking, their fractions are parameterized to be jointly controlled by various environmental factors, such as soil properties, SM, vapor pressure deficit, and vegetation parameters (e.g., Leaf Area Index, LAI, and Normalized Difference Vegetation Index, NDVI), and vary greatly among models due to differences and uncertainties of model parameterizations and a lack of observation-based constraints [4]. Taking transpiration (the largest overall contributor to terrestrial ET; [87,88]) as an example, Jasechko et al. [89] pointed out that 90% of terrestrial ET was cycled via vegetation transpiration based on isotope techniques. However, conclusions from Miralles et al. [90] stated that the ratio of transpiration to terrestrial ET from the GLEAM3.0a product was 76% for the whole landmass. This implies that fractions of transpiration have larger discrepancies among models, which possibly propagate into the ET products; the same applies for the ratio of soil or interception evaporation, despite the values being generally smaller [90]. It is particularly noteworthy that even the estimated interception precipitation from the most popular applied approach of the Gash analytical model may produce substantial errors, e.g., an annual overestimation of 39.8 mm in a subtropical evergreen forest of Central-South China [91]; thus, this causes considerable uncertainties in interception evaporation. Therefore, if inaccurate and even incorrect functions for constraining each ET component are used by the models, questionable ET may be provided [4].

As an aside, errors within the estimated ET originate from the neglect of some components of ET, such as night transpiration [92,93]. Based on the assumption that plant stomata is closed at night, and thus transpiration stops, night ET can commonly be ignored for the terrestrial ecosystem; but recent observations have provided evidence that night transpiration is of significance across a wide range of biomes and climate regimes [92–97]. For example, Novick et al. [92] reviewed previous studies and pointed out that the percentage of night transpiration accounting for the daily total was basically 10–30%; however, this varied among plant functional groups (i.e., C3 and C4; [93]) and SM conditions [94]. Despite there being no agreement on the mechanisms of night transpiration, it is generally believed that the processes are closely related to W, vapor deficit, SM, and circadian regulation of stomatal conductance [93,97]. Hence, models with no or insufficient considerations of night transpiration processes may lead to systematically underestimated ET.

4.1.2. Model Inputs

If a given ET model is ideally full-biophysical and, thus, can comprehensively describe the ET processes, errors in the ET estimates and differences among the ET products are mainly dependent on various inputs, especially for precipitation and radiation [36,54,60,75,76,79]. Studies have been extensively performed to evaluate different precipitation products (e.g., gauge-based, and reanalysis and remote sensing-related datasets) over the globe [98–102]. For example, Nair and Indu [99] noted that the MSWEP products (input for GLEAM3.0a) in India showed large errors in higher precipitation (i.e., >75th and >95th quantiles), which was confirmed by Alijanian et al. [98] in Iran. Sun et al. [102] found that the CPC-U precipitation (input for MERRA-Land) averaged over the world was underestimated for each season and correspondingly led to the annual value being the smallest compared to other datasets. Moreover, because of relatively limited gauge observations, the CPC-U dataset has overall potential to smooth the precipitation structure and miss local heavy precipitation events [103]. Based on gauge data over the Adige Basin of Italy, the PUMFD precipitation (used by GLDAS2.0-Noah) was assessed by Duan et al. [104], and the conclusions showed that the performance of this precipitation product was the worst relative to others, with biases in the occurrence frequency of daily precipitation for some intensity ranges and higher errors in winter. By comparing different daily precipitation products over Canada, Wong et al. [105] suggested that the skills of the WFDEI (used by EartH2Observe-En) dataset differed from region to region, with underestimation in the northern and eastern parts and overestimation in the west. As shown in Table 1, net radiation used by the four ET models came from different datasets, including ERA-Interim, MERRA version 1.0, PUMFD, and WFDEI for GLEAM3.0a, MERRA-Land, GLDAS2.0-Noah, and EartH2Observe-En, respectively. With respect to the radiation datasets, assessments have been conducted across the world [4,66,79, 106–112]. In Boilly and Wald [109], ERA-Interim radiation was overestimated overall to some degree in Europe, Africa, and the Atlantic Ocean, whereas clear and cloudy sky conditions, respectively, corresponded to overestimation and underestimation. Regarding MERRA version 1.0, it showed significantly overestimated net radiation at the twenty-three EC sites and aggregated over the whole of China; moreover, the net radiation was almost 2.8 times the Global Energy and Water Exchanges (GEWEX) value [79], which might be caused by the overestimation of the occurrence of clear sky conditions [107,109]. Tory and Wood [106] compared and evaluated gridded radiation products across northern Eurasia and found that there were smaller biases for the PUMFD dataset on an annual scale, but larger errors on a seasonal scale. For the WFDEI dataset, the downwelling shortwave radiation is higher in northern Africa but lower in northern South America, despite the effects of interannual changes in the atmospheric aerosol optical depths being considered [110]; thus, net radiation would be overestimated. Apart from precipitation and radiation, other meteorological forcings (e.g., T, W, Q, and SP) are also different for MERRA-Land, GLDAS2.0-Noah, and EartH2Observe-En (Table 1), integrated with different accuracies [6,26,66]. These studies indicated evident discrepancies among the existing meteorological datasets in both magnitude and variability on daily to annual scales (e.g., owing to the number and spatial coverage of surface stations, satellite algorithms, and data assimilation systems); meanwhile, their capabilities to capture meteorological conditions differed from region to region.

It is well known that descriptions of vegetation processes, definitions of land use/cover (LUC) and relevant vegetation character parameters (e.g., NDVI, LAI, and/or VOD) are needed; thus, their differences and uncertainties potentially propagate into the ET estimates [113]. There are a number of available LUC (e.g., Table S3) and NDVI/LAI/VOD products derived from different data sources (e.g., various satellite images), algorithms, and classification schemes [114,115]. It should be noted, however, that these datasets were produced for specific purposes and applications, including analyses of LUC and vegetation changes and their impacts on the climate, hydrology, and ecosystem, and the developments of various geo-scientific models; thus, obvious discrepancies and even errors in these products have been reported, especially at the regional scale [115–125]. Therefore, without considering the suitability of LUC and NDVI/LAI/VOD products, biases originating from raw data and inconsistencies among the selected products and uncertainties owing to product selection and processing can be of the same magnitude as those from the representation of the processes under investigation [113,121,126–129]. For example, Branger et al. [126] investigated the impact of different LUC datasets on the long-term water balance of the Yzeron peri-urban catchment of France and stated that most water quantities (including ET) were sensitive to LUC selections. Liu et al. [113] quantified uncertainties of simulated water fluxes using MODIS (MOD15), GLASS, and the Four-Scale Geometric Optical Model (FSGOM)-based LAI, and concluded that LAI products could lead to substantial uncertainties in the ET estimates. For these selected ET products, different LUCs and vegetation character parameters are used and cause differences in performances and uncertainties of the ET estimates.

4.2. Uncertainties of EC ET

Since the EC technique was first applied [130], it has been used extensively to directly measure terrestrial carbon, water, and energy cycles, and taken as ground truth values for evaluating various ET products [4,131]. Nevertheless, there are still uncertainties regarding EC observations. Especially problematic is energy imbalance at EC sites, mainly characterized by the energy closure ratio (i.e., the sum of observed latent and sensible heat divided by the difference of net radiation and ground heat flux), not being equal to one [132]. Based on numerous previous conclusions, energy balance non-closure can generally be attributed to the missed very low and/or high-frequency fluctuations of fluxes, measurement errors associated with sensor separation, interference from tower or instrument-mounting structures, not fully considering the storage term (e.g., canopy and photosynthesis storage), mismatch between the scales of energy balance components, large-eddy transport, or secondary circulations not captured by the EC technique [37,133]. It is reported that, in general, the sum of observed latent and sensible heat is 10–30% smaller than the difference between net radiation and ground heat flux at EC sites [32,132]; moreover, the closure error can vary seasonally and inter-annually and from biome to biome [32,133–135]. Scholars have often suggested that underestimation of latent heat has largely contributed to this energy non-closure of the EC technique [136–138]. For instance, Finkelstein and Sims [136] indicated that the normalized errors for sensible and latent heat were 10% and 25–30%, respectively.

In addition, the spatial context of the EC measurement is limited and defined within the footprint of a turbulent flux measurement [131,139]. For a deployed turbulent flux sensor, its detected signals reflect influences of the underlying surface on the turbulent exchange. Over a homogeneous surface with enough spatial extent (i.e., at least ~1 km; [111]), the measured fluxes from all parts of the surface are, by definition, equal. However, the surface is typically inhomogeneous; the EC measurement is dependent on which part of the surface exerts the strongest impact on the sensor and consequently on the location and size of its footprint [139]. To reduce the influence from the inhomogeneous surface and, thus, enhance the spatial representation, many footprint models have been developed and used to identify and parameterize the footprint of each EC site [139–141]. Despite that, the measured signals in most cases involve influences from the untargeted surfaces within the footprint, indicating that the observations at the EC site cannot perfectly reflect energy and gas fluxes from the targeted surface. Notably, the spatial extent of the footprint is not unchangeable, but can vary with W and its direction, stability, and measurement heights [4,142,143]; therefore, the fixed parameterization of the footprint can also introduce uncertainties into the EC observations. Besides the energy imbalance and limited spatial representativeness, errors of EC ET can result from missing data post-processing, which are attributed to instrument failure, poor maintenance, instances of bad weather, and data rejection [131].

4.3. Other Factors Influencing Validation Results

In addition to uncertainties from the ET models and the EC observations, impacts from other factors (e.g., spatial scale problems among ET estimates and necessary inputs for running models, and data aggregation) on the validation results should be considered. In this study, the selected four ET products have different spatial resolutions and corresponding spatial extents much larger than the footprint of the EC site. Given the larger grid, greater potential exists for spatial heterogeneity in surface characteristics (e.g., LUC, vegetation parameters, and elevation) and meteorological inputs for estimating ET. The estimated ET value by the models actually reflects the combination of influences from different landscapes rather than any single landscape. By contrast, the EC measurement corresponds to a relatively homogenous footprint (even though it is not perfectly uniform) and represents the ET from a given landscape to a great extent. As a result, not considering impacts from the spatial scale mismatch, conducting a direct comparison between the ET products and the EC measurements is likely to influence the validation results [23,76,144]. To qualitatively compare the impacts of different LUCs, we have collected most (i.e., GLEAM3.0a, MERRA-Land, GLDAS2.0-Noah, and seven models within EartH2Observe-En) of the LUC maps used by these ET products, including MODIS (i.e., MOD12Q1, MCD12Q1, and MOD44B), and the Global Land Cover Characterization (GLCC) Version 2 and GlobCover 2009 v2.3 products, which are produced at different spatial resolutions and classification systems (i.e., IGBP, Simple Biosphere 2 Model, and GlobCover legends). Corresponding LUC types at 12 EC sites are identified (Table S3). As depicted in Figure 8b, GLEAM3.0a ET outperforms other products in each validation criterion. This result may be related to the reasonable treatment on vegetation types at Yc and Sx sites in GLEAM3.0a [i.e., dominant type of low vegetation (e.g., grassland) versus IGBP CRO]. The EC sites with MF, ENF, or EBF correspond to the dominant types of tall vegetation (except for Qyz) in GLEAM3.0a (Table S3). However, based on *RME* and *RRMSE* (which can partly remove regional differences), the performance of GLEAM3.0a ET is better than MERRA-Land, with smaller differences among the forest sites. This may be associated with the GLEAM3.0a ET algorithms for tall vegetation. As another example, WET at the Has site is simply specified as low vegetation, agriculture, or C3 grassland, and GRA by the ET products (Table S3). As a result, the ETs are underestimated at this site due to large discrepancies of ET mechanisms between WET and other vegetation types (i.e., generally there are no water limits for evaporation and transpiration in WET). Because of the lack of detailed descriptions of the digital elevation model (DEM) datasets used by some ET models, we would like to discuss the impacts of mismatch in elevation levels from several popular DEM datasets (Table S4) on validation results. Obviously, there is perfect agreement on elevation levels based on grid mean elevations over all the EC sites. However, we found that within the $0.25° \times 0.25°$ grid at the Qyz and Dhs sites, there exists larger spatial variability in elevations for each DEM dataset compared to the corresponding grid mean values (the ratio between mean and spatial variability is less than 2.5); this suggests that the representativeness of the topography at these two sites is lower, and consequently influences the evaluation results at low elevation levels. Generally, we found that elevations from EC metadata have limited impacts on the results at moderate and high elevation levels; however, future studies should examine to what magnitude the higher spatial variability of the elevation at the Qyz and Dhs sites impacts validations at low elevation levels.

To make a comparison and evaluation possible, all of the ET products were aggregated to the same spatial ($0.25° \times 0.25°$) and temporal (monthly) resolutions, and the EC measurements were integrated into monthly values. The aggregations of the ET products and EC ET can impact the comparisons and often reduce the confidence in any subsequent model performance ranking [75,76,145]. Among the ET models, we can find different spatial resolutions for the driving factors, which are dependent on the specified requirements of the ET model. Several studies have examined the impacts of spatial resolutions of inputs on the estimated ET [32,146,147]. McCabe and Wood [147] calculated the ET based on the Surface Energy Balance method and necessary inputs derived from three satellite platforms with different spatial resolutions, and compared the results with the flux tower ET on the Walnut Creek watershed in Iowa. They found that despite the comparable accuracy of the regional mean

MODIS-based ET relative to the other two higher resolution estimates, the MODIS-based retrievals could not effectively reproduce the flux tower ET, mainly because the MODIS inputs were unable to discriminate the influence of land surface heterogeneity at field scales. Thus, the influences of the different spatial scales of the inputs for driving models would be reflected in the ET products.

5. Conclusions

In this study, we conduct point-scale evaluations of four ET global products [one remote sensing-based product (GLEAM3.0a), two reanalysis-based product (MERRA-Land and GLDAS2.0-Noah), and one LSM ensemble dataset (EartH2Observe-En)] at 12 EC sites across China, focusing on the bias, error, and overall performance of the datasets, as well as their capabilities in capturing the spatio-temporal variability of ET. The major results are summarized below:

- Validation using all of 12 EC sites: Generally, these products reproduce intra-annual ET fluctuations but perform differently in view of each validation criterion. GLDAS2.0-Noah (GLEAM3.0a) shows minimum monthly biases (annual errors). The highest monthly and annual *Rs* (*TS* values) occur in GLEAM3.0a and MERRA-Land (GLEAM3.0a and EartH2Observe-En), respectively. The metrics vary among all 12 months.
- Validation by biome: ETs in MF, ENF, and EBF are generally overestimated, but underestimated in GRA and WET. In CRO, MERRA-Land, and GLDAS2.0a-Noah (remaining two products) overestimate (underestimate) ET. Except for GLEAM3.0a and MERRA-Land in ENF and EBF, and GLDAS2.0-Noah and EartH2Observe-En in WET, a comparable error exists among the six biomes. Relative to EBF, the products in the remaining biomes (excluding GLDAS2.0-Noah and EartH2Observe-En in WET) show higher *Rs* and *TS* values.
- Validation by elevation level: All products underestimated and overestimated ET, respectively, for high and medium/low elevations (excluding EartH2Observe-En for moderate elevations). Each product showed comparable error, except for the *RMES* values of MERRA-Land for low and moderate elevations and errors of GLDAS2.0-Noah and EartH2Observe-En for high elevations. Compared to low elevation levels, *Rs* for medium and high elevation levels were slightly larger. Larger *TS* values were found in all elevation levels, except for GLDAS2.0-Noah and EartH2Observe-En for high elevation levels.
- Validation by climate regime: ETs in wet (dry) regions were always overestimated (underestimated). In wet regions, GLEAM3.0a and MERRA-Land (remaining two products) show larger (smaller) errors, in contrast to dry regions. Excluding GLDAS2.0-Noah and EartH2Observe-En in dry regions (MERRA-Land and EartH2Observe-En in wet and dry regions, respectively), *Rs* (*TS* values) are larger for each climate regime.
- OETs: Overall, the OETs varied among stratification classifications (the selected six criteria). In other words, no product always performed best in terms of the validation criteria.

Supplementary Materials: The following are available online at http://www.mdpi.com/2072-4292/10/11/1692/s1, Table S1: Members of Eearth2Observe-En ET product, with their PET schemes and references. Table S2: Monthly comparisons of the estimated EC ET with the constant and the variable λ at 11 sites. Table S3: Comparisons of LUC types used by the four ET products at EC sites. Table S4: Comparisons of elevation levels from several popular digital elevation model (DEM) datasets at EC sites.

Author Contributions: S.L., G.W. and S.S. conceived and designed this study. S.L., S.S. and P.B. were the main authors, whose works included data collection and analysis, interpretation of the results and manuscript preparation. G.W. and H.C. played a supervisory role. S.Z. and Y.H. contributed by processing data and providing EC observations at Shouxian site, respectively. All authors discussed the results and revised the manuscript.

Funding: This work was jointly supported by the National Key Research and Development Program of China (Grant NO. 2017YFA0603701), the Natural Science Foundation of China (Grant NOs. 41875094, 41605042, and 41501029), the Natural Science Foundation of Jiangsu Province, China (Grant NOs. BK20151525 and BK20150922), the Qinglan Project of Jiangsu Province of China and the Priority Academic Program Development (PAPD) of Jiangsu Higher Education Institutions, China.

Acknowledgments: This work used eddy covariance data acquired and shared by the FLUXNET community, including these networks: AmeriFlux, AfriFlux, AsiaFlux, CarboAfrica, CarboEuropeIP, CarboItaly, CarboMont, ChinaFlux, Fluxnet-Canada, GreenGrass, ICOS, KoFlux, LBA, NECC, OzFlux-TERN, TCOS-Siberia, and USCCC. The FLUXNET eddy covariance data processing and harmonization was carried out by the European Fluxes Database Cluster, the AmeriFlux Management Project, and the Fluxdata project of FLUXNET, with the support of the CDIAC and the ICOS Ecosystem Thematic Center, and the OzFlux, ChinaFlux and AsiaFlux offices. Following the fair use policy (see website at http://fluxnet.fluxdata.org/data/data-policy/), these EC measurements at FLUXNET (8 sites) and ChinaFlux datasets (3sites) are open and free for scientific and educational purpose. Notably, the EC measurements at Shouxian site and the meteorological variables at 1211 weather sites are not available for the public, but which can be obtained and used through cooperation with the CMA. The four ET global products are downloaded from different websites, i.e., GLEAM3.0a from https://www.gleam.eu/, MERRA-Land from https://disc.gsfc.nasa.gov/datasets?keywords=merra-land&page=1, GLDAS2.0-Noah from https://disc.gsfc.nasa.gov/datasets?keywords=GLDAS, and EartH2Observe-En from http://www.earth2observe.eu/. The land cover/use products of MODIS (e.g., MOD44B, MOD12Q1 and MCD12Q1; https://lpdaac.usgs.gov/dataset_discovery/modis/modis_products_table/), GLCC v2 (https://lta.cr.usgs.gov/glcc) and GLobCover2009 (http://due.esrin.esa.int/page_globcover.php), and the DEM datasets of GTOPO30 (https://lta.cr.usgs.gov/GTOPO30), HYDRO1k (https://lta.cr.usgs.gov/HYDRO1K), GMTED2010 (https://topotools.cr.usgs.gov/gmted_viewer/) and SRTM3 (http://srtm.csi.cgiar.org/index.asp) are available according to the specific data use and citation policies. We thank all the data developers and their managers and funding agencies, whose work and support were essential for obtaining the datasets, without which the analyses conducted in this study would not be possible. The authors appreciate the very constructive suggestions and comments from two anonymous reviewers, and our colleague of Ph.D Li Qingqing at NUIST. In addition, the source codes for conducting the evaluations are available from the authors upon request (gwang@nuist.edu.cn and ppsunsanlei@126.com).

Conflicts of Interest: The authors declare no conflict of interest.

References

1. Wild, M.; Grieser, J.; Schär, C. Combined surface solar brightening and increasing greenhouse effect support recent intensification of the global land-based hydrological cycle. *Geophys. Res. Lett.* **2008**, *35*, L17706. [CrossRef]

2. Sun, S.L.; Chen, H.S.; Ju, W.M.; Yu, M.; Hua, W.J.; Yin, Y. On the attribution of the changing hydrological cycle in Poyang Lake Basin, China. *J. Hydrol.* **2014**, *514*, 214–225. [CrossRef]

3. Miralles, D.G.; van den Berg, M.J.; Gash, J.H.; Parinussa, R.M.; De Jeu, R.A.M.; Beck, H.E.; Holmes, T.R.H.; Jiménez, C.; Verhoest, N.E.C.; Dorigo, W.A.; et al. El Niño-La Niña cycle and recent trends in continental evaporation. *Nat. Clim. Chang.* **2014**, *4*, 122–126. [CrossRef]

4. Wang, K.C.; Dickinson, R.E. A review of global terrestrial evapotranspiration: Observation, modeling, climatology, and climatic variability. *Rev. Geophys.* **2012**, *50*, RG2005. [CrossRef]

5. Sörensson, A.A.; Ruscica, R.C. Intercomparison and uncertainty assessment of nine evapotranspiration estimates over South America. *Water Resour. Res.* **2018**, *54*, 2891–2908. [CrossRef]

6. Sheffield, J.; Wood, E.F.; Roderick, M.L. Little change in global drought over the past 60 years. *Nature* **2012**, *491*, 435–438. [CrossRef] [PubMed]

7. Mueller, B.; Hirschi, M.; Jimenez, C.; Ciais, P.; Dirmeyer, P.A.; Dolman, A.J.; Fisher, J.B.; Jung, M.; Ludwig, F.; Maignan, F.; et al. Benchmark products for land evapotranspiration: LandFlux-EVAL multi-data set synthesis. *Hydrol. Earth Syst. Sci.* **2013**, *17*, 3707–3720. [CrossRef]

8. Majozi, N.P.; Mannaerts, C.M.; Ramoelo, A.; Mathieu, R.; Mudau, A.E.; Verhoef, W. An intercomparison of satellite-based daily evapotranspiration estimates under different eco-climatic regions in South Africa. *Remote Sens.* **2017**, *9*, 307. [CrossRef]

9. Sun, S.L.; Chen, H.S.; Ju, W.M.; Wang, G.J.; Sun, G.; Huang, J.; Ma, H.D.; Hua, W.J.; Yan, G.X. On the coupling between precipitation and potential evapotranspiration: Contributions to decadal drought anomalies in the Southwest China. *Clim. Dyn.* **2017**, *48*, 3779–3797. [CrossRef]

10. Kousari, M.R.; Ahani, H. An investigation on reference crop evapotranspiration trend from 1975 to 2005 in Iran. *Int. J. Climatol.* **2012**, *32*, 2387–2402. [CrossRef]

11. Andam-Akorful, S.A.; Ferreira, V.G.; Awange, J.L.; Forootan, E.; He, X.F. Multi-model and multi-sensor estimations of evapotranspiration over the Volta Basin, West Africa. *Int. J. Climatol.* **2015**, *35*, 3132–3145. [CrossRef]

12. Zhang, K.; Kimball, J.S.; Running, S.W. A review of remote sensing based actual evapotranspiration estimation. *Wiley Interdiscip. Rev. Water* **2016**, *3*, 834–853. [CrossRef]

13. Jian, D.; Li, X.; Sun, H.; Tao, H.; Jiang, T.; Su, B.; Hartmann, H. Estimation of actual evapotranspiration by the complementary theory-based advection-aridity model in the Tarim River Basin, China. *J. Hydrometeorol.* **2018**, *19*, 289–303. [CrossRef]

14. Allen, R.G.; Howell, T.A.; Pruitt, W.O.; Walter, I.A.; Jensen, M.E. (Eds.) *Lysimeters for Evapotranspiration and Environmental Measurements*; ASCE Publication: Reston, VA, USA, 1991; p. 444.

15. Bowen, I.S. The ratio of heat loss by conduction and by evaporation from any water surface. *Phys. Rev.* **1926**, *27*, 779–787. [CrossRef]

16. Monteith, J.L.; Unsworth, M.H. *Principles of Environmental Physics*; Edword Arnold: London, UK, 1990.

17. Everson, C.S.; Clulow, A.; Mengitsu, M. *Feasibility Study on the Determination of Riparian Evaporation in Non-Perennial Systems*; WRC Report No. TT 424/09; Water Research Commission: Pretoria, South Africa, 2009.

18. Xu, C.-Y.; Singh, V.P. Evaluation of three complementary relationship evapotranspiration models by water balance approach to estimate actual regional evapotranspiration in different climatic regions. *J. Hydrol.* **2005**, *308*, 105–121. [CrossRef]

19. Su, Z. The surface energy balance system SEBS for estimation of turbulent heat fluxes. *Hydrol. Earth Syst. Sci.* **2002**, *6*, 85–99. [CrossRef]

20. Fisher, J.B.; Tu, K.P.; Baldocchi, D.D. Global estimates of the land-atmosphere water flux based on monthly AVHRR and ISLSCP-II data, validated at 16 FLUXNET sites. *Remote Sens. Environ.* **2008**, *112*, 901–919. [CrossRef]

21. Miralles, D.G.; de Jeu, R.A.M.; Gash, J.H.; Holmes, T.R.H.; Dolman, A.J. Magnitude and variability of land evaporation and its components at the global scale. *Hydrol. Earth Syst. Sci.* **2011**, *15*, 967–981. [CrossRef]

22. Miralles, D.G.; Holmes, T.R.H.; de Jeu, R.A.M.; Gash, J.H.; Meesters, A.G.C.A.; Dolman, A.J. Global land-surface evaporation estimated from satellite-based observations. *Hydrol. Earth Syst. Sci.* **2011**, *15*, 453–469. [CrossRef]

23. Mu, Q.; Zhao, M.; Running, S.W. Improvements to a MODIS Global Terrestrial Evapotranspiration Algorithm. *Remote Sens. Environ.* **2011**, *115*, 1781–1800. [CrossRef]

24. Onogi, K.; Tsutsui, J.; Koide, H.; Sakamoto, M.; Kobayashi, S.; Hatsushika, H.; Matsumoto, T.; Yamazaki, N.; Kamahori, H.; Takahashi, K.; et al. The JRA-25 Reanalysis. *J. Meteorol. Soc. Jpn.* **2007**, *85*, 369–432. [CrossRef]

25. Dee, D.P.; Uppala, S.M.; Simmons, A.J.; Berrisford, P.; Poli, P.; Kobayashi, S.; Andrae, U.; Balmaseda, M.A.; Balsamo, G.; Bauer, P.; et al. The ERA-Interim reanalysis: Configuration and performance of the data assimilation system. *Q. J. R. Meteorol. Soc.* **2011**, *137*, 553–597. [CrossRef]

26. Reichle, R.H.; Draper, C.S.; Liu, Q.; Girotto, M.; Mahanama, S.P.P.; Koster, R.D.; de Lannoy, G.J.M. Assessment of MERRA-2 land surface hydrology estimates. *J. Clim.* **2017**, *30*, 2937–2960. [CrossRef]

27. Bosilovich, M.G.; Robertson, F.R.; Takacs, L.; Molod, A.; Mocko, D. Atmospheric water balance and variability in the MERRA-2 reanalysis. *J. Clim.* **2017**, *30*, 1177–1196. [CrossRef]

28. Rodell, M.; Houser, P.R.; Jambor, U.; Gottschalck, J.; Mitchell, K.; Meng, C.-J.; Arsenault, K.; Cosgrove, B.; Radakovich, J.; Bosilovich, M.; et al. The global land data assimilation system. *Bull. Am. Meteorol. Soc.* **2004**, *85*, 381–394. [CrossRef]

29. Haddeland, I.; Clark, D.B.; Franssen, W.; Ludwig, F.; Voß, F.; Arnell, N.W.; Bertrand, N.; Best, M.; Folwell, S.; Gerten, D.; et al. Multimodel estimate of the global terrestrial water balance: Setup and first results. *J. Hydrometeorol.* **2011**, *12*, 869–884. [CrossRef]

30. Schellekens, J.; Dutra, E.; Martínez-de la Torre, A.; Balsamo, G.; van Dijk, A.; Sperna Weiland, F.; Minvielle, M.; Calvet, J.-C.; Decharme, B.; Eisner, S.; et al. A global water resources ensemble of hydrological models: The eartH2Observe Tier-1 dataset. *Earth Syst. Sci. Data* **2017**, *9*, 389–413. [CrossRef]

31. Jung, M.; Reichstein, M.; Bondeau, A. Towards global empirical upscaling of FLUXNET eddy covariance observations: Validation of a model tree ensemble approach using a biosphere model. *Biogeosciences* **2009**, *6*, 2001–2013.

32. Glenn, E.P.; Huete, A.R.; Nagler, P.L.; Hirschboeck, K.K.; Brown, P. Integrating remote sensing and ground methods to estimate evapotranspiration. *Crit. Rev. Plant Sci.* **2007**, *26*, 139–168. [CrossRef]

33. Kalma, J.D.; McVicar, T.R.; McCabe, M.F. Estimating land surface evaporation: A Review of methods using remotely sensed surface temperature data. *Surv. Geophys.* **2008**, *29*, 421–469. [CrossRef]

34. Ershadi, A.; McCabe, M.; Evans, J.; Chaney, N.; Wood, E. Multi-site evaluation of terrestrial evaporation models using FLUXNET data. *Agric. For. Meteorol.* **2014**, *187*, 46–61. [CrossRef]

35. Allen, R.G.; Pereira, L.S.; Howell, T.A.; Jensen, M.E. Evapotranspiration information reporting: I. Factors governing measurement accuracy. *Agric. Water Manag.* **2011**, *98*, 899–920. [CrossRef]
36. Fisher, J.B.; Melton, F.; Middleton, E.; Hain, C.; Anderson, M.; Allen, R.; McCabe, M.; Hook, S.; Baldocchi, D.; Townsend, P.A.; et al. The Future of Evapotranspiration: Global requirements for ecosystem functioning, carbon and climate feedbacks, agricultural management, and water resources. *Water Resour. Res.* **2017**, *53*, 2618–2626. [CrossRef]
37. Foken, T. The energy balance closure problem: An overview. *Ecol. Appl.* **2008**, *18*, 1351–1367. [CrossRef] [PubMed]
38. Kim, H.W.; Hwang, K.; Mu, Q.; Lee, S.O.; Choi, M. Validation of MODIS 16 global terrestrial evapotranspiration products in various climates and land cover types in Asia. *KSCE J. Civ. Eng.* **2012**, *16*, 229–238. [CrossRef]
39. Zhang, F.; Zhou, G.; Wang, Y.; Yang, F.; Nilsson, C. Evapotranspiration and crop coefficient for a temperate desert steppe ecosystem using eddy covariance in Inner Mongolia, China. *Hydrol. Process.* **2012**, *26*, 379–386. [CrossRef]
40. Michel, D.; Jiménez, C.; Miralles, D.G.; Jung, M.; Hirschi, M.; Ershadi, A.; Martens, B.; McCabe, M.F.; Fisher, J.B.; Mu, Q.; et al. The WACMOS-ET project-Part 1: Tower-scale evaluation of four remote-sensing-based evapotranspiration algorithms. *Hydrol. Earth Syst. Sci.* **2016**, *20*, 803–822. [CrossRef]
41. Numata, I.; Khand, K.; Kjaersgaard, J.; Cochrance, M.; Silva, S. Evaluation of Landsat-based METRIC modeling to provide high-spatial resolution evapotranspiration estimates for Amazonian forests. *Remote Sens.* **2017**, *9*, 46. [CrossRef]
42. Yang, X.; Yong, B.; Ren, L.; Zhang, Y.; Long, D. Multi-scale validation of GLEAM evapotranspiration products over China via ChinaFLUX ET measurements. *Int. J. Remote Sens.* **2017**, *38*, 5688–5709. [CrossRef]
43. Fu, G.; Yu, J.; Zhang, Y.; Hu, S.; Ouyang, R.; Liu, W. Temporal variation of wind speed in China for 1961–2007. *Theor. Appl. Climatol.* **2010**, *104*, 313–324. [CrossRef]
44. Xie, B.; Zhang, Q.; Ying, Y. Trends in precipitable water and relative humidity in China: 1979–2005. *J. Clim.* **2011**, *50*, 1985–1994. [CrossRef]
45. Huang, J.; Sun, S.; Xue, Y.; Li, J.; Zhang, J. Spatial and temporal variability of precipitation and dryness/wetness during 1961–2008 in Sichuan Province, West China. *Water Resour. Manag.* **2014**, *28*, 1655–1670. [CrossRef]
46. Huang, J.; Sun, S.; Xue, Y.; Zhang, J. Changing characteristics of precipitation during 1960–2012 in Inner Mongolia, northern China. *Meteorol. Atmos. Phys.* **2015**, *127*, 257–271. [CrossRef]
47. Huang, J.; Sun, S.; Zhang, J. Detection of trends in precipitation during 1960–2008 in Jiangxi province, southeast China. *Theor. Appl. Climatol.* **2013**, *114*, 237–251. [CrossRef]
48. Liao, W.; Wang, X.; Fan, Q.; Zhou, S.; Chang, M.; Wang, Z.; Wang, Y.; Tu, Q. Long-term atmospheric visibility, sunshine duration and precipitation trends in South China. *Atmos. Environ.* **2015**, *107*, 204–216. [CrossRef]
49. Wang, K.C.; Ma, Q.; Li, Z.; Wang, J. Decadal variability of surface incident solar radiation over China: Observations, satellite retrievals, and reanalyses. *J. Geophys. Res.* **2015**, *120*, 6500–6514. [CrossRef]
50. Cao, L.; Yan, Z.; Zhao, P.; Zhu, Y.; Yu, Y.; Tang, G.; Jones, P. Climatic warming in China during 1901–2015 based on an extended dataset of instrumental temperature records. *Environ. Res. Lett.* **2017**, *12*, 064005. [CrossRef]
51. Feng, F.; Wang, K. Merging satellite retrievals and reanalyses to produce global long-term and consistent surface incident solar radiation datasets. *Remote Sens.* **2018**, *10*, 115. [CrossRef]
52. Schneider, K.; Ketzer, B.; Breuer, L.; Vaché, K.B.; Bernhofer, C.; Frede, H.G. Evaluation of evapotranspiration methods for model validation in a semi-arid watershed in northern China. *Adv. Geosci.* **2007**, *11*, 37–42. [CrossRef]
53. Xiao, J.; Sun, G.; Chen, J.; Chen, H.; Chen, S.; Dong, G.; Gao, S.; Guo, H.; Guo, J.; Han, S.; et al. Carbon fluxes, evapotranspiration, and water use efficiency of terrestrial ecosystems in China. *Agric. For. Meteorol.* **2013**, *182–183*, 76–90. [CrossRef]
54. Chen, Y.; Xia, J.; Liang, S.; Feng, J.; Fisher, J.B.; Li, X.; Li, X.; Liu, S.; Ma, Z.; Miyata, A.; et al. Comparison of satellite-based evapotranspiration models over terrestrial ecosystems in China. *Remote Sens. Environ.* **2014**, *140*, 279–293. [CrossRef]

55. Tang, R.; Shao, K.; Li, Z.L.; Wu, H.; Tang, B.-H.; Zhou, G.; Zhang, L. Multiscale validation of the 8-day MOD16 evapotranspiration product using flux data collected in China. *IEEE J. Sel. Top. Appl. Earth Obs. Remote Sens.* **2015**, *8*, 1478–1486. [CrossRef]

56. Yang, Y.; Long, D.; Guan, H.; Liang, W.; Simmons, C.; Batelaan, O. Comparison of three dual-source remote sensing evapotranspiration models during the MUSOEXE-12 campaign: Revisit of model physics. *Water Resour. Res.* **2015**, *51*, 3145–3165. [CrossRef]

57. Zhou, Y.; Li, X.; Yang, K.; Zhou, J. Assessing the impacts of an ecological water diversion project on water consumption through high-resolution estimations of actual evapotranspiration in the downstream regions of the Heihe River Basin, China. *Agric. For. Meteorol.* **2018**, *249*, 210–227. [CrossRef]

58. Martens, B.; Miralles, D.G.; Lievens, H.; van der Schalie, R.; de Jeu, R.A.M.; Fernández-Prieto, D.; Beck, H.E.; Dorigo, W.A.; Verhoest, N.E.C. GLEAM v3: Satellite-based land evaporation and root-zone soil moisture. *Geosci. Model Dev.* **2017**, *10*, 1903–1925. [CrossRef]

59. Liu, Y.Y.; de Jeu, R.A.M.; McCabe, M.F.; Evans, J.P.; van Dijk, A.I.J.M. Global long-term passive microwave satellite based retrievals of vegetation optical depth. *Geophys. Res. Lett.* **2011**, *38*, L18402. [CrossRef]

60. Reichle, R.H.; Koster, R.D.; de Lannoy, G.J.M.; Forman, B.A.; Liu, Q.; Mahanama, S.P.P.; Touré, A. Assessment and enhancement of MERRA land surface hydrology estimates. *J. Clim.* **2011**, *24*, 6322–6338. [CrossRef]

61. Dutra, E.; Balsamo, G.; Calvet, J.-C.; Minvielle, M.; Eisner, S.; Fink, G.; Pessenteiner, S.; Orth, R.; Burke, S.; van Dijk, A.I.J.M.; et al. Report on the current state-of-the-art Water Resources Reanalysis. Available online: http://earth2observe.eu/files/PublicDeliverables/D5.1_ReportontheWRR1tier1.pdf (accessed on 31 March 2015).

62. Rienecker, M.M.; Suarez, M.J.; Gelaro, R.; Todling, R.; Bacmeister, J.; Liu, E.; Bosilovich, M.G.; Schubert, S.D.; Takacs, L.; Kim, G.-K.; et al. MERRA: NASA's Modern-Era Retrospective Analysis for research and applications. *J. Clim.* **2011**, *24*, 3624–3648. [CrossRef]

63. Koster, R.D.; Suarez, M.J.; Ducharne, A.; Stieglitz, M.; Kumar, P. A catchment-based approach to modeling land surface processes in a general circulation model: 1. Model structure. *J Geophys. Res.* **2000**, *105*, 24809–24822. [CrossRef]

64. Wang, W.; Cui, W.; Wang, X.; Chen, X. Evaluation of GLDAS-1 and GLDAS-2 forcing data and Noah model simulations over China at the monthly scale. *J. Hydrometeorol.* **2016**, *17*, 2815–2833. [CrossRef]

65. Sheffield, J.; Goteti, G.; Wood, E.F. Development of a 50-year high-resolution global dataset of meteorological forcings for land surface modeling. *J. Clim.* **2006**, *19*, 3088–3111. [CrossRef]

66. Weedon, G.P.; Balsamo, G.; Bellouin, N.; Gomes, S.; Best, M.J.; Viterbo, P. The WFDEI meteorological forcing data set: WATCH Forcing Data methodology applied to ERA-Interim reanalysis data. *Water Resour. Res.* **2015**, *50*, 7505–7514. [CrossRef]

67. Gosling, S.N.; Bretherton, D.; Haines, K.; Arnell, N.W. Global hydrology modelling and uncertainty: Running multiple ensembles with a campus grid. *Philos. Trans. R. Soc. A* **2010**, *368*, 4005–4021. [CrossRef] [PubMed]

68. Vuichard, N.; Papale, D. Filling the gaps in meteorological continuous data measured at FLUXNET sites with ERA-Interim reanalysis. *Earth Syst. Sci. Data* **2015**, *7*, 157–171. [CrossRef]

69. Reichstein, M.; Falge, E.; Baldocchi, D.; Papale, D.; Aubinet, M.; Berbigier, P.; Bernhofer, C.; Buchmann, N.; Gilmanov, T.; Granier, A.; et al. On the separation of net ecosystem exchange into assimilation and ecosystem respiration: Review and improved algorithm. *Glob. Chang. Biol.* **2005**, *11*, 1424–1439. [CrossRef]

70. Huang, J.; Ji, M.; Xie, Y.; Wang, S.; He, Y.; Ran, J. Global semi-arid climate change over last 60 years. *Clim. Dyn.* **2016**, *46*, 1131–1150. [CrossRef]

71. Allen, R.G.; Pereira, L.S.; Raes, D.; Smith, M. *Crop Evapotranspiration: Guidelines for Computing Crop Requirements, Irrigation and Drainage Paper 56*; FAO: Roma, Italia, 1998.

72. Senay, G. Modeling landscape evapotranspiration by integrating land surface phenology and a water balance algorithm. *Algorithms* **2008**, *1*, 52–68. [CrossRef]

73. Henderson-Sellers, B. A new formula for latent heat of vaporization of water as a function of temperature. *Q. J. R. Meteorol. Soc.* **1984**, *110*, 1186–1190. [CrossRef]

74. Taylor, K. Summarizing multiple aspects of model performance in a single diagram. *J. Geophys. Res.* **2001**, *106*, 7183–7192. [CrossRef]

75. Jiménez, C.; Prigent, C.; Mueller, B.; Seneviratne, S.I.; McCabe, M.F.; Wood, E.F.; Rossow, W.B.; Balsamo, G.; Betts, A.K.; Dirmeyer, P.A.; et al. Global intercomparison of 12 land surface heat flux estimates. *J. Geophys. Res.* **2011**, *116*, 3–25. [CrossRef]

76. McCabe, M.F.; Ershadi, A.; Jimenez, C.; Miralles, D.G.; Michel, D.; Wood, E.F. The GEWEX LandFlux project: Evaluation of model evaporation using tower-based and globally-gridded forcing data. *Geosci. Model Dev.* **2016**, *9*, 283–305. [CrossRef]

77. Xue, B.-L.; Wang, L.; Li, X.; Yang, K.; Chen, D.; Sun, L. Evaluation of evapotranspiration estimates for two river basins on the Tibetan Plateau by a water balance method. *J. Hydrol.* **2013**, *492*, 290–297. [CrossRef]

78. Long, D.; Longuevergne, L.; Scanlon, B.R. Uncertainty in evapotranspiration from land surface modeling, remote sensing, and GRACE satellites. *Water Resour. Res.* **2014**, *50*, 1131–1151. [CrossRef]

79. Badgley, G.; Fisher, J.B.; Jiménez, C.; Tu, K.P.; Vinukollu, R. On uncertainty in global terrestrial evapotranspiration estimates from choice of input forcing datasets. *J. Hydrometeorol.* **2015**, *16*, 1449–1455. [CrossRef]

80. Purdy, A.J.; Fisher, J.B.; Goulden, M.L.; Famiglietti, J.S. Ground heat flux: An analytical review of 6 models evaluated at 88 sites and globally. *J. Geophys. Res.* **2016**, *121*, 3045–3059. [CrossRef]

81. Deardorff, J.W. Efficient prediction of ground surface temperature and moisture, with inclusion of a layer of vegetation. *J. Geophys. Res.* **1978**, *83*, 1889–1903. [CrossRef]

82. Beven, K. Sensitivity analysis of the Penman-Monteith actual evapotranspiration estimates. *J. Hydrol.* **1979**, *44*, 169–190. [CrossRef]

83. Ball, J.T.; Woodrow, I.E.; Berry, J.A. A model predicting stomatal conductance and its contribution to control of photosynthesis under different environmental conditions. In *Progress in Photosynthesis Research*; Biggins, J., Ed.; Martinus Nijhof: Zoetermeer, The Netherlands, 1987; pp. 221–234.

84. Stewart, J.B. Modelling surface conductance of pine forest. *Agric. For. Meteorol.* **1988**, *43*, 19–35. [CrossRef]

85. Alves, I.; Pereira, L.S. Modelling surface resistance from climatic variables. *Agric. Water Manag.* **2000**, *42*, 371–385. [CrossRef]

86. Komatsu, H. Forest categorization according to dry-canopy evaporation rates in the growing season: Comparison of the Priestley-Taylor coefficient values from various observation sites. *Hydrol. Process.* **2005**, *19*, 3873–3896. [CrossRef]

87. Dirmeyer, P.A.; Gao, X.; Zhao, M.; Guo, Z.; Oki, T.; Hanasaki, N. GSWP-2: Multimodel analysis and implications for our perception of the land surface. *Bull. Am. Meteorol. Soc.* **2006**, *87*, 1381–1397. [CrossRef]

88. Lawrence, D.M.; Thornton, P.E.; Oleson, K.W.; Bonan, G.B. The partitioning of evapotranspiration into transpiration, soil evaporation, and canopy evaporation in a GCM: Impacts on land-atmosphere interaction. *J. Hydrometeorol.* **2007**, *8*, 862–880. [CrossRef]

89. Jasechko, S.; Sharp, Z.D.; Gibson, J.J.; Birks, S.J.; Yi, Y.; Fawcett, P.J. Terrestrial water fluxes dominated by transpiration. *Nature* **2013**, *496*, 347–350. [CrossRef] [PubMed]

90. Miralles, D.G.; Jiménez, C.; Jung, M.; Michel, D.; Ershadi, A.; McCabe, M.F.; Hirschi, M.; Martens, B.; Dolman, A.J.; Fisher, J.B.; et al. The WACMOS-ET project. Part 2: Evaluation of global terrestrial evaporation data sets. *Hydrol. Earth Syst. Sci.* **2016**, *20*, 823–842. [CrossRef]

91. Zhang, G.; Zeng, G.M.; Jiang, Y.M.; Huang, G.H.; Li, J.B.; Yao, J.M.; Tan, W.; Xiang, R.; Zhang, X.L. Modelling and measurement of two-layer-canopy interception losses in a subtropical evergreen forest of central-south China. *Hydrol. Earth Syst. Sci.* **2006**, *10*, 65–77. [CrossRef]

92. Novick, K.A.; Oren, R.; Stoy, P.C.; Siqueira, M.B.S.; Katul, G.G. Nocturnal evapotranspiration in eddy-covariance records from three co-located ecosystems in the Southeastern U.S.: Implications for annual fluxes. *Agric. For. Meteorol.* **2009**, *149*, 1491–1504. [CrossRef]

93. O'Keefe, K.; Nippert, J.B. Drivers of nocturnal water flux in a tallgrass prairie. *Funct. Ecol.* **2018**. [CrossRef]

94. Snyder, K.A.; Richards, J.H.; Donovan, L.A. Night-time conductance in C3 and C4 species: Do plants lose water at night? *J. Exp. Bot.* **2003**, *54*, 861–865. [CrossRef] [PubMed]

95. Fisher, J.B.; Baldocchi, D.D.; Misson, L.; Dawson, T.E.; Goldstein, A.H. What the towers don't see at night: Nocturnal sap flow in trees and shrubs at two AmeriFlux sites in California. *Tree Physiol.* **2007**, *27*, 597–610. [CrossRef] [PubMed]

96. Zippel, M.; Tissue, D.; Macinnis-Ng, C.; Eamus, D. Rates of nocturnal transpiration in two evergreen temperate woodland species with differing water-use strategies. *Tree Physiol.* **2010**, *30*, 988–1000. [CrossRef] [PubMed]

97. De Dios, V.R.; Roy, J.; Ferrio, J.P.; Alday, J.G.; Landais, D.; Milcu, A.; Gessler, A. Processes driving nocturnal transpiration and implications for estimating land evapotranspiration. *Sci. Rep.* **2015**, *5*, 10975. [CrossRef] [PubMed]

98. Alijanian, M.; Rakhshandehroo, G.R.; Mishra, A.K.; Dehghani, M. Evaluation of satellite rainfall climatology using CMORPH, PERSIANN-CDR, PERSIANN, TRMM, MSWEP over Iran. *Int. J. Climatol.* **2017**, *37*, 4896–4914. [CrossRef]

99. Nair, A.; Indu, J. Performance Assessment of multi-source weighted-ensemble precipitation (MSWEP) product over India. *Climate* **2017**, *5*, 2. [CrossRef]

100. Beck, H.E.; van Dijk, A.I.J.M.; Levizzani, V.; Schellekens, J.; Miralles, D.G.; Martens, B.; de Roo, A. MSWEP: 3-hourly 0.25 degrees global gridded precipitation (1979–2015) by merging gauge, satellite, and reanalysis data. *Hydrol. Earth Syst. Sci.* **2017**, *21*, 589–615. [CrossRef]

101. Beck, H.E.; Vergopolan, N.; Pan, M.; Levizzani, V.; van Dijk, A.I.J.M.; Weedon, G.P.; Brocca, L.; Pappenberger, F.; Huffman, G.J.; Wood, E.F. Global-scale evaluation of 22 precipitation datasets using gauge observations and hydrological modeling. *Hydrol. Earth Syst. Sci.* **2017**, *21*, 6201–6217. [CrossRef]

102. Sun, Q.; Miao, C.; Duan, Q.; Ashouri, H.; Sorooshian, S.; Hsu, K.-L. A review of global precipitation datasets: Data sources, estimation, and intercomparisons. *Rev. Geophys.* **2018**, *56*, 79–107. [CrossRef]

103. Shen, Y.; Xiong, A. Validation and comparison of a new gauge-based precipitation analysis over mainland China. *Int. J. Climatol.* **2016**, *36*, 252–265. [CrossRef]

104. Duan, Z.; Liu, J.; Tuo, Y.; Chiogna, G.; Disse, M. Evaluation of eight high spatial resolution gridded precipitation products in Adige Basin (Italy) at multiple temporal and spatial scales. *Sci. Total Environ.* **2016**, *573*, 1536–1553. [CrossRef] [PubMed]

105. Wong, J.S.; Razavi, S.; Bonsal, B.R.; Wheater, H.S.; Asong, Z.E. Inter-comparison of daily precipitation products for large-scale hydro-climatic applications over Canada. *Hydrol. Earth Syst. Sci.* **2017**, *21*, 2163–2185. [CrossRef]

106. Tory, T.J.; Wood, E.F. Comparison and evaluation of gridded radiation products across northern Eurasia. *Environ. Res. Lett.* **2009**, *4*, 045008. [CrossRef]

107. Bosilovich, M.G.; Robertson, F.R.; Chen, J. Global energy and water budgets in MERRA. *J. Clim.* **2011**, *24*, 5721–5739. [CrossRef]

108. Zhao, L.; Lee, X.; Liu, S. Correcting surface solar radiation of two data assimilation systems against FLUXNET observations in North America. *J. Geophys. Res.* **2013**, *118*, 9552–9564. [CrossRef]

109. Boilley, A.; Wald, L. Comparison between meteorological re-analyses from ERA-Interim and MERRA and measurements of daily solar irradiation at surface. *Renew. Energy* **2015**, *75*, 135–143. [CrossRef]

110. Schmied, H.M.; Müller, R.; Sanchez-Lorenzo, A.; Ahrens, B.; Wild, M. Evaluation of radiation components in a global freshwater model with station-based observations. *Water* **2016**, *8*, 450. [CrossRef]

111. Draper, C.S.; Reichle, R.H.; Koster, R.D. Assessment of MERRA-2 Land Surface Energy Flux Estimates. *J. Clim.* **2018**, *31*, 671–691. [CrossRef]

112. Jia, A.; Liang, S.; Jiang, B.; Zhang, X.; Wang, G. Comprehensive assessment of global surface net radiation products and uncertainty analysis. *J. Geophys. Res.* **2018**, *123*, 1970–1989. [CrossRef]

113. Liu, Y.; Xiao, J.; Ju, W.; Zhu, G.; Wu, X.; Fan, W.; Li, D.; Zhou, Y. Satellite-derived LAI products exhibit large discrepancies and can lead to substantial uncertainty in simulated carbon and water fluxes. *Remote Sens. Environ.* **2018**, *206*, 174–188. [CrossRef]

114. McCallum, I.; Obersteiner, M.; Nilsson, S.; Shvidenko, A. A spatial comparison of four satellite derived 1 km global land cover datasets. *Int. J. Appl. Earth Obs. Geoinf.* **2006**, *8*, 246–255. [CrossRef]

115. Kaptué, T.A.T.; Roujean, J.-L.; de Jong, S.M. Comparison and relative quality assessment of the GLC2000, GLOBCOVER, MODIS and ECOCLIMAP land cover data sets at the African continental scale. *Int. J. Appl. Earth Obs. Geoinf.* **2011**, *13*, 207–219. [CrossRef]

116. Liu, R.; Liang, S.; Liu, J.; Zhuang, D. Continuous tree distribution in China: A comparison of two estimates from Moderate-Resolution Imaging Spectroradiometer and Landsat data. *J. Geophys. Res.* **2006**, *111*, D08101. [CrossRef]

117. Fritz, S.; See, L. Identifying and quantifying uncertainty and spatial disagreement in the comparison of Global Land Cover for different applications. *Glob. Chang. Biol.* **2008**, *14*, 1057–1075. [CrossRef]

118. Herold, M.; Mayaux, P.; Woodcock, C.E.; Schmullius, C. Some challenges in global land cover mapping: An assessment of agreement and accuracy in existing 1 km datasets. *Remote Sens. Environ.* **2008**, *112*, 2538–2556. [CrossRef]

119. Fensholt, R.; Rasmussen, K.; Nielsen, T.T.; Mbow, C. Evaluation of earth observation based long term vegetation trends: Intercomparing NDVI time series trend analysis consistency of Sahel from AVHRR GIMMS, Terra MODIS and SPOT VGT data. *Remote Sens. Environ.* **2009**, *113*, 1886–1898. [CrossRef]

120. Beck, H.E.; McVicar, T.R.; van Dijk, A.I.J.M.; Schellekens, J.; de Jeu, R.A.M.; Bruijnzeel, L.A. Global evaluation of four AVHRR-NDVI data sets: Intercomparison and assessment against Landsat imagery. *Remote Sens. Environ.* **2011**, *115*, 2547–2563. [CrossRef]

121. Verburg, P.H.; Neumann, K.; Nol, L. Challenges in using land use and land cover data for global change studies. *Glob. Chang. Biol.* **2011**, *17*, 974–989. [CrossRef]

122. Camacho, F.; Cernicharo, J.; Lacaze, R.; Baret, F.; Weiss, M. GEOV1: LAI, FAPAR essential climate variables and FCOVER global time series capitalizing over existing products. Part 2: Validation and intercomparison with reference products. *Remote Sens. Environ.* **2013**, *137*, 310–329. [CrossRef]

123. Xiao, Z.; Liang, S.; Jiang, B. Evaluation of four long time-series global leaf area index products. *Agric. For. Meteorol.* **2017**, *246*, 218–230. [CrossRef]

124. Yang, Y.; Xiao, P.; Feng, X.; Li, H. Accuracy assessment of seven global land cover datasets over China. *ISPRS J. Photogramm. Remote Sens.* **2017**, *125*, 156–173. [CrossRef]

125. Li, X.; Lu, H.; Yu, L.; Yang, K. Comparison of the spatial characteristics of four remotely sensed leaf area index products over China: Direct validation and relative uncertainties. *Remote Sens.* **2018**, *10*, 148. [CrossRef]

126. Branger, F.; Kermadi, S.; Jacqueminet, C.; Michel, K.; Labbs, M.; Krause, M.; Kralisch, S.; Braud, I. Assessment of the influence of land use data on the water balance components of a peri-urban catchment using a distributed modelling approach. *J. Hydrol.* **2013**, *505*, 312–325. [CrossRef]

127. Polhamus, A.; Fisher, J.B.; Tu, K.P. What controls the error structure in evapotranspiration models? *Agric. For. Meteorol.* **2013**, *169*, 12–24. [CrossRef]

128. Ghilain, N.; Gellensmeulenberghs, F. Assessing the impact of land cover map resolution and geolocation accuracy on evapotranspiration simulations by a land surface model. *Remote Sens. Lett.* **2014**, *5*, 491–499. [CrossRef]

129. Madhusoodhanan, C.G.; Sreeja, K.G.; Eldho, T.I. Assessment of uncertainties in global land cover products for hydro-climate modeling in India. *Water Resour. Res.* **2017**, *53*, 1713–1734. [CrossRef]

130. Högström, U.; Bergström, H. Organized turbulence structures in the near-neutral atmospheric surface layer. *J. Atmos. Sci.* **1996**, *53*, 2452–2464. [CrossRef]

131. Baldocchi, D.; Falge, E.; Gu, L.; Olson, R.; Hollinger, D.; Running, S.; Anthoni, P.; Bernhofer, C.; Davis, K.; Evans, R.; et al. FLUXNET: A new tool to study the temporal and spatial variability of ecosystem scale carbon dioxide, water vapor, and energy flux densities. *Bull. Am. Meteorol. Soc.* **2001**, *82*, 2415–2434. [CrossRef]

132. Wilson, K.; Goldstein, A.; Falge, E.; Aubinet, M.; Baldocchi, D.; Berbigier, P.; Bernhofer, C.; Ceulemans, R.; Dolman, H.; Field, C.; et al. Energy balance closure at FLUXNET sites. *Agric. For. Meteorol.* **2002**, *113*, 223–243. [CrossRef]

133. Xu, Z.; Ma, Y.; Liu, S.; Shi, W.; Wang, J. Assessment of the energy balance closure under advective conditions and its impact using remote sensing data. *J. Appl. Meteorol. Climatol.* **2017**, *56*, 127–140. [CrossRef]

134. Moderow, U.; Aubinet, M.; Feigenwinter, C.; Kolle, O.; Lindroth, A.; Molder, M.; Montagnani, L.; Rebmann, C.; Bernhofer, C. Available energy and energy balance closure at four coniferous forest sites across Europe. *Theor. Appl. Climatol.* **2009**, *98*, 397–412. [CrossRef]

135. Sánchez, J.M.; Caselles, V.; Rubio, E.M. Analysis of the energy balance closure over a FLUXNET boreal forest in Finland. *Hydrol. Earth Syst. Sci.* **2010**, *14*, 1487–1497. [CrossRef]

136. Finkelstein, P.L.; Sims, P.F. Sampling error in eddy correlation flux measurements. *J. Geophys. Res. Atmos.* **2001**, *106*, 3503–3509. [CrossRef]

137. Castellvi, F.; Martinez-Cob, A.; Perez-Coveta, O. Estimating sensible and latent heat fluxes over rice using surface renewal. *Agric. For. Meteorol.* **2006**, *139*, 164–169. [CrossRef]

138. Castellvi, F.; Snyder, R.L.; Baldocchi, D.D. Surface energy-balance closure over rangeland grass using the eddy covariance method and surface renewal analysis. *Agric. For. Meteorol.* **2008**, *148*, 1147–1160. [CrossRef]

139. Schmid, H.P. Footprint modeling for vegetation atmosphere exchange studies: A review and perspective. *Agric. For. Meteorol.* **2002**, *113*, 159–183. [CrossRef]

140. Kormann, R.; Meixner, F.X. An Analytical footprint model for non-neutral stratification. *Bound. Layer Meteorol.* **2001**, *99*, 207–224. [CrossRef]

141. Kljun, N.; Calanca, P.; Rotach, M.W.; Schmid, H.P. The simple two-dimensional parameterisation for Flux Footprint Predictions FFP. *Geosci. Model Dev.* **2015**, *8*, 3695–3713. [CrossRef]
142. Kanemasu, E.T.; Verma, S.B.; Smith, E.A.; Fritschen, L.J.; Wesely, M.; Field, R.T.; Kustas, W.P.; Weaver, H.; Stewart, J.B.; Gurney, R.; et al. Surface flux measurements in FIFE: An overview. *J. Geophys. Res.* **1992**, *97*, 18547–18555. [CrossRef]
143. Castelli, M.; Anderson, M.C.; Yang, Y.; Wohlfahrt, G.; Bertoldi, G.; Niedrist, G.; Hammerle, A.; Zhao, P.; Zebisch, M.; et al. Two-source energy balance modeling of evapotranspiration in Alpine grasslands. *Remote Sens. Environ.* **2018**, *209*, 327–342. [CrossRef]
144. Ramoelo, A.; Majozi, N.; Mathieu, R.; Jovanovic, N.; Nickless, A.; Dzikiti, S. Validation of global evapotranspiration product (MOD16) using flux tower data in the African savanna, South Africa. *Remote Sens.* **2014**, *6*, 942–945. [CrossRef]
145. Ershadi, A.; Mccabe, M.F.; Evans, J.P.; Walker, J.P. Effects of spatial aggregation on the multi-scale estimation of evapotranspiration. *Remote Sens. Environ.* **2013**, *131*, 51–62. [CrossRef]
146. Hall, F.; Huemmrich, K.; Goetz, S.; Sellers, P.; Nickeson, J. Satellite remote sensing of surface energy balance: Success, failures and unresolved issues in FIFE. *J. Geophys. Res.* **1992**, *97*, 19061–19090. [CrossRef]
147. McCabe, M.F.; Wood, E.F. Scale influences on the remote estimation of evapotranspiration using multiple satellite sensors. *Remote Sens. Environ.* **2006**, *105*, 271–285. [CrossRef]

Article

Continuous Daily Evapotranspiration Estimation at the Field-Scale over Heterogeneous Agricultural Areas by Fusing ASTER and MODIS Data

Zhenyan Yi, Hongli Zhao * and Yunzhong Jiang

Department of Water Resources, China Institute of Water Resources and Hydropower Research, Beijing 100038, China; daisy7303@yeah.net (Z.Y.); lark@iwhr.com (Y.J.)
* Correspondence: zhaohl@iwhr.com; Tel.: +86-139-1190-0329

Received: 23 August 2018; Accepted: 24 October 2018; Published: 26 October 2018

Abstract: Continuous daily evapotranspiration (ET) monitoring at the field-scale is crucial for water resource management in irrigated agricultural areas in arid regions. Here, an integrated framework for daily ET, with the required spatiotemporal resolution, is described. Multi-scale surface energy balance algorithm evaluations and a data fusion algorithm are combined to optimally exploit the spatial and temporal characteristics of image datasets, collected by the advanced space-borne thermal emission reflectance radiometer (ASTER) and the moderate resolution imaging spectroradiometer (MODIS). Through combination with a linear unmixing-based method, the spatial and temporal adaptive reflectance fusion model (STARFM) is modified to generate high-resolution ET estimates for heterogeneous areas. The performance of this methodology was evaluated for irrigated agricultural fields in arid and semiarid areas of Northwest China. Compared with the original STARFM, a significant improvement in daily ET estimation accuracy was obtained by the modified STARFM (overall mean absolute percentage error (MAP): 12.9% vs. 17.2%; root mean square error (RMSE): 0.7 mm d^{-1} vs. 1.2 mm d^{-1}). The modified STARFM additionally preserved more spatial details than the original STARFM for heterogeneous agricultural fields, and provided field-to-field variability in water use. Improvements were further evident in the continuous daily ET, where the day-to-day dynamics of ET estimates were captured. ET data fusion provides a unique means of monitoring continuous daily crop ET values at the field-scale in agricultural areas, and may have value in supporting operational water management decisions.

Keywords: evapotranspiration; field-scale; STARFM; unmixing-based method; MPDI-integrated SEBS

1. Introduction

Evapotranspiration (ET)—the sum of land surface evaporation, vegetation transpiration, and evaporation of water intercepted by plant canopies—is a major component of the water cycle and energy exchange in the soil-plant-atmosphere-climate system [1]. Continuous daily ET monitoring at the field-scale can provide detailed information about crop water use and soil moisture status. This information has long been a critical requirement for a wide range of applications, including irrigation scheduling, increasing the efficiency of crop water use, and assessing the impacts of drought on crop yields [2,3]. The spatiotemporal variation of ET in irrigated fields is particularly relevant in arid and semiarid areas where water resources are scarce, such as in northern and western China [4–6].

Remote sensing (RS) has provided a suitable alternative for obtaining spatially distributed estimates of the temporal evolution of ET over the growing season. During the last few decades, various RS-based methods have been proposed to estimate ET. RS-based approaches now generally include empirical and semi-empirical methods [7], surface energy balance models (e.g., the surface energy balance algorithm for land (SEBAL), the surface energy balance system (SEBS), the mapping ET with internalized calibration (METRIC), the two-source energy balance model (TSEB) [8–11], the vegetation index combined with the Penman-Monteith (PM) or Priestley-Taylor (PT) method [12,13], and data assimilation combined with land surface models and hydrological models [14,15]. These approaches have been developed and applied from local to global scales, at both the satellite overpass time and daily time scales, achieving relative errors of 10–30% for different ecosystems around the world [16–18].

Local water management usually requires temporally continuous ET measurements at the field-scale in order to understand the ET variation of crops, which is related to spatial heterogeneity in crop type, phenological stage, meteorological conditions, and soil moisture conditions [19]. However, the estimation of ET values for croplands using remote sensing is particularly challenging in heterogeneous landscapes, where agricultural plots are small. The 100 m resolution characteristic of thermal infrared (TIR) imagery (e.g., Landsat series, ASTER (the advanced space-borne thermal emission reflectance radiometer)), which we define as "moderate resolution" to distinguish from high-resolution (meter scale) shortwave imagery, has proven critical in providing detailed information about vegetation status, soil moisture, and surface temperature [20,21]. Moderate-resolution RS can be used for ET estimates, because it typically enables the discrimination of individual agricultural fields. However, the long revisit interval makes its application problematic in areas with high cloud cover and dynamic land cover [22]. Although geostationary satellites (e.g., moderate resolution imaging spectroradiometer (MODIS)) provide sources of ET information on a daily basis, their resolution is too coarse (km-scale or above) to resolve individual fields in most irrigation districts [23]. Therefore, exploiting the complementary spatial and temporal characteristics of different satellite sensors to provide daily field-scale ET estimates could be significant for water resource management.

Some efforts have focused on combining imagery from different platforms to generate high-resolution ET maps. Downscaling converts RS-based ET from a low- to a high-spatial resolution. Singh et al. (2014) developed a linear regression method with zero intercept for downscaling MODIS-based monthly ET maps to Landsat-scale ones [24]. Ke et al. (2016) used machine learning algorithms, including support vector regression (SVR), cubist, and random forest (RF) to model the relationship between Landsat indices and MODIS eight-day, one kilometer ET products, and then predicted 30 m ET based on the model using Landsat-8 indices [25]. Although these downscaling methods provide useful means to improve the coarse-resolution ET data to finer spatial resolutions, they cannot simultaneously enhance their temporal resolution. Gao et al. (2006) proposed a data fusion technique named the "spatial and temporal adaptive reflectance fusion model" (STARFM), that simultaneously integrates the temporal advantage of MODIS-like images and the spatial advantage of Landsat-like images [26]. This data fusion approach has lately received much attention, since it can provide high-resolution vegetation indices and land surface temperature estimates [27–30]. This approach additionally appears to hold great utility for high-resolution TIR data for ET mapping. Cammalleri et al. (2013) described the initial implementation and evaluation of the fusion of Landsat and MODIS ET measurements using the STARFM model over corn and soybean fields in the Walnut Creek watershed [31]. It was subsequently tested for corn and cotton fields in both rain-fed and irrigated areas [32], a forest [33], and vineyards [34]. Additionally, Li et al. (2017) applied the STARFM algorithm to fuse ASTER and MODIS images for continuous daily ET at the field-scale over irrigated agricultural areas [35].

While the STARFM algorithm provides a useful tool to generate high-resolution daily ET estimates, several limitations should be noted. If changes are transient and not recorded in at least one of the base fine resolution images, it may not be possible to predict them in fine resolution by this algorithm [26,29,36]. STARFM depends on the temporal information from pure homogenous pixels ("similar pixel") in the MODIS image [37,38]. And the predicted results can be misleading when the homogeneous coarse-resolution pixels cannot be found in the search window, which happens in heterogeneous landscapes such as agricultural areas. Many spatial-temporal fusion algorithms have been developed for improving STARFM. For example, Hilker et al. (2009) developed the spatial-temporal adaptive algorithm for mapping reflectance changes (STAARCH) based on STARFM, to detect the date on which land-cover change occurs, and to record this information in a Landsat image to improve the final predicted result of the original STARFM approach [36]. Zhu et al. (2010) proposed an enhanced STARFM (ESTARFM) algorithm, by introducing a conversion coefficient, retrieved from the ratio of change between the MODIS pixels and Landsat endmembers, into STARFM to enhance the prediction accuracy for heterogeneous landscapes [37]. Fu et al. (2013) modified the procedure of similar pixel selection in ESTARFM, according to the standard deviation of the reflectance and the number of land cover types within a local moving window [38]. Considering heterogeneous landscapes with complex vegetation types, soil water, and meteorological conditions in small-scale agricultural irrigation areas, Bai et al. (2017) applied ESTARFM to produce daily field-scale ET estimates based on Landsat and MODIS images for different crops [39]. However, ESTARFM is more computationally intensive, and requires at least two pairs of fine- and coarse-resolution images acquired at the same date between two base dates, which increases the difficulty of data acquisition and limits its applicability.

The goal of this work was to obtain accurate continuous daily ET values at the field-scale for heterogeneous agricultural areas. We developed a multiresolution modeling framework by combing STARFM with a linear unmixing model (u-STARFM) to resolve the difficulties that STARFM presents from the mixed pixel of MODIS in heterogeneous areas. The fusion methodology was applied to MODIS and ASTER images acquired during a period of crop development from June to September, 2012, for an oasis-desert agricultural region in the middle reaches of the Heihe River Basin. Instantaneous flux retrievals and daily ET estimates were estimated by using an enhanced SEBS model, namely MPDI (modified perpendicular drought index)-integrated SEBS, as proposed by Yi et al. (2018), thereby allowing an improved mapping of ET under water-limited conditions [40]. The results were validated for multiple land-cover types, including cropland, residential areas, woodland, water, desert, desert steppe, and wetlands, using in situ observations from eddy covariance (EC) systems.

2. Experimental Region and Data

2.1. Study Area

The study area (97.1°E–102.0°E and 37.7°N–42.7°N) was located in an oasis-desert agricultural region in the middle reaches of the Heihe River Basin (HRB) (Figure 1). The climate was continental, with an average annual temperature of 7.5 °C, an average annual rainfall of 136.8 mm, and an average annual potential evaporation of 1840 mm [41]. The dominant landscapes in the region were an artificial oasis, the Gobi Desert, and transitional zones between oasis and desert. The central part of the study area was a typical irrigated crop ecosystem, mainly covered by maize, vegetables, orchards, and wheat. The agriculture in the HRB depended heavily on irrigation water extracted from the Heihe River or from an aquifer. During the period of crop growth, the croplands were irrigated using flood irrigation at an interval between 20 days and one month.

The Multi-Scale Observation Experiment on Evapotranspiration (MUSOEXE) for heterogeneous land surfaces was conducted in the study area from May to September, 2012. This was the first thematic experiment in the Heihe Watershed Allied Telemetry Experimental Research (HiWATER) project [42]. It involved a flux observation network composed of two nested matrices; one large experimental area (30 km × 30 km) and one kernel experimental area (5.5 km × 5.5 km) (Figure 1). The accurate quantification of ET in an irrigated oasis using remote sensing is one of the prime targets of the artificial oasis experiment.

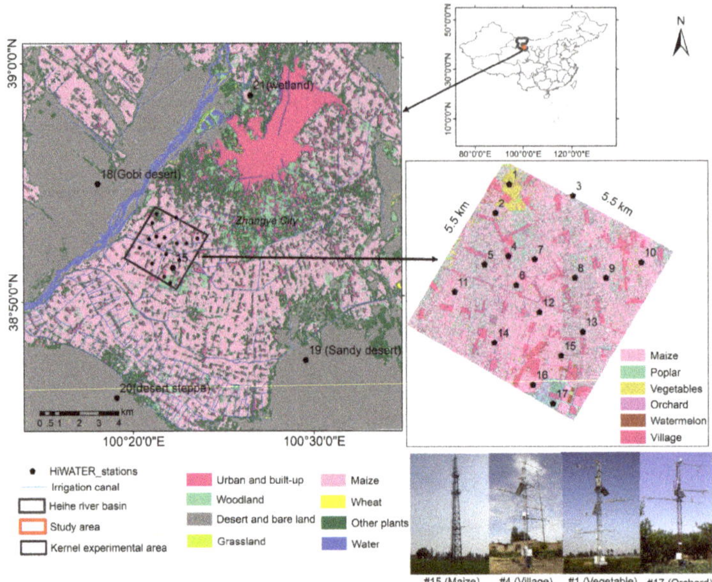

Figure 1. The experimental region and field observation system.

2.2. Field Measurements

The large experimental area contained five stations that had differing underlying surfaces, being the Gobi Desert (station #18), sandy desert (station #19), desert steppe (station #20), wetland (station #21) and maize (station #15). The kernel experimental area contained 16 stations, 13 of which were installed in seed maize plots, and the remaining three of which were installed in a vegetable plot (station #1), a village (station #4), and an orchard plot (station #17). The field observation system of the HiWATER mainly included an automatic weather station (AWS), an eddy covariance (EC), a soil moisture and temperature measurement system with wireless sensor network (WSN). The details of the meteorological and flux sites can be found in Xu et al. (2013) [43].

Each AWS was equipped with sensors to collect data including air temperature, wind speed and direction, air pressure, the relative air humidity, precipitation, soil moisture profile, solar radiation, four-component radiation, soil heat flux, and infrared temperature, every 10 min. The accurate quantification of ET estimates depends on the reliability of input meteorological variables. In this sense, the integrity and quality of these meteorological data were assessed through several quality control tests, including range (fixed or dynamic) test, step test, internal consistency test, and persistence test [44,45]. All variables were interpolated into a map, with a resolution of 90 m, covering the study area, using Kriging method [46]. The turbulent fluxes were measured at 10 Hz sampling frequency by EC. The fluxes of water vapor, latent heat, and sensible heat were processed into a half hour interval. The Gaussian fitting method was used to interpolate the missing data [47]. The energy balance closure was enforced into the EC system observations by using the Bowen ratio closure method [48].

2.3. Satellite Data

ASTER satellite images cover three visible to near-infrared (VNIR) bands with a resolution of 15 m, six shortwave infrared (SWIR) bands with a resolution of 30 m, and five thermal infrared (TIR) bands with a resolution of 90 m. During HiWATER-MUSOEXE, a total of eight ASTER Level-1 (ASTER L1A) images with minimum cloud cover (<5% within the study area), taken between May and September, 2012 (DOY 151, 167, 176, 192, 215, 231, 240, and 247), were collected from the online EARTHDATA database (https://earthdata.nasa.gov/). All of these images were subjected to radiative, atmospheric, and geometric corrections. The VNIR and SWIR bands of the chosen ASTER image were resampled to a resolution of 90 m, consistent with the TIR resolution. The ASTER land surface temperature (LST) was retrieved using the temperature and emissivity separation method [45].

The MODIS surface reflectance products (MOD09GA) and the LST products (MOD11A1) from 30 May (DOY 151) to 3 September (DOY 247), 2012, were collected from the online EARTHDATA database. Cloud mask products (MOD35) were used to detect clear-sky pixels, assuming a threshold at the 99% confidence interval. As a result of this data screening, the images for 54 days were considered to be cloudy on the experimental fields, with an average actual average revisit interval of two days. All of these MODIS products were projected to the UTM projection and resampled to a spatial resolution of 90 m, using the MODIS reprojection tool (MRT). The monthly land-cover classification products (30 m) for the HRB were acquired from the Ecological and Environmental Science Data Center for Western China (http://westdc.westgis.ac.cn). The leaf area index (LAI) was additionally calculated from ASTER images, using a look-up table constructed by a unified model developed by Zhao et al. (2015) [49].

3. Model Descriptions

A schematic overview of the framework for retrieving continuous daily ET values at the field-scale is illustrated in Figure 2, including inputs and image processing steps. The process consists of three steps. The first step is to apply the MPDI-integrated SEBS model to retrieve ET values on the day when satellites overpass for ASTER and MODIS, respectively. The coarse-resolution ET image is then unmixed, based on a linear spectral mixture model, to obtain the ET of each land-cover endmember. Finally, the unmixing of ET images is performed for the directly resampled MODIS ET data, and STARFM is run to predict ASTER-like ET images. The daily ET, retrieved from the MPDI-integrated SEBS and the fusion model, were all validated using ground measurements.

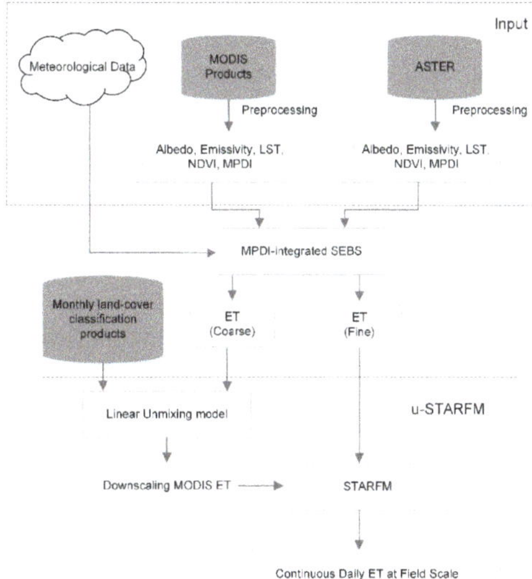

Figure 2. Schematic overview of the inputs and processing steps in the evapotranspiration (ET) data fusion system. (NDVI: normalized difference vegetation index; LST: land surface temperature; MPDI: the modified perpendicular drought index; MODIS: the moderate resolution imaging spectroradiometer; ASTER: the advanced space-borne thermal emission reflectance radiometer; STARFM: spatial and temporal adaptive reflectance fusion model; u-STARFM: the modified STARFM model)

3.1. A Brief Description of the MPDI-Integrated SEBS Model

The SEBS model is a single source model, and takes latent heat flux as the residual of the surface energy balance. It is commonly used to estimate atmospheric turbulent fluxes on the basis of remote sensing and meteorological data, which mainly consists of the estimation of land surface physical parameters, the roughness length for heat transfer, the sensible heat flux (H), and the latent heat flux (λET) [9]. The surface energy balance is normally written as:

$$\lambda ET = R_n - G - H \tag{1}$$

$$R_n = (1 - \alpha)R_{swd} + \varepsilon R_{lwd} - \varepsilon \delta T_s^4 \tag{2}$$

$$G = R_n[\Gamma_c + (1 - f_c)(\Gamma_s - \Gamma_c)] \tag{3}$$

where λET, R_n, G, and H are the latent heat flux, net radiation flux, soil heat flux, and sensible heat flux (W m^{-2}), respectively; R_{swd} and R_{lwd} are the downward solar radiation and longwave radiation (W m^{-2}), respectively; α and ε are the albedo and emissivity of land surface, respectively; δ is the Stefan–Boltzmann constant (W m^{-2}K^{-4}); T_s is the land surface temperature (K); f_c is the fraction of canopy cover; and $\Gamma_c = 0.05$ for the full vegetation canopy and $\Gamma_s = 0.315$ for bare soil [50]. Su (2002) detailed a set of equations for the estimation of the land surface physical parameters and variables. H is derived from similarity theory along with a dynamic roughness height formulation for the heat transfer. In the atmospheric surface layer (ASL), the similarity relationship for the profiles of mean wind speed (u) and mean temperature ($T_0 - T_a$) are usually described as:

$$u = \frac{u^*}{k}\left[\ln\left(\frac{z - d_0}{z_{0m}}\right) - \Psi_m\left(\frac{z - d_0}{L}\right) + \Psi_m\left(\frac{z_{0m}}{L}\right)\right] \tag{4}$$

$$T_0 - T_a = \frac{H}{ku^*\rho C_p}\left[\ln\left(\frac{z - d_0}{z_{0h}}\right) - \Psi_h\left(\frac{z - d_0}{L}\right) + \Psi_h\left(\frac{z_{0h}}{L}\right)\right] \tag{5}$$

where u is the mean wind speed, u^* is the friction velocity (m s^{-1}), k is the von Karman constant (0.4), z is the height above the land surface where the meteorological observations were made (m), C_p is the specific heat of air at constant pressure (J kg^{-1}K^{-1}), d_0 is the zero plane displacement height (m), z_{0h} and z_{0m} are the surface roughness heights for heat and momentum transport (m), Ψ_h and Ψ_m are the stability correction functions for heat and momentum transport, L is the Obukhov length (m), T_0 and T_a are the temperature of the land surface and air (K), and ρ is the density of air (kg m^{-3}). A set of equations for the estimation of land surface physical parameters and variables are detailed in Su (2002).

The SEBS algorithm assumes that the information on the ratio of actual to potential evaporation is implicitly embedded in the land surface temperature. This assumption is usually appropriate where available energy is the limiting factor for ET, but there is a problem when water availability becomes limiting for ET. Some recent studies have reported that SEBS can underestimate the sensible heat flux and overestimate ET in arid and semiarid regions [51–53]. To reduce the overestimation of ET for vegetation under water-limited conditions, Yi et al. (2018) proposed an enhanced SEBS model, namely MPDI-integrated SEBS, that integrated MPDI, as an indicator of soil moisture, into SEBS through a modified definition of the dimensionless k B^{-1} parameter. Here, we provide a brief description of the MPDI-integrated SEBS model. The k B^{-1} value is modified using a scaling factor (SF) represented by a reverse sigmoid function in the MPDI-integrated SEBS model:

$$k\,B_u^{-1} = SF \times k\,B^{-1} \tag{6}$$

$$SF = \left[a + \frac{1}{(1 + \exp(b - c/MPDI))}\right] \tag{7}$$

where k B_u^{-1} is the updated k B^{-1}; and a, b, and c are the coefficients of the reverse sigmoid function. MPDI represents the modified perpendicular drought index, which is based on the spatial distribution features of soil moisture in NIR-Red spectral space [54]. To obtain the coefficients of the sigmoid function (Equation (7)), we calibrated MPDI-integrated SEBS using an optimization by reducing the error between the simulated sensible heat flux and EC measurements in 2015 [40]. The parameters a, b, and c were determined to be 0.024, 3.1, and 1.6, respectively.

3.2. Description of the Scheme for Fusion of Daily ET Values at Different Resolutions

A sketch map of the fusion algorithm is shown in Figure 3. A pair of ASTER–MODIS ET for t_s (the basis date) and a MODIS ET for t_p (the predicted date) were used as the inputs of the fusion model. Firstly, the land cover components and their abundances were produced from a monthly high-resolution classification map. A downscaled MODIS ET was then retrieved by resolving the linear mixture model combined with the abundance of each land cover based on a least square method. Finally, the downscaled ET maps for t_s and t_p, along with ASTER ET for t_s, were inputted into STARFM to produce the ASTER-like ET on the predicted date.

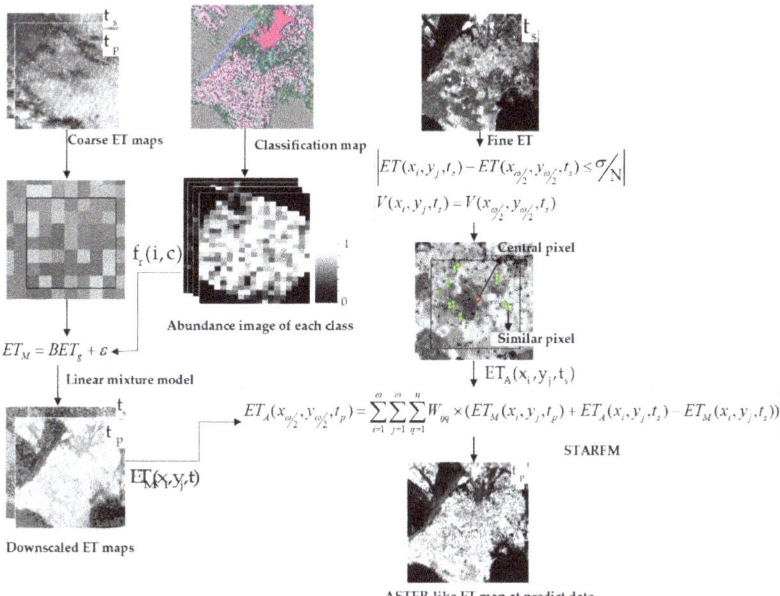

Figure 3. Schematic description of the data fusion procedure.

3.2.1. Unmixing the Coarse-Resolution Images

Due to the complexity of the land surface, most of the pixels in coarse-resolution images were mixed pixels (i.e., covered by multiple classes of land cover). A mixed pixel is typically modeled as a linear combination of endmembers and abundances [55,56]. Here, the ET value of a mixed pixel was calculated as the sum of the mean ET values of different land-cover types within the pixel, weighted by the corresponding abundance. A sliding window of [$\omega \times \omega$] MODIS pixels was applied to the land-cover image to record the endmember abundance matrix B, a [$\omega^2 \times g$] matrix with ω^2 rows (one for each pixel within the neighborhood) and g columns (one for each endmember). Abundances (i.e., land cover fractions) were the ratio of each endmember within the coarse-resolution pixel. The abundance was defined as:

$$f_r(e,\ g) = \frac{Q}{N},\ f_r(e,\ g) \geq 0, \sum f_r(e,\ g) = 1 \tag{8}$$

where $f_r(e,\ g)$ denotes the fraction of the gth endmember (i.e., the gth land cover type) in the coarse pixel e, Q denotes the total number of fine pixels for each land cover class in the pixel e, and N is the total number of fine pixels in the coarse pixel e. The numbers $f_r(e,\ g)$ should strictly be non-negative, and add to unity. The abundance of each land cover type was calculated every month from the monthly land-cover classification products in the HRB between May and September, 2012. The aim of unmixing was to solve for ET_g, a [g $\times$ 1] column vector that contained each daily ET values for each land cover type. ET_M was a [$\omega^2 \times 1$] column vector containing the daily ET values of each MODIS pixel in the sliding window which was currently being unmixed. This was achieved by minimizing the residual error (ε) of the linear model (Equation (9)) with an ordinary least square technique [57]:

$$\mathbf{ET_M = BET_g + \varepsilon} \tag{9}$$

3.2.2. STARFM

STARFM was originally designed by Gao et al. (2006) to fuse the surface reflectance of high-spatial-resolution Landsat and high-temporal-resolution MODIS images. The algorithm is based on the premise that both Landsat and MODIS imagery observe the same reflectance on the same day, biased by an error. This error is constant, assuming that the land cover type and system errors from satellites do not change over short time intervals. Hence, the error can be calculated if a base Landsat-MODIS image pair is available for the same date (t_s). The Landsat-like prediction images can then be obtained from the MODIS image on the predicted day and the error. Here, we applied STARFM to retrieve high-spatial-resolution ET maps at daily timesteps combining the spatiotemporal characteristics of MODIS and ASTER.

STARFM data processing included three major processing steps. Firstly, a sliding window of size $\omega \times \omega$ was applied to identify similar neighboring pixels. We added the monthly land cover classification maps as an auxiliary for searching for similar neighboring pixels. The similar neighbor pixels were determined by using the threshold of the standard deviation and the condition that the candidate pixel had the same land cover type (V) as the central pixel of the sliding window for t_s. This rule can be described as follows:

$$\begin{cases} \left| ET(x_i, y_i,\ t_s) - ET(x_{\omega/2},\ y_{\omega/2},\ t_s) \right| \leq \sigma/N \\ V(x_i, y_i,\ t_s) = V(x_{\omega/2},\ y_{\omega/2},\ t_s) \end{cases} \tag{10}$$

where (x_i, y_i) is the location of the similar pixel; $(x_{\omega/2},\ y_{\omega/2})$ is the location of the central pixel; σ is the standard deviation of ET within the sliding window; and N presents the total number of land cover types within the sliding window. Secondly, a normalized reverse distance (W_{ijq}) was used as the weight function combining three distances in Equation (11): (1) The ET difference between the base MODIS-ASTER pair (S_{ijq}); (2) the temporal difference between ET values of the MODIS images for date t_s and t_p (T_{ijq}); and (3) the spatial Euclidean distance between the central pixel ($x_{\omega/2},\ y_{\omega/2}$) and candidate pixel ($x_i, y_i$) for date t_p (D_{ijq}). The final step was to calculate the ET value of the central pixel for t_p. The algorithm is characterized in Equation (12):

$$W_{ijq} = \frac{\frac{1}{S_{ijq} \times T_{ijq} \times D_{ijq}}}{\sum_{i=1}^{\omega} \sum_{j=1}^{\omega} \sum_{q=1}^{n} \frac{1}{S_{ijq} \times T_{ijq} \times D_{ijq}}} \tag{11}$$

$$ET_A\left(x_{\omega/2},\ y_{\omega/2},\ t_p\right) = \sum_{i=1}^{\omega} \sum_{j=1}^{\omega} \sum_{q=1}^{n} W_{ijq} \times \left(ET_M\left(x_i,\ y_j,\ t_p\right) + ET_A\left(x_i, y_j, t_s\right) - ET_M(x_i, y_i, t_s) \right) \tag{12}$$

where ET_A and ET_M denote the fine-resolution (ASTER) and coarse-resolution (MODIS) ET, respectively, and n is the total number of coarse pixels within the predefined window. For our simulation, the sliding window size was set at 7×7 pixels for MODIS images. A more detailed description of the STARFM algorithm can be found in Gao et al. (2006).

4. Results

4.1. Validating the Quality of Meteorological Data

Metrological variables that were validated included: Daily solar radiation (R_s), daily mean, maximum and minimum air temperatures (T_m, T_{max}, and T_{min}, respectively), daily mean, maximum and minimum relative humidity of the air (RH_m, RH_{max}, and RH_{min}, respectively), and daily mean and maximum wind speed data (U_m and U_{max}, respectively). The percentage of flagged meteorological data (discarding data) for each variable on all the AWS stations are summarized in Table 1. On average, all the air temperature flagged data by the quality tests were below 2.0%. The persistence test for solar radiation showed the average and maximum percentages of data flagged were 10.25% and 60.91%, respectively. This was mainly due to constant values reported when there was a continuous sensor failure in station #4. The highest percentage of data flagged by persistence test was obtained for RH_{max}, with an average of 7.03%, and the highest fraction of 26.08% was detected in station #6. The maximum fraction (36.82%) of data flagged for U_m was detected in station #14.

Table 1. Percentage of flagged meteorological data by quality control tests (Max = maximum; Avg = average; stdv. = standard deviation of total automatic weather station (AWS) data). The percentages on each cell correspond to Max/Avg (stdv.).

Variables		Range Test (Fixed)	Range Test (Dynamic)	Step Test	Internal Consistency Test	Persistency Test
Air temperature	T_m	1.01/0.08 (0.30)	0/0 (0)	1.23/0.40 (0.45)	0.99/0.20 (0.25)	0.31/0.05 (0.24)
	T_{max}	0.12/0.02 (0.03)	0/0 (0)	2.01/0.30 (0.36)	0.99/0.20 (0.25)	0.31/0.05 (0.24)
	T_{min}	0.06/0.01 (0.02)	0/0 (0)		0.50/0.21 (0.12)	0.42/0.09 (0.12)
Solar radiation	R_s	2.8/0.30 (0.58)	56.02/13.29 (23.65)			60.91/10.25 (18.93)
Relative humidity	RH_m	0/0 (0)			1.52/0.21 (0.38)	17.26/2.02 (4.18)
	RH_{max}	0.10/0.01 (0.03)			1.03/0.05 (0.21)	26.08/7.03 (8.54)
	RH_{min}	0.82/0.03 (0.20)			1.35/0.18 (0.32)	15.28/6.10 (4.20)
Wind speed	U_m	1.23/0.08 (0.31)		1.34/0.22 (0.30)	36.82/9.05 (8.72)	34.18/8.03 (7.20)
	U_{max}	0.13/0.05 (0.04)			0.24/0.03 (0.05)	30.27/4.89 (5.65)

4.2. Evaluating the Performance of the MPDI-Integrated SEBS Model

The MPDI-integrated SEBS model was validated with flux data from all 21 EC stations. Note that some flux towers did not generate usable data for all the MODIS and ASTER dates. The scatter plots in Figure 4 show comparisons between the modeled values of sensible heat flux and latent heat flux from MPDI-integrated SEBS, and those observed from EC measurements. The results were further corroborated by the statistical performance metrics derived for each flux, as summarized in Table 2. The statistical metrics include mean bias error (Bias), root-mean-square error (RMSE), mean absolute bias error (MAE), and mean absolute percentage error (MAP; i.e., the MAE divided by the observed flux).

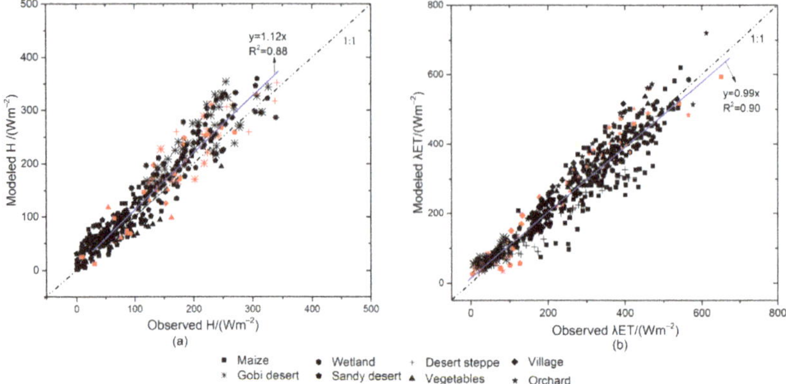

Figure 4. A comparison of the modeled and observed surface heat fluxes at the eight experimental sites from 30 May to 12 September, 2012. (**a**) Shows scatter plots of the modeled and observed sensible heat fluxes (H); and (**b**) shows scatter plots of the modeled and observed latent heat fluxes (λET). Red and black symbols represent ASTER and MODIS data, respectively.

Table 2. The statistical differences between surface heat fluxes modeled by MPDI-integrated SEBS, and observed surface heat fluxes.

Land Use Type	H (W m⁻²)				λET (W m⁻²)			
	Bias	**MAE**	**MAP**	**RMSE**	**Bias**	**MAE**	**MAP**	**RMSE**
	(W m⁻²)	(W m⁻²)	(%)	(W m⁻²)	(W m⁻²)	(W m⁻²)	(%)	(W m⁻²)
Overall	12.7	39.5	16.0	34.3	−4.8	40.9	10.9	40.0
Maize	10.1	43.7	15.6	38.5	−5.2	45.5	10.7	46.9
Orchard	−0.5	27.0	17.5	18.7	−20.4	48.2	12.4	42.4
Vegetable	7.7	21.5	16.1	19.2	−12.1	35.5	11.6	15.3
Gobi Desert	30.4	41.2	14.0	39.0	−2.8	28.2	14.1	20.6
Sandy desert	22.9	35.7	13.3	29.2	−5.2	24.6	13.2	20.5
Desert steppe	31.2	43.5	14.5	36.0	−7.1	27.5	14.8	23.8
Village	28.2	32.3	25.7	25.2	23.1	30.0	18.9	31.1
Wetland	4.9	17.5	15.8	14.9	−3.3	28.3	10.2	31.9

H: Sensible heat flux; λET: Latent heat flux; Bias: Mean difference between modeled and observed heat fluxes; MAE: Mean absolute bias error; MAP: The MAE divided by the observed flux; RMSE: Root-mean-square error.

The instantaneous sensible heat flux observations at satellite overpass times were compared with the sensible heat flux values estimated by MPDI-integrated SEBS, using MODIS and ASTER data, as shown in Figure 4a. The coefficient of determination (R^2) was 0.88, indicating a strong correlation between the modeled and measured values of H. The performance of MPDI-integrated SEBS for H estimation was good for all land types, with 16.0% of overall MAP (32% in Li et al., 2017). Compared to an earlier study (Huang et al., 2015), the accuracy of H estimation based on the MPDI-integrated SEBS model was higher than that based on the original SEBS model, with an RMSE of 34.3 W m⁻² and a Bias of 12.7 W m⁻² in the study area (the RMSE and Bias for H were 84.1 W m⁻² and −24.9 W m⁻², respectively, in Huang et al., 2015). The difference between the modeled and observed values of H was large at the village site, with Bias, MAE, RMSE, and MAP values of 28.2 W m⁻², 32.3 W m⁻², 25.7 W m⁻², and 25.2%, respectively. The representativeness of the satellite pixels for the land surface and vegetation parameters were likely responsible for the large errors at the village site, since the village had a significant spatial variability with croplands adjacent to the village site.

A comparison between the latent heat flux observed from EC systems, and the values modeled using the MPDI-integrated SEBS model using MODIS and ASTER data, is shown in Figure 4b. The coefficient of determination (R^2) was 0.90, which indicates a strong correlation between the modeled and the measured values of λET. Other than the village site, almost all of the modeled values of λET agree well with the observed values for all land-cover types, with MAP values ranging from

10% to 15%. Compared to the original SEBS, the overestimation of λET was improved when MPDI was integrated into SEBS, with an RMSE of 40.0 W m^{-2}, a Bias of -4.8 W m^{-2}, and a MAP of 10.9%, respectively (an RMSE of 69.4 W m^{-2}, a Bias of 117.8 W m^{-2}, and a MAP of 14% in Huang et al., 2015). Overall, the MPDI–SEBS integration model tended to outperform the original SEBS model, particularly in arid and semiarid regions where water is limited.

4.3. Assessing the Performance of the Fusion Approach on Daily ET Retrievals over Heterogeneous Regions

The daily ET estimates by the u-STARFM and the original SEBS were compared with EC measurements from the 21 observation stations in Figure 5, along with a statistical evaluation listed in Table 3. Both of the data fusion algorithms successfully predicted ASTER-like daily ET values from MODIS observations in homogeneous areas such as desert steppe, sandy desert, and the Gobi Desert (Figure 5a–c). The performances of u-STARFM and STARFM in the prediction of daily ET were consistent, with scatter points around the 1:1 line. They also had similar values of Bias, MAE, MAP, and RMSE (e.g., for sandy desert, Bias: 0.04 mm d^{-1} vs. 0.05 mm d^{-1}; MAE: 0.1 mm d^{-1} vs. 0.2 mm d^{-1}; MAP: 13.2% vs. 13.5%; RMSE: 0.2 mm d^{-1} vs. 0.3 mm d^{-1}). As for the heterogeneous areas, such as agricultural land and wetland, the predicted daily ET by u-STARFM more closely matched the actual observations (1:1 line) than that of STARFM (Figure 5d–h). The prediction errors of u-STARFM were lower than those of STARFM for all the EC sites covered by maize, orchard, vegetables, village, the Gobi Desert, sandy desert, desert steppe, and wetland, with lower MAP and RMSE (Table 3). Taking the maize plots for example, the RMSE and MAP decreased by 0.6 mm d^{-1} and 5.3%, respectively. When the unmixed fraction data took place of the directly resampled MODIS data for STARFM, the synthetic ASTER-like daily ET values had a higher accuracy than STARFM, with the overall MAP value of 12.9% vs. 17.2%. This was due to the fact that the unmixed fraction data as basis for u-STARFM were able to incorporate prior heterogeneity information, whereas the resampled MODIS data only offer the STARM uniform spot without reflecting the heterogeneous land cover. Overall, the modified STARFM had better performance than the STARFM in predicting high-resolution daily ET, particularly in heterogeneous areas.

Table 3. The statistical differences between the fused daily ET values, modeled by u-STARFM and STARFM, and the observed daily ET values.

Cover Type	u-STARFM				STARFM			
	Bias	MAE	MAP	RMSE	Bias	MAE	MAP	RMSE
	(mm d^{-1})	(mm d^{-1})	(%)	(mm d^{-1})	(mm d^{-1})	(mm d^{-1})	(%)	(mm d^{-1})
Maize	0.4	0.5	12.5	0.8	-0.6	0.8	17.8	1.4
Orchard	-0.5	0.8	14.0	0.9	0.2	1.0	19.1	1.2
Vegetable	-0.4	0.6	12.6	0.7	-0.1	0.9	17.5	1.1
Gobi Desert	-0.05	0.2	13.7	0.2	0.02	0.3	14.1	0.3
Sandy desert	0.04	0.1	13.2	0.2	0.05	0.2	13.5	0.3
Desert steppe	-0.03	0.1	12.6	0.2	-0.04	0.1	12.1	0.2
Village	0.3	0.8	20.4	1.0	-0.2	0.9	21.8	1.1
Wetland	-0.2	0.4	9.8	0.5	-0.1	0.6	12.0	0.6
Overall	0.3	0.5	12.9	0.7	-0.4	0.7	17.2	1.2

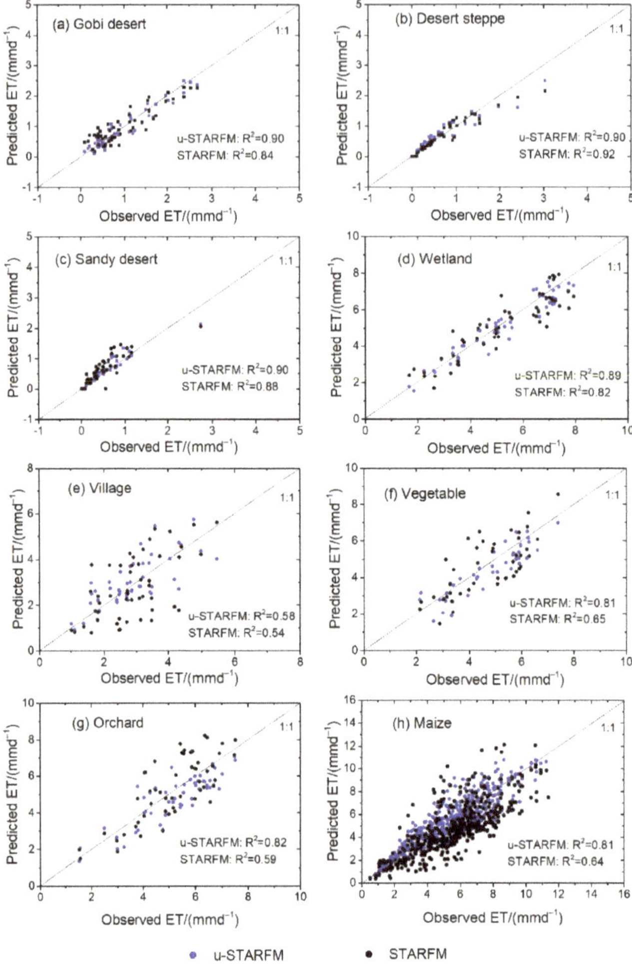

Figure 5. Scatter plots of observed and simulated daily ET, by u-STARFM (blue dots) and STARFM (black dots), for the Gobi Desert (**a**); desert steppe (**b**); sandy desert (**c**); wetland (**d**); village (**e**); vegetable (**f**); orchard (**g**); and maize (**h**).

4.4. Spatial Patterns in Daily ET

The synthetic image at ASTER spatial resolution was simulated by one pair of adjacent MODIS and ASTER images for the base date, and one MODIS image for the simulated date. The data fusion model used the temporally-closest image base pair for the predictions. For example, the synthetic ET images from 15 June (DOY 167), shown in Figure 6, were computed with input image pairs from 30 May (DOY 151), as well as the MODIS image acquired on 15 June. The synthetic ET images from 2 August (DOY 215), shown in Figure 8, were computed with input image pairs from 10 July (DOY 192), as well as the MODIS image acquired on 2 August. The performance was assessed by the ASTER ET images from 15 June and 2 August. The mean and standard deviation (stdv.) of the daily ET values are also marked. The mean represents the average strength of ET, and the standard deviation illustrates the degree of the spatial variability of ET values in the study region.

Figure 6 and Figure 8 show the prediction results in comparison with the inputs for: (a) MODIS ET image; (b) unmixed MODIS ET image; (c) ASTER ET image; (d) predicted ET image based on STARFM; and (e) predicted ET image based on u-STARFM, on 15 June and 2 August, respectively. The theoretical basis of the algorithms caused STARFM and u-STARFM to produce ASTER-like ET values, and caused the unmixing-based method to produce MODIS-like ET values. MODIS ET values remained relatively high and had a uniform spatial ET distribution with a high mean value (15 June: 4.2 mm d^{-1}; 2 August: 5.6 mm d^{-1}) and low standard deviation (15 June: 1.4 mm d^{-1}; 2 August: 1.7 mm d^{-1}). However, the unmixed MODIS ET provided relatively more spatial details, and mainly retained the characteristics of the MODIS image (e.g., mean value: 4.2 mm d^{-1} vs. 3.9 mm d^{-1}; stdv.: 1.4 mm d^{-1} vs 1.6 mm d^{-1}, on 15 June). The STARFM and u-STARFM algorithms accurately preserved most of the fine spatial detail in the ASTER ET maps (i.e., stdv. for STARFM: 1.8 mm d^{-1} vs. 2.3 mm d^{-1}; stdv. for u-STARFM: 2.1 mm d^{-1} vs. 2.3 mm d^{-1}). Compared to STARFM, the prediction of u-STARFM had a closer value to that of ASTER. The frequency of the difference between the predicted ASTER-like LST obtained from the u-STARFM and ASTER ET values distributed in $(-1, 1)$ was highest (Figure 7). Although there was a larger time span for the predicted date of 2 August, the prediction of u-STARFM was also able to describe the detailed ET information (mean: 4.8 mm d^{-1} vs 4.9 mm d^{-1}; stdv.: 2.4 mm d^{-1} vs. 2.8 mm d^{-1}) (Figure 8). Most of the difference, on 2 August, between the predicted ASTER-like ET values based on the u-STARFM and the ASTER ET values, ranged from -1 mm d^{-1} to 1 mm d^{-1} (Figure 9d). The results of the fused ET indicate that u-STARFM can correctly predict the daily ET values of small objects.

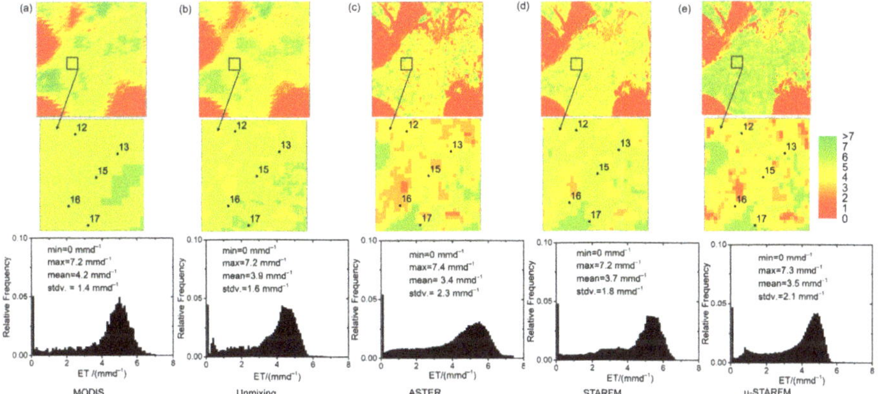

Figure 6. Spatial pattern, magnified view, and frequency distribution of ET maps, on 15 June, from: (**a**) MODIS; (**b**) unmixing method; (**c**) ASTER; (**d**) fusion using STARFM; and (**e**) u-STARFM model. Field boundaries were overlaid to compare inter- and intra-field variability in ET.

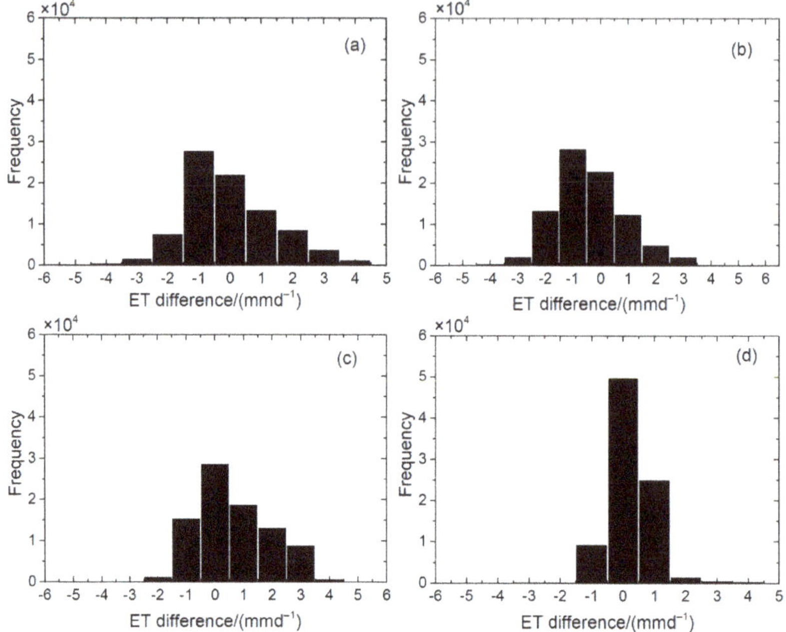

Figure 7. Histograms for: (**a**) MODIS ET and ASTER ET; (**b**) unmixing ET; (**c**) ASTER-like ET based on the STARFM model and ASTER ET; and (**d**) ASTER-like ET based on the u-STARFM model and ASTER ET, on 15 June.

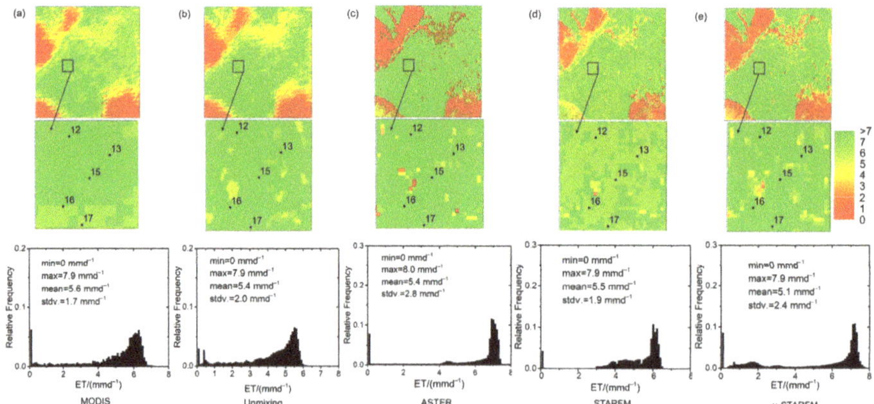

Figure 8. Spatial pattern, magnified view and frequency distribution of ET maps, on 2 August, from: (**a**) MODIS; (**b**) unmixing; (**c**) ASTER; (**d**) fusion using the STARFM model; and (**e**) fusion using the u-STARFM model. Field boundaries were overlaid to compare inter- and intra-field variability in ET.

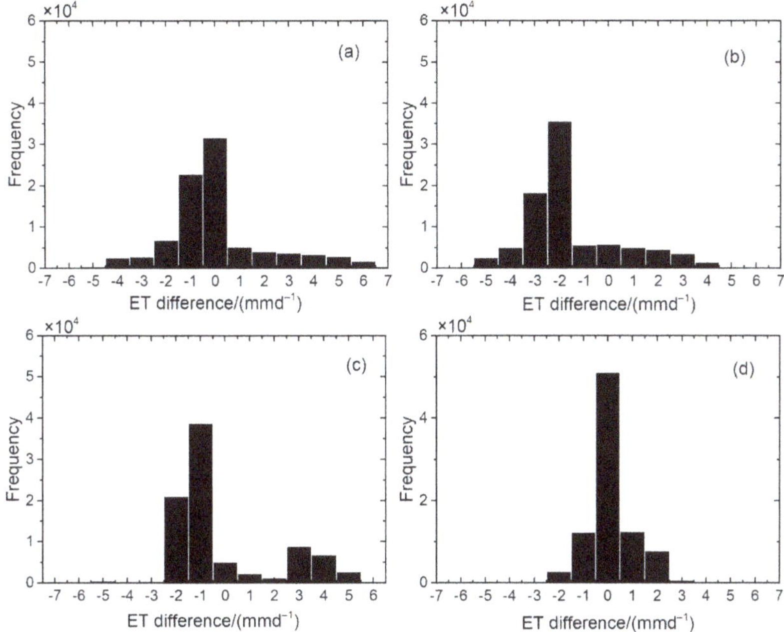

Figure 9. Histograms for: (**a**) MODIS ET and ASTER ET; (**b**) unmixing ET; (**c**) ASTER-like ET based on the STARFM model and ASTER ET; and (**d**) ASTER-like ET based on the u-STARFM model and ASTER ET, on 2 August.

The synthetic ASTER-like ET values based on u-STARFM also demonstrated that the spatial ET patterns in the agricultural fields over the course of the growing season were varied and complex, which could be the result of different crop types, irrigation practices, or soil characteristics, such as texture and water holding capacity. As is shown in the magnified views in Figures 6 and 8, ET was distributed relatively non-uniformly over the area. On 15 June, maize (i.e., stations #12, #13, #15, and #16) was in the tillering or jointing stage, with a mean LAI < 2, and the ET of maize was relatively low at around 3 mm d^{-1}, whereas orchard (i.e., #17), with a LAI > 3, had higher ET values of around 5mm d^{-1}. When all the crops were in the vigorous growth stage, on 2 August, with LAI > 5, ET was relatively uniformly distributed over the study area, with high ET values (ET > 5 mm d^{-1}) when the background crops were in the vigorous growth stage.

4.5. Temporal Patterns in Daily ET

Both the ASTER and synthetic ASTER-like daily ET streams were gap-filled to be continuous using a cubic spline interpolation. All ASTER-like results were compared with the observed ET in Figure 10. The results indicate that a good agreement was obtained between interpolated daily ET and observed daily ET, with RMSE, MAE, and MAP of 0.8 mm d^{-1}, 1.1 mm d^{-1}, and 13.4%, respectively. Figure 11 shows the temporal variation of ASTER-only and ASTER-like results at each of the eight flux tower sites. Clearly, ASTER, with a revisit period of 7–16 days, could hardly capture the temporal variation of daily ET values and was expected to result in large uncertainties. The role played by the MODIS ET in guiding the temporal interpolation of daily ET values between ASTER overpasses was evident from the fused time series. The complete time series of ASTER-like ET estimates generally produced reasonably accurate ET trends for all land-cover types. The day-to-day fluctuations in the ET data stream emphasized the detailed behavior of water stress, crop growth dynamics, and meteorological conditions at the different sites. For the sparsely vegetated sites (i.e., the Gobi Desert, sandy desert, and desert steppe), ET values were less than 1 mm d^{-1}, except for on rainy days. Additionally, the sporadic peaks of daily ET always occurred after rainfall events at these three sites, since the proportion of evaporation from bare soil was high. The fluctuations of daily ET values in vegetated areas covered by maize, vegetables, orchard, wetland, and village were clearly highly complex. Figure 6a–e highlights some significant differences among these sites for different vegetated surfaces, where the individual effects of soil evaporation, soil moisture storage, stomatal regulation, transpiration, and interception storage were all implicitly incorporated into the resulting surface evapotranspiration [52]. The u-STARFM is better able to capture the response of ET to rainfall and irrigation events. Spikes in ET usually occurred after rainfall or on days of irrigation. When precipitation occurred, the daily values of ET at all sites were relatively low, whereas they were relatively high after rainfall (DOY 174, 183, 193, 207). Additionally, peaks in ET values always occurred on days of irrigation (DOY 176, 205, 223, 238).

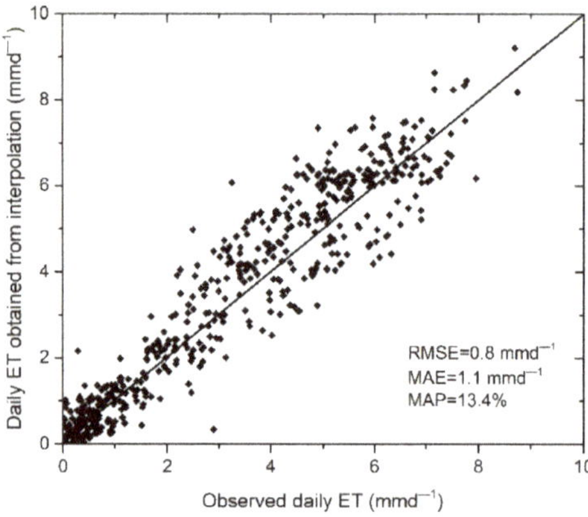

Figure 10. A comparison of the continuous daily ET by interpolation and observed daily ET.

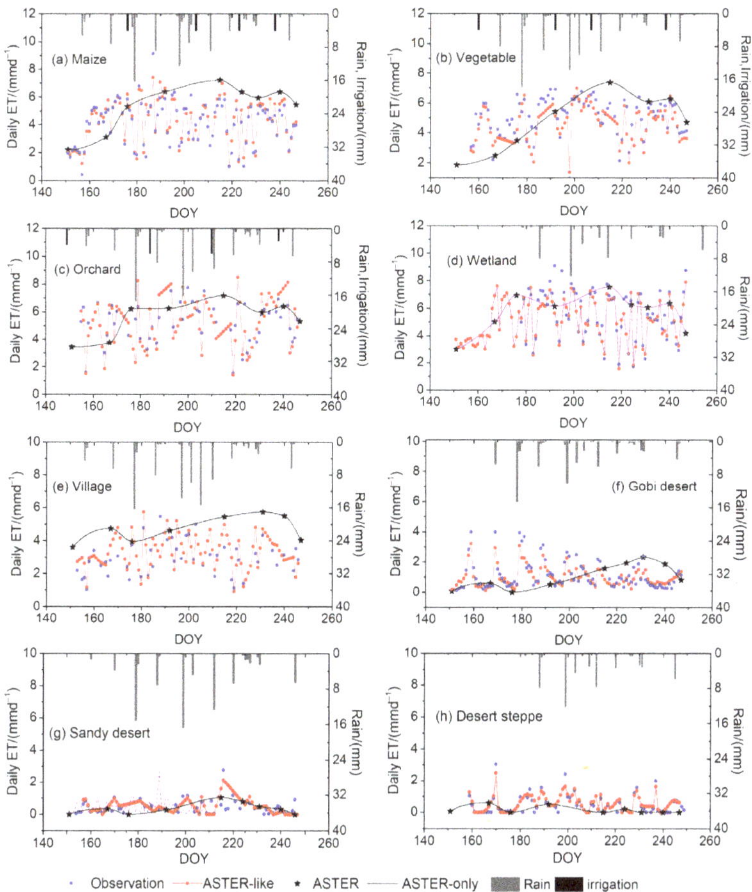

Figure 11. Time series of observed ET (blue dots), daily ET estimates from u-STARFM (red line with dots), and ASTER ET (black pentagram and line) from May to September, 2012. Rainfall and irrigation events are shown as blue and grey bars, respectively.

5. Conclusions

Measurements of continuous daily evapotranspiration (ET) at the field-scale are of substantial benefit for agricultural water management worldwide. The current lack of concurrent high spatial- and temporal-resolution remote sensing (RS) data significantly limits the applicability of RS-based methodologies. We proposed a data fusion framework for predicting continuous daily ET values at the field-scale by optimally exploiting the combined datasets of MODIS and ASTER. The STARFM algorithm, which was devised for producing Landsat surface reflectance time series, was modified and improved using a linear unmixing method (u-STARFM) for generating continuous daily ET estimates at the field-scale for heterogenous agricultural areas. In this study, an experiment was performed from July to September, 2012, in the middle reaches of the Heihe River Basin, Northwest China. Compared with the observed flux data, the synthetic fine-resolution daily ET estimates produced by the u-STARFM model had higher accuracies than those produced by STARFM, especially for heterogeneous agricultural areas.

Although the applications of data fusion to ET products from different satellite sensors is not straightforward, previous research has emphasized the potential advantages of STARFM for generating daily ET products with a high spatial and temporal resolution. However, STARFM often relies on the existence of "homogeneous" pixels at both fine- and coarse-resolution, which is more problematic for ET estimation than for reflectance or vegetation indices. The u-STARFM model presents several improvements over previous models. The most significant improvement is the ability to unmix coarse-resolution MODIS ET maps based on the abundance of land-cover types, and to replace the resampled MODIS ET values in the STARFM. The unmixed fraction of the unmixing data, which incorporates prior information about the MODIS ET values of each endmember in heterogeneous areas, can increase the probability of searching the "homogeneous" pixels and guarantee fusion accuracy. And the u-STARFM tends to outperform the original STARFM in irrigated heterogeneous agricultural areas, with overall MAP of 12.9% vs. 17.2%. Although many agricultural lands are smaller than one MODIS pixel, u-STARFM can adjust the daily ET changes of mixed MODIS pixels to the changes of internal ASTER pixels. Secondly, the land cover data was used as auxiliary information for the selection of similar pixels, which allowed more accurate similar pixels to be obtained. Thirdly, unlike ESTARFM, which requires at least two pairs of fine- and coarse-resolution images acquired on the same day, the u-STARFM needs just one input pair. Accordingly, in some cloudy regions, where it is difficult to acquire two high-quality input pairs simultaneously, u-STARFM may be more appropriate.

It should be noted that u-STARFM also contains a few limitations and constraints. Firstly, it assumes that the land-cover type is unchanged during the predicted period. Similar to STARFM and ESTARFM, u-STARFM cannot accurately predict transient changes that are not recorded in any of the base fine-resolution images. Therefore, combining u-STARFM with the STAARCH algorithm may be a feasible approach. Secondly, the size of the sliding window plays an important role in searching for similar pixels, and directly affects their weight distribution. The determination of sliding window size is not strictly regulated in the u-STARFM, being defined with an empirical or semi-quantitative method. Different study areas may require setting up different values, and it is necessary to perform a sensitivity analysis of different window sizes before the modeling.

In conclusion, u-STARFM improved the capability for producing daily ET products with both high-spatial resolution and frequent coverage from multi-source satellite data, which capture field-to-field variability in water usage. This information, if generated operationally, could be of utility for irrigation managers at the scale of individual fields, as well as for the regional monitoring of water use toward allocation and conservation efforts. Although the current study applied data fusion to ASTER and MODIS images, the techniques described here could be easily applied to other sensors. Further research could continue to analyze algorithm performance in various situations.

Author Contributions: Z.Y., H.Z., and Y.J. conceived and designed the experiments; Z.Y performed the experiments; Z.Y and H.Z. analyzed the data; and Z.Y. wrote the paper.

Funding: This research was funded by the National Key Research Project of Stereoscopic Monitoring System for Water Resources and the Application of Remote Sensing Technology, grant number [2017YFC0405803]; the Dynamic Assessment and Prediction of Monthly Water Storage and Water Demand in China, grant number [WR0145B012017]; the Special Fund for the Commercialization of Research Findings from China Institute of Water Resources and Hydropower Research—Actual Irrigated Area Estimation Using Remote Sensing, grant number [WR1003A112016]; and the Framework Design of National Intelligent Water Network Project, grant number [WR0145B272016].

Conflicts of Interest: The authors declare no conflicts of interest.

References

1. Katul, G.G.; Oren, R.; Manzoni, S.; Higgins, C.; Parlange, M.B. Evapotranspiration: A process driving mass transport and energy exchange in the soil-plant-atmosphere-climate system. *Rev. Geophys.* **2012**, *50*, RG3002. [CrossRef]
2. Anderson, M.C.; Allen, R.G.; Morse, A.; Kustas, W.P. Use of Landsat thermal imagery in monitoring evapotranspiration and managing water resources. *Remote Sens. Environ.* **2012**, *122*, 50–65. [CrossRef]

3. Magali, O.L.; Isidro, C.; Christopher, M.U.N.; Samuel, O.F.; Carlos, P.E.; Claudio, B.; Alfonso, C. Estimating Evapotranspiration of an Apple Orchard Using a Remote Sensing-Based Soil Water Balance. *Remote Sens.* **2016**, *8*, 253. [CrossRef]

4. Wu, X.J.; Zhou, J.; Wang, H.J.; Li, Y.; Zhong, B. Evaluation of irrigation water use efficiency using remote sensing in the middle reach of the Heihe river, in the semi-arid Northwestern China. *Hydrol. Process.* **2015**, *29*, 2243–2257. [CrossRef]

5. Yang, Y.T.; Shang, S.H.; Jiang, L. Remote sensing temporal and spatial patterns of evapotranspiration and the responses to water management in a large irrigation district of North China. *Agric. For. Meteorol.* **2012**, *164*, 112–122. [CrossRef]

6. Hu, G.C.; Jia, L. Monitoring of Evapotranspiration in a Semi-Arid Inland River Basin by Combining Microwave and Optical Remote Sensing Observations. *Remote Sens.* **2015**, *7*, 3056–3087. [CrossRef]

7. Nagler, P.L.; Morino, K.; Murray, R.S.; Osterberg, J.; Glenn, E.P. An empirical algorithm for estimating agricultural and riparian evapotranspiration using MODIS enhanced vegetation index and ground measurements of ET. I. Description of method. *Remote Sens.* **2009**, *1*, 1273–1297. [CrossRef]

8. Bastiaanssen, W.G.M.; Menenti, M.; Feddes, R.A.; Holtslag, A.A.M. A remote sensing surface energy balance algorithm for land (SEBAL)1. Formulation. *J. Hydrol.* **1998**, *212–213*, 198–212. [CrossRef]

9. Su, Z. The Surface Energy Balance System (SEBS) for estimation of turbulent heat fluxes. *Hydrol. Earth Syst. Sci.* **2002**, *6*, 85–99. [CrossRef]

10. Allen, R.G.; Tasumi, M.; Trezza, R. Satellite-based energy balance for mapping evapotranspiration with internalized calibration (METRIC)-model. *J. Irrig. Drain. Eng.* **2007**, *133*, 380–394. [CrossRef]

11. Kustas, W.P.; Norman, J.M. A two-source approach for estimating turbulent fluxes using multiple angle thermal infrared observations. *Water Resour. Res.* **1997**, *33*, 1495–1508. [CrossRef]

12. Leuning, R.; Zhang, Y.Q.; Rajaud, A.; Cleugh, H.; Tu, K. A simple surface conductance model to estimate regional evaporation using MODIS leaf area index and the Penman-Monteith equation. *Water Resour. Res.* **2008**, *44*, 652–655. [CrossRef]

13. Yao, Y.J.; Liang, S.L.; Li, X.L.; Chen, J.Q.; Wang, K.C.; Jia, K. A satellite-based hybrid algorithm to determine the Priestley-Taylor parameter for global terrestrial latent heat flux estimation across multiple biomes. *Remote Sens. Environ.* **2015**, *165*, 216–233. [CrossRef]

14. Huang, C.L.; Chen, W.J.; Li, Y.; Shen, H.F.; Li, X. Assimilating multi-source data into land surface model to simultaneously improve estimations of soil moisture, soil temperature, and surface turbulent fluxes in irrigated fields. *Agric. Forest Meteorol.* **2016**, *230–231*, 142–156. [CrossRef]

15. Hartanto, I.M.; van der Kwast, J.; Alexandridis, T.K.; Almeida, W.; Song, Y.; van Andel, S.J.; Solomatine, D.P. Data assimilation of satellite-based actual evapotranspiration in a distributed hydrological model of a controlled water system. *Int. J. Appl. Earth Observ. Geoinf.* **2017**, *57*, 123–135. [CrossRef]

16. Yang, Y.T. Evapotranspiration over Heterogeneous Vegetated Surfaces. Ph.D. Thesis, Tsinghua University, Beijing, China, 2015.

17. Wang, K.; Dickinson, R.E. A review of global terrestrial evapotranspiration: Observation, modelling, climatology, and climatic variability. *Rev. Geophys.* **2011**, *50*, 1–54. [CrossRef]

18. Verstraeten, W.W.; Veroustraete, F.; Feyen, J. Assessment of Evapotranspiration and Soil Moisture Content across Different Scales of Observation. *Sensors* **2008**, *8*, 70–117. [CrossRef] [PubMed]

19. Evett, S.R.; Kustas, W.P.; Gowda, P.H.; Anderson, M.C.; Prueger, J.H.; Howell, T.A. Overview of the Bushland Evapotranspiration and Agricultural Remote sensing EXperiment 2008 (BEAREX08): A field experiment evaluating methods for quantifying ET at multiple scales. *Adv. Water Resour.* **2012**, *50*, 4–19. [CrossRef]

20. Kustas, W.P.; Li, F.; Jackson, T.J.; Prueger, J.H.; MacPherson, J.I.; Wolde, M. Effects of remote sensing pixel resolution on modeled energy flux variability of croplands in Iowa. *Remote Sens. Environ.* **2004**, *92*, 535–547. [CrossRef]

21. He, R.Y.; Jin, Y.F.; Kandelous, M.M.; Zaccaria, D.; Sanden, B.L.; Snyder, R.L.; Jiang, J.B.; Hopmans, J.W. Evapotranspiration Estimate over an Almond Orchard Using Landsat Satellite Observations. *Remote Sens.* **2017**, *9*, 436. [CrossRef]

22. Senay, G.B.; Friedrichs, M.; Singh, R.K.; Velpuri, N.M. Evaluating Landsat 8 evapotranspiration for water use mapping in the Colorado River Basin. *Remote Sens. Environ.* **2016**, *185*, 171–185. [CrossRef]

23. McCabe, M.F.; Wood, E.F. Scale Influences on the Remote Estimation of Evapotranspiration Using Multiple Satellite Sensors. *Remote Sens. Environ.* **2006**, *105*, 271–285. [CrossRef]

24. Singh, R.K.; Senay, G.B.; Velpuri, N.M.; Bohms, S.; Verdin, J.P. On the Downscaling of Actual Evapotranspiration Maps Based on Combination of MODIS and Landsat-Based Actual Evapotranspiration Estimates. *Remote Sens.* **2014**, *6*, 10483–10509. [CrossRef]

25. Ke, Y.H.; Im, J.H.; Park, S.; Gong, H.L. Downscaling of MODIS One Kilometer Evapotranspiration Using Landsat-8 Data and Machine Learning Approaches. *Remote Sens.* **2016**, *8*, 215. [CrossRef]

26. Gao, F.; Masek, J.; Schwaller, M.; Hall, F. On the Blending of the Landsat and MODIS Surface Reflectance: Predicting Daily Landsat Surface Reflectance. *IEEE Trans. Geosci. Remote Sens.* **2006**, *44*, 2207–2218.

27. Bhandari, S.; Phinn, S.; Gill, T. Preparing Landsat Image Time Series (LITS) for Monitoring Changes in Vegetation Phenology in Queensland, Australia. *Remote Sens.* **2012**, *4*, 1856–1886. [CrossRef]

28. Li, X.D.; Ling, F.; Foody, M.G.; Ge, Y.; Zhang, Y.H.; Du, Y. Generating a series of fine spatial and temporal resolution land cover maps by fusing coarse spatial resolution remotely sensed images and fine spatial resolution land cover maps. *Remote Sens. Environ.* **2017**, *196*, 293–311. [CrossRef]

29. Weng, Q.H.; Fu, P.; Gao, F. Generating Daily Land Surface Temperature at Landsat Resolution by Fusing Landsat and MODIS data. *Remote Sens. Environ.* **2014**, *45*, 55–67. [CrossRef]

30. Yang, G.J.; Weng, Q.H.; Pu, R.L.; Gao, F.; Sun, C.H.; Li, H.; Zhao, C.J. Evaluation of ASTER-Like Daily Land Surface Temperature by Fusing ASTER and MODIS Data during the HiWATER-MUSOEXE. *Remote Sens.* **2016**, *8*, 75. [CrossRef]

31. Cammalleri, C.; Anderson, M.C.; Gao, F.; Hain, C.R.; Kustas, W.P. A Data Fusion Approach for Mapping Daily Evapotranspiration at Field Scale. *Water Resour. Res.* **2013**, *49*, 4672–4686. [CrossRef]

32. Cammalleri, C.; Anderson, M.C.; Gao, F.; Hain, C.R.; Kustas, W.P. Mapping Daily Evapotranspiration at Field Scales over Rainfed and Irrigated Agricultural Areas Using Remote Sensing Data Fusion. *Agric. For. Meteorol.* **2014**, *186*, 1–11. [CrossRef]

33. Yang, Y.; Anderson, M.C.; Gao, F.; Hain, C.R.; Semmens, K.A.; Kustas, W.P.; Noormets, A.; Wynne, R.H.; Thomas, V.A.; Sun, G. Daily Landsat-scale Evapotranspiration Estimation over a Forested Landscape in North Carolina, USA Using Multi-satellite Data Fusion. *Hydrol. Earth Syst. Sci.* **2016**, *21*, 1017–1037. [CrossRef]

34. Semmens, K.A.; Anderson, M.C.; Kustas, W.P.; Gao, F.; Alfieri, J.G.; McKee, L.; Prueger, J.H. Monitoring Daily Evapotranspiration over Two California Vineyards Using Landsat 8 in a Multi-sensor Data Fusion Approach. *Remote Sens. Environ.* **2016**, *185*, 155–170. [CrossRef]

35. Li, Y.; Huang, C.L.; Hou, J.L.; Gu, J.; Zhu, G.F.; Li, X. Mapping Daily Evapotranspiration Based on Spatiotemporal Fusion of ASTER and MODIS Images over Irrigated Agricultural Areas in the Heihe River Basin, Northwest China. *Agric. For. Meteorol.* **2017**, *244–245*, 82–97. [CrossRef]

36. Hilker, T.; Wulder, M.A.; Coops, N.C.; Linke, J.; McDermid, G.; Masek, J.G.; Gao, F.; White, J.C. A new data fusion model for high spatial- and temporal-resolution mapping of forest disturbance based on Landsat and MODIS. *Remote Sens. Environ.* **2009**, *113*, 1613–1627. [CrossRef]

37. Zhu, X.L.; Chen, J.; Gao, F.; Chen, X.H.; Masek, J.G. An Enhanced Spatial and Temporal Adaptive Reflectance Fusion Model for Complex Heterogeneous Regions. *Remote Sens. Environ.* **2010**, *114*, 2610–2623. [CrossRef]

38. Fu, D.J.; Chen, B.Z.; Wang, J.; Zhu, X.L.; Hilker, T. An Improved Image Fusion Approach Based on Enhanced Spatial and Temporal the Adaptive Reflectance Fusion Model. *Remote Sens.* **2013**, *5*, 6346–6360. [CrossRef]

39. Bai, L.L.; Cai, J.B.; Liu, Y.; Chen, H.; Zhang, B.Z.; Huang, L.X. Responses of field evapotranspiration to the changes of cropping pattern and groundwater depth in large irrigation district of Yellow River basin. *Agric. Water Manag.* **2017**, *188*, 1–11. [CrossRef]

40. Yi, Z.Y.; Zhao, H.L.; Jiang, Y.Z.; Yan, H.W.; Cao, Y.; Huang, Y.Y.; Hao, Z. Daily Evapotranspiration Estimation at the Field Scale: Using the Modified SEBS Model and HJ-1 Data in a Desert-Oasis Area, Northwestern China. *Water* **2018**, *10*, 640. [CrossRef]

41. Zhang, M.M.; Wang, S.; Fu, B.J.; Gao, G.Y.; Shen, Q. Ecological effects and potential risks of the water diversion project in the Heihe River Basin. *Sci. Total Environ.* **2018**, *619–620*, 794–803. [CrossRef] [PubMed]

42. Li, X.; Liu, S.M.; Xiao, Q.; Ma, M.G.; Jin, R.; Che, T.; Wang, W.Z.; Hu, X.L.; Xu, Z.W.; Wen, J.G.; et al. A multiscale dataset for understanding complex eco-hydrological processes in a heterogeneous oasis system. *Sci. Data* **2017**, *4*, 170083. [CrossRef] [PubMed]

43. Xu, Z.W.; Liu, S.M.; Li, X.; Shi, S.J.; Wang, J.M.; Zhu, Z.L.; Xu, T.R.; Wang, W.; Ma, M. Intercomparison of surface energy flux measurement systems used during the HiWATER-MUSOEXE. *J. Geophys. Res. Atmos.* **2013**, *118*, 13140–13157. [CrossRef]

44. Estévez, J.; Gavilán, P.; Giráldez, J.V. Guidelines on validation procedures for meteorological data from automatic weather stations. *J. Hydrol.* **2011**, *402*, 144–154. [CrossRef]
45. Estévez, J.; García-Marín, A.P.; Morábito, J.A.; Cavagnaro, M. Quality assurance procedures for validating meteorological input variables of reference evapotranspiration in mendoza province (Argentina). *Agric. Water Manag.* **2016**, *172*, 96–109. [CrossRef]
46. Xu, Z.W.; Ma, Y.F.; Liu, S.M.; Shi, W.J.; Wang, J.M. Assessment of the Energy balance closure under advective conditions and its impact using remote sensing data. *J. Appl. Meteorol. Climatol.* **2017**, *56*, 127–140. [CrossRef]
47. Li, H.; Sun, D.L.; Yu, Y.Y.; Wang, H.Y.; Liu, Y.L.; Liu, Q.H.; Du, Y.M.; Wang, H.; Cao, B. Evaluation of the VIIRS and MODIS LST Products in an Arid Area of Northwest China. *Remote Sens. Environ.* **2014**, *142*, 111–121. [CrossRef]
48. Zhong, B.; Yang, A.X.; Nie, A.H.; Yao, Y.J.; Zhang, H.; Wu, S.L.; Liu, Q.H. Finer Resolution Land-Cover Mapping Using Multiple Classifiers and Multisource Remotely Sensed Data in the Heihe River Basin. *IEEE J. Sel. Top. Appl. Earth Obs. Remote Sens.* **2015**, *99*, 1–20. [CrossRef]
49. Zhao, J.; Li, J.; Liu, Q.H.; Fan, W.J.; Zhong, B.F.; Wu, S.L.; Yang, L.; Zeng, Y.L.; Xu, B.D.; Yin, G.F. Leaf Area Index Retrieval Combining HJ1/CCD and Landsat8/OLI Data in the Heihe River Basin, China. *Remote Sens.* **2015**, *7*, 6862–6885. [CrossRef]
50. Kustas, W.P.; Daughtry, C.S.T. Estimation of the soil heat flux/net radiation ratio from spectral data. *Agric. For. Meteorol.* **1990**, *49*, 205–223. [CrossRef]
51. Gokmen, M.; Vekerdy, Z.; Verhoef, A.; Verhoef, W.; Batelaan, O.; van der Tol, C. Integration of soil moisture in SEBS for improving evapotranspiration estimation under water stress conditions. *Remote Sens. Environ.* **2012**, *121*, 261–274. [CrossRef]
52. Huang, C.L.; Li, Y.; Gu, J.; Lu, L.; Li, X. Improving Estimation of Evapotranspiration under Water-Limited Conditions Based on SEBS and MODIS Data in Arid Regions. *Remote Sens.* **2015**, *7*, 16795–16814. [CrossRef]
53. Li, Y.; Zhou, J.; Wang, H.J.; Li, D.Z.; Jin, R.; Zhou, Y.Z.; Zhou, Q.G. Integrating soil moisture retrieved from L-band microwave radiation into an energy balance model to improve evapotranspiration estimation on the irrigated oases of arid regions in northwest China. *Agric. Forest Meteorol.* **2015**, *214–215*, 306–318. [CrossRef]
54. Ghulam, A.; Qin, Q.M.; Teyip, T.; Li, Z.L. Modified perpendicular drought index (MPDI): A real-time drought monitoring method. *ISPRS J. Photogramm. Remote Sens.* **2007**, *62*, 150–164. [CrossRef]
55. Settle, J.J.; Drake, N. Linear Mixing and the Estimation of Ground Cover Proportions. *Int. J. Remote Sens.* **1993**, *14*, 1159–1177. [CrossRef]
56. Zhang, Y.; Li, L.; Chen, L.Q.; Liao, Z.H.; Wang, Y.C.; Wang, B.Y.; Yang, X.Y. A Modified Multi-Source Parallel Model for Estimating Urban Surface Evapotranspiration Based on ASTER Thermal Infrared Data. *Remote Sens.* **2017**, *9*, 1029. [CrossRef]
57. Xie, D.F.; Zhang, J.S.; Zhu, X.F.; Pan, Y.Z.; Liu, H.L.; Yuan, Z.M.Q.; Yun, Y. An Improved STARFM with Help of an Unmixing-Based Method to Generate High Spatial and Temporal Resolution Remote Sensing Data in Complex Heterogeneous Regions. *Sensors* **2016**, *16*, 207. [CrossRef] [PubMed]

Article

Estimating Calibration Variability in Evapotranspiration Derived from a Satellite-Based Energy Balance Model

Sulochan Dhungel * and Michael E. Barber

Department of Civil and Environmental Engineering, University of Utah, 110 S. Central Campus Drive, Salt Lake City, UT 84112, USA; michael.barber@utah.edu
* Correspondence: sulochandhungel@gmail.com or sulochan.dhungel@utah.edu

Received: 21 September 2018; Accepted: 24 October 2018; Published: 26 October 2018

Abstract: Computing evapotranspiration (ET) with satellite-based energy balance models such as METRIC (Mapping EvapoTranspiration at high Resolution with Internalized Calibration) requires internal calibration of sensible heat flux using anchor pixels ("hot" and "cold" pixels). Despite the development of automated anchor pixel selection methods that classify a pool of candidate pixels using the amount of vegetation (normalized difference vegetation index, NDVI) and surface temperature (T_s), final pixel selection still relies heavily on operator experience. Yet, differences in final ET estimates resulting from subjectivity in selecting the final "hot" and "cold" pixel pair (from within the candidate pixel pool) have not yet been investigated. This is likely because surface properties of these candidate pixels, as quantified by NDVI and surface temperature, are generally assumed to have low variability that can be attributed to random noise. In this study, we test the assumption of low variability by first applying an automated calibration pixel selection process to 42 nearly cloud-free Landsat images of the San Joaquin area in California taken between 2013 and 2015. We then compute T_s (vertical near-surface temperature differences) vs. T_s relationships at all pixels that could potentially be used for model calibration in order to explore ET variance between the results from multiple calibration schemes where NDVI and T_s variability is intrinsically negligible. Our results show significant variability in ET (ranging from 5% to 20%) and a high—and statistically consistent—variability in dT values, indicating that there are additional surface properties affecting the calibration process not captured when using only NDVI and T_s. Our findings further highlight the potential for calibration improvements by showing that the dT vs. T_s calibration relationship between the cold anchor pixel (with lowest dT) and the hot anchor pixel (with highest dT) consistently provides the best daily ET estimates. This approach of quantifying ET variability based on candidate pixel selection and the accompanying results illustrate an approach to quantify the biases inadvertently introduced by user subjectivity and can be used to inform improvements on model usability and performance.

Keywords: remote sensing; surface energy balance model; calibration; METRIC; Google Earth Engine

---

1. Introduction

Evapotranspiration (ET) is an essential part of any hydrologic investigation as it represents the flux of water between soils, plants, and biosphere [1]. Since 75% of the world's freshwater withdrawals support agriculture [2], estimating ET as consumptive use accurately over a watershed is crucial for efficient water resources planning [3]. ET can be assessed at different scales of observation with multiple approaches and methods [4] such as the field-scale water balance approach using weighing lysimeter [5], the field-scale energy balance approach using the Eddy Covariance (EC) technique [6] and scintillometers [7], and the leaf- or plant-scale energy balance approach using

Sap-flow techniques [8]. These methods vary in size of effective area measured, cost, complexity, and accuracy. However, the footprint of field ET measured by these methods is smaller than the watershed area, usually by orders of magnitude, which often creates significant discrepancies in ET estimates at spatial scales larger than plot or field scales [9]. Most commonly, ET over an agricultural region is measured by multiplying a weather-based estimate of reference ET by crop coefficients (which can vary with crop type and crop growth stage) [10]. This method of estimating ET is popular since it can be applied with ease. However, since crop growth and vegetative stage can vary significantly even for the same crop in a watershed due to spatial variability in soil, fertilization and irrigation practice, slope, basin orientation, temperature, and numerous other factors [11], this method of estimating watershed ET requires multiple assumptions and increases uncertainty in the results. Uncertainties can propagate directly or indirectly (i.e., through additional hydrologic models) into decision-making processes of water right managers, water resource planners, ecologists, and researchers, potentially causing issues such as inaccurate allocation of water rights, higher redundancies or deficiencies in water infrastructure, and inaccurate estimates of ecological flows [12–14].

To overcome these problems, remotely sensed data have been used in conjunction with surface energy balance models to estimate ET [15]. Of the many surface-energy-balance-based remote sensing models available, this work uses the Mapping EvapoTranspiration at high Resolution using Internalized Calibration (METRIC) model since it was specifically designed for agricultural areas and parameterized according to the factors most relevant to the characteristics of our study area. It was also chosen since it uses freely available Landsat satellite imageries that are taken every 16 days, thereby capturing the vegetation growth stages within a watershed.

ET is estimated in METRIC by taking the energy consumed by the ET process as a residual of the surface energy balance equation [16]. This energy is called latent heat flux (LE) [W/m^2] and is represented by

$$LE = R_n - G - H \tag{1}$$

where the net radiation, R_n, is the balance of all incoming and outgoing short-wave and long-wave radiation [W/m^2], G is the soil heat flux [W/m^2], and H is the sensible heat flux [W/m^2].

Although METRIC has been proven to be a powerful tool in estimating evapotranspiration for larger areas, the accuracy and final results depend on user selection of parameters, submodels for calculation of components in energy balance, and calibration pixels [16]. Acknowledging this, Allen et al. [17] states that anchor pixel selection used in the calibration process (called CIMEC—Calibration using Inverse Modeling at Extreme Conditions) can correct for lingering biases in parameter selection associated with estimation of albedo, atmospheric correction, land surface temperature, emissivity, soil heat flux, and net radiation. This makes calibration of METRIC one of the most important steps in obtaining accurate ET estimates. The calibration process in METRIC requires identifying a "hot" and "cold" anchor pixel pair based on extreme conditions within a given image. In such an image, a well-irrigated pixel with maximum possible evapotranspiration is considered to be "cold", and a dry field with almost no evapotranspiration is "hot". Selecting these pixels, however, requires careful consideration of properties other than surface temperature (T_s), including normalized difference vegetation index (NDVI), surface albedo, and land use type. In the past, calibration pixel selection was completely dependent on expert judgement, but researchers have recently begun creating processes to automatically identify a subset of pixels from which the final selection is made [18,19]. Although METRIC has been used by water resource managers for a variety of applications such as water rights management, water allocation, and estimation of water yield, the need for expert judgement in the calibration process makes METRIC a complicated tool for operational use. A calibration process that facilitates a completely automatic application of METRIC would significantly expand its use as an operational model rather than primarily as a research tool.

One of these automatic pixel selection algorithms, a statistics-based procedure by Allen et al. [19], simplifies the techniques traditionally requiring extensive knowledge of remote sensing processes and vegetation indices. Although their procedure significantly reduces the number of potential anchor pixels, the final pair must still be selected using expert judgement. Furthermore, while the METRIC model itself has been validated (e.g., Bhattarai et al. [20] using a process similar to the procedure explained in Allen et al. [19]), no attempt has been made to evaluate ET results from the model calibrated using pixels from automated calibration. Rather, previous evaluations of the automated calibration pixel selection processes [18,19] have only compared the temperature of the anchor pixels selected by an expert to the temperature of the pixels selected by the algorithm.

A study by Morton et al. [21] has, to a certain extent, evaluated calibration uncertainty in METRIC ET results by applying a calibration process similar to that used by Allen et al. [19], and concluded that automatic calibration processes can produce ET estimates very similar to time-intensive manual efforts. Morton's study used a distribution of reference ET fraction (ETrF) computed for the specific study area using METRIC results from five trained users. Although Morton's method would help in automating the calibration of METRIC, two major limitations prevent the uptake of the method by a general METRIC user: (1) it involves a computationally intensive procedure (Monte Carlo simulation), and (2) it requires a generic cumulative distribution function (CDF) of ETrF for the study regions (which is usually lacking).

In this paper, with an objective of estimating the calibration variability in METRIC ET, we examine the properties of anchor pixels selected by automatic procedures and identify additional properties other than NDVI and T_s that affect calibration processes. To do so, five sets of calibration pixel pairs, encompassing all possible combinations, were prepared using all available calibration pixel sets. These calibration sets were then used to find the variation in ET. Previous studies suggest that, based on the nature of the automatic calibration pixel selection process, only small and random variations of final ET should occur when using this approach. However, our results indicated significant spatial differences in ET across the study area. Furthermore, one of the five sets we investigated provided consistently better results across multiple years, showing that NDVI and surface temperature alone do not capture all relevant surface properties required for the calibration pixel selection process. These results will be useful in operational aspects of METRIC calibration, allowing the user to be aware of the variability in ET results caused by calibration pixel selection.

2. Materials and Methods

2.1. Study Area

This study focused on an agricultural area in San Joaquin—Sacramento River Delta area (Figure 1 and Table 1). A U.S. AmeriFlux site with an EC flux tower in the area was used to compare and quantify the variability in ET due to the calibration pixel selection process in METRIC. The farm plot (Twitchell Alfalfa) has been used to farm alfalfa, with crop rotation occurring every 5–6 years. The alfalfa at the site is harvested by mowing and bailing several times per year and is irrigated periodically by flooding ditches around the field [22]. Based on the recommended procedure developed by Allen et al. [19], we selected calibration pixels from a study area within a 30 km radius around the available weather station (Figure 1).

Table 1. Description of sites used in this study.

Site Descriptors	Twitchell Island 140	US-Tw3
Latitude (N)	38.1217	38.1159
Longitude (W)	121.6745	121.6467
Elevation (m)	0	−9
Cover type	Large Pasture	Alfalfa
Data availability period	1998–current	2013–2015
Mean Annual temperature (°C)	15	15
Mean Annual precipitation (mm)	382	421
Temporal resolution provided at (min)	60	30
Source of data	CIMIS	Baldocchi [1]

[1] http://dx.doi.org/10.17190/AMF/1246149 (last assessed: 10 Octobe 2017).

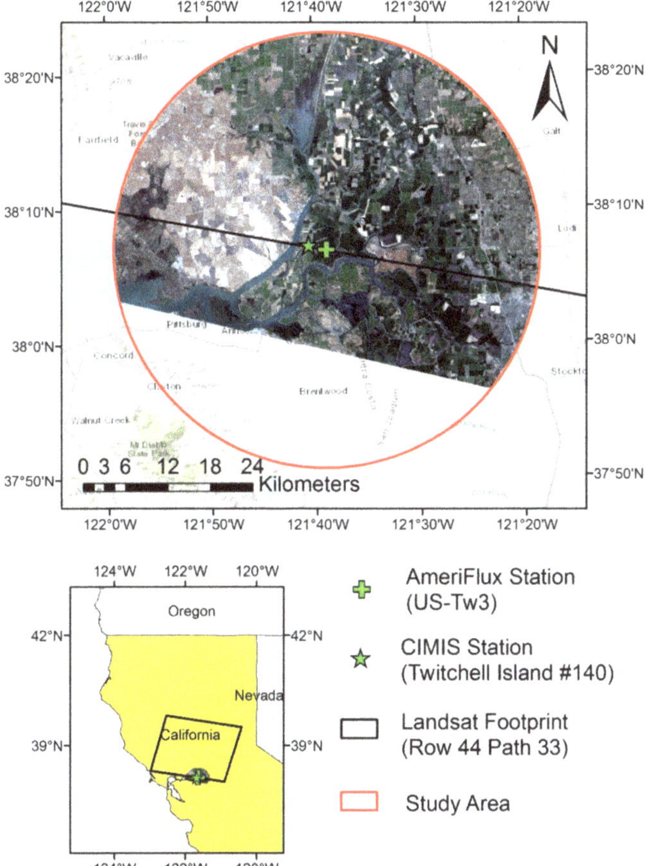

Figure 1. Location of study area with AmeriFlux site and nearby California Irrigation Management Irrigation System (CIMIS) weather station in California. A Red Green Blue (RGB) Landsat 8 image for 24 July 2014 is clipped and overlaid over the study area.

2.1.1. Measured ET Data

An Eddy Covariance (EC) system with GILL WM 1590 sonic anemometers and a LICOR LI-7500A CO_2/H_2O gas analyzer was used at the flux site. Daily flux data from this site was postprocessed and provided as 30 min average datasets [23]. The EC instruments were mounted at a height of 2.8 m above the ground.

2.1.2. Landsat Images

We performed our analysis with Landsat images since this allows ET to be mapped at field scales. Additionally, ET measured using an EC tower has a fetch of about 100 times [24] the station height, so METRIC results using Landsat will give an accurate representation of the area around the station.

We used all available nearly cloud-free Landsat 7 (With Enhanced Thematic Mapper Plus or ETM+ sensor) and Landsat 8 (with Operational Land Imager or OLI sensor) images covering the station within the ET measurement period (Table 2). Even though the satellite images are captured using different sensors, Kilic et al. [25] did not find significant differences in NDVI and ETrF computed from these imageries. Therefore, we did not treat these images differently in our study. The cloud mask created using the CFMask algorithm [26], provided within the Landsat Collection 1 Level-1 Quality Assessment (QA) band, was used to determine cloudy pixels. Only images with less than 20% cloud coverage (by area) were used. To remove cloudy images not identified by CFMask and QA bands, the weather station and validation site location in these images were visually inspected using a contrast enhancement function provided by Google Earth Engine (GEE) [27] (see Section 2.3). This function uses a Styled Layer Descriptor (SLD) to render red, green, and blue bands that produce images with equal brightness levels. Rather than download Landsat imageries, we used GEE to access and analyze Landsat data from the public catalog GEE provides.

Table 2. List of Landsat images used in this study. Path/row of the image was 044/033.

YearMonthDay (YYYYMMDD)	Area Cloud Covered (%)	YearMonthDay (YYYYMMDD)	Area Cloud Covered (%)
2013			
Landsat 7		Landsat 8	
20130713	2%	20130603	5%
20130729	4%	20130619	4%
20130814	2%	20130705	7%
20130830	3%	20130721	7%
20130915	3%	20130822	7%
20131001	2%	20130907	7%
20131017	3%	20130923	7%
		20131009	7%
		20131025	8%
2014			
Landsat 7		Landsat 8	
20140614	4%	20140505	13%
20141004	2%	20140606	9%
		20140622	7%
		20140724	7%
		20140809	7%
		20140825	9%
		20141012	8%
2015			
Landsat 7		Landsat 8	
20150414	2%	20150422	8%
20150430	2%	20150508	12%
20150516	3%	20150524	8%
20150703	4%	20150625	8%
20150719	5%	20150711	7%
20150820	2%	20150727	7%
20150905	2%	20150812	7%
20150921	3%	20151031	8%
20151007	3%		

2.1.3. Weather Data

Freely available hourly meteorological data from the California Irrigation Management Information System (CIMIS) (last accessed 17/11/2017) was used for this study. Specifically, we used CIMIS Station Number 140 (Twitchell Island) in Sacramento, CA since this is the only site that has data for the complete study period.

2.1.4. Land Cover Data

Land cover data available in the GEE catalog from the USGS National Land Cover Database 2011 were used to identify the agricultural land use pixels and pasture land use pixels required for this study.

2.2. METRIC Model

METRIC is a satellite-based image processing model that computes ET as the residual of energy balance as given in Equation (1) above. METRIC first computes available energy as $(R_n - G)$, which is then divided into *LE* and *H*. In our work, the components of energy balance needed to compute R_n were estimated using the formulations provided in Allen et al. [16] and Allen et al. [17]. In these studies, the formulations have been provided in detail for Landsat 5 and 7. However, since we also included Landsat 8 for our studies, there were a few components which we had to formulate using related newer studies.

These components include surface albedo, α, which was computed for Landsat 8 using weighing coefficients based on Ke et al. [28]. They used the Simple Model of Atmospheric Radiative Transfer of Sunshine (SMARTS) radiative transfer model to determine the weighing coefficients for Landsat 8 OLI sensors. This is similar to the procedure used in Tasumi et al. [29] to compute surface albedo for Landsat 5 TM and Landsat 7 ETM+ sensors. Due to the consistency between the two methods of calculating albedo using weights for each band of similar wavelengths across different satellite sensors, significant biases in the reflected solar radiation were not introduced into the model [30].

Surface temperature was computed based on the single channel method outlined in Sobrino et al. [31] for Landsat 5, Jimenez-Munoz et al. [32] for Landsat 7, and Yu et al. [33] for Landsat 8. Of the two bands in Landsat 7 that can be used to compute radiometric surface temperature, we used the high gain band (VCID_2) since our analysis required better estimates of temperature for vegetated areas. Similarly, from the two Thermal Infrared Sensor (TIRS) bands of Landsat 8, band 10 was used since band 11 is known to have larger calibration uncertainty [34]. In addition to thermal bands, this method requires land surface emissivity (ε) and water vapor content (w) to compute surface temperature. Water vapor content (w) was used from the weather station data and emissivity (ε) was computed using the NDVI threshold method.

The remaining components of energy balance for the model were computed using Allen et al. [17]. Of the multiple soil heat flux relationships available, our study used the following relation which uses net radiation (R_n) to compute soil heat flux [16,35,36]:

$$\frac{G}{R_n} = (T_s - 273.15)(0.0038 + 0.0074\,\alpha)\left(1 - 0.98\,NDVI^4\right). \tag{2}$$

Sensible heat flux (*H*), which is one of the key components of METRIC, is estimated using a one-dimensional bulk aerodynamic temperature-gradient-based method as

$$H = \frac{\rho\,c_p dT}{r_{ah\,1,2}} \tag{3}$$

where ρ is the air density (kg/m^3), c_p is the air specific heat at constant pressure (J/kg K), $r_{ah1,2}$ is the aerodynamic resistance (s/m) between two heights Z_1 and Z_2 above the zero-plane displacement height of vegetation (z_{om}) where endpoints of dT are defined, and dT is the vertical near-surface temperature difference (K). The near-surface heights Z_1 and Z_2 are assumed to be 0.1 m and 2 m above the zero-plane displacement height (z_{om}).

The value of $r_{ah1,2}$ is calculated using wind speed extrapolated from some blending height above the ground surface (typically 200 m) and using an iterative stability correction scheme based on Monin–Obhukhov functions [37,38]. An iterative solution is needed because this aerodynamic resistance ($r_{ah1,2}$) is influenced by the buoyancy caused by surface heating which is driven in turn by the sensible heat Flux (H), and both $r_{ah1,2}$ and H are unknown at a given pixel.

The parameter dT, which is the difference in temperature at two near-surface heights (usually $T_{z1} = 0.1 + z_{om}$ and $T_{z2} = 2.0 + z_{om}$), is difficult to compute directly because the temperatures at these two levels estimated from satellite sensors will have uncertainties greater than their differences [16]. To overcome this issue, METRIC uses an approach developed by Bastiaanssen [35] that approximates dT as a linear function of T_s given by

$$dT = a + bT_s \tag{4}$$

where a and b are regression coefficients. These coefficients are determined using a procedure that first selects two extreme condition pixels in the image, one which represents maximum LE (cold pixel) and the other which represents minimum or zero LE (hot pixel), then implements Equations (1) and (3) on these pixels. Bastiaanssen [39] assumed that, at a cold pixel, all available energy is used up for latent heating, so $H \approx 0$. At a hot pixel, conversely, all available energy is used up for sensible heat flux, so $LE \approx 0$. For the METRIC application, it is assumed that the latent heat flux corresponds to the ET value that is 5% more than the ASCE (American Society of Civil Engineers) standardized alfalfa-based reference ET at a selected cold pixel. Similarly, it is assumed that, at a selected hot pixel, the latent heat flux corresponds to the ET value that is bare soil ET due to prior precipitation events calculated using the FAO-56 soil water evapotranspiration model [10,17]. An iterative solution approach is then applied at these two "anchor" pixel locations to determine dT and, subsequently, the relationship between dT and T_s. Readers are referred to Allen et al. [17] for more details on the iterative process.

2.3. Google Earth Engine for METRIC

We used Google Earth Engine (GEE) [27] and the associated Python API to develop and implement METRIC in an entirely web-based environment that does not require download of any raster data (Landsat, Digital Elevation Model (DEM), and land use) or use of a personal computational engine. GEE is a cloud-based geospatial processing platform developed by Google Inc. and combines large (petabyte scale) storage capabilities, which contain archived remotely sensed imageries, with a powerful computational infrastructure optimized for parallel processing of large geospatial datasets. It also supports charting, dynamic mapping, and table/image export features, making it easier to visualize intermediate results without needing to download anything onto a personal computer. Additionally, most of the GEE's algorithms use per-pixel algebraic functions, making it ideal to apply irrespective of the region or scale (provided that the data are available for the region).

2.4. Selection of Calibration Pixels

An overview of the modeling process with selection of calibration pixels for this study is presented in Figure 2. We used the process outlined in Allen et al. [19] and Bhattarai et al. [18] to select the sets of candidate calibration pixels. The selection of candidate calibration pixels using the method outlined by Bhattarai et al. [18] provided a smaller subset from within those pixels selected using Allen et al. [19], because we iteratively increased the window size in small increments for NDVI and T_s as suggested by the authors. This resulted in a final window size which was much smaller for the former method than the latter one. Although the selection method outlined by Bhattarai et al. [18] provides a smaller set of calibration pixels and has been shown to perform with comparable accuracy in the final ET results when compared to the method used by Allen et al. [19], we used the latter method since our goal was to understand the variability in dT between candidate pixels due to small variations in T_s and NDVI. Additionally, the calibration pixel selection process developed by METRIC developers (i.e., Allen et al. [16]) is considered a standard method and has been recently used in Earth Engine Evapotranspiration Flux (EEFlux) [40].

The study area for this work was selected following Allen et al. [19] who suggest an area with a radius of 20–30 km around the weather station to be used for calibration of the model for the given image. Since the elevation around our weather station does not change drastically, we assume that the elevation adjusted temperature, determined using a customized lapse rate (T_{sDEM}), and wind speed across the terrain [41], which is used for calibration, do not change drastically. Due to the minimal elevation differences, we used an area with the maximum possible radius of 30 km suggested by Allen et al. [19]. The corrections for temperature with elevation (T_{sDEM}) and wind speed were, however, performed for this study using Allen et al. [17] and Allen et al. [41]. If there are significant changes in elevation which could affect T_{sDEM} or wind speed, the area used for calibration needs to be divided into regions with similar elevation and climate, and terrain corrections applied.

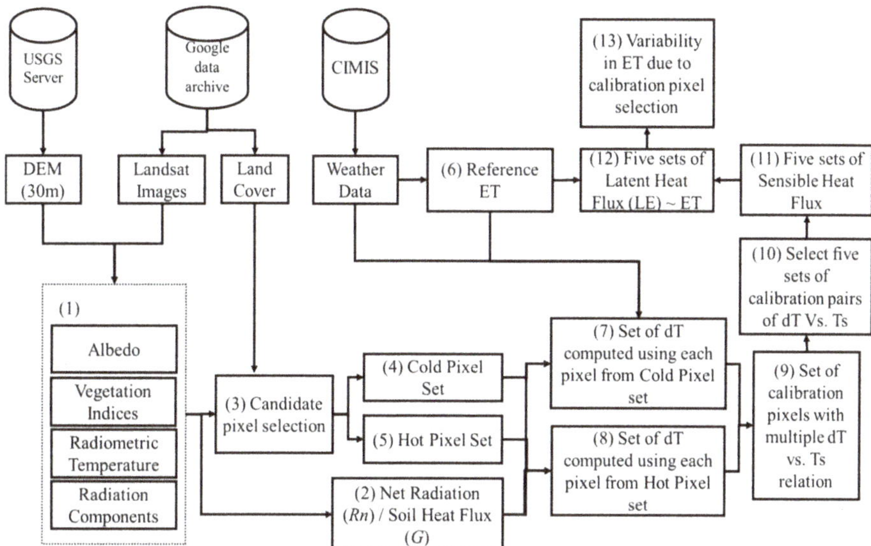

Figure 2. Methodology for computing variability in Mapping EvapoTranspiration at high Resolution with Internalized Calibration (METRIC) evapotranspiration (ET) due to calibration pixel selection.

Within the 30 km radius area, we used pixels with land use (LU) classified as Pasture/Hay, Cultivated Crops, and Grassland (LU codes 81, 82, and 71, respectively). The selected pixels were filtered as recommended by Allen et. al [19] to eliminate effects of clouds, field edges for thermal pixels, and T_s vs. NDVI relationships. For homogeneity in the thermal pixels, we filtered the image using a 7×7 pixel cluster and rejected any pixels having a standard deviation greater than 1.5 K, as recommended by Bhattarai et al. [18]. Similarly, to avoid too much variation in NDVI within calibration polygons, we rejected pixels with coefficients of variation (Standard Deviation/Mean) greater than 0.15 (per [19]). Some images had clouds not identified by the F-mask algorithm in QA bands; these were manually delineated before calibration.

2.4.1. Hot and Cold Pixel Set Identification

Allen et al. [19] states that a cold pixel candidate can been found to lie in the coolest, most vegetated 1% of areas within the image (the coldest 20% of the 5% highest vegetated pixels). Similarly, a hot pixel candidate can be found to lie in the hottest, least vegetated 2% of the image (the warmest 20% of the 10% lowest pixels). So, we followed Allen et al. [19], which has been used successfully for finding calibration pixels, so that the method of the calibration pixel selection process would be consistent with other similar applications of METRIC [21,42]. Additionally, since the model is used for estimating consumptive use in agricultural watersheds, we assume that there will be ample agricultural pixels during the growing season other than dry bare pixels, so a larger subset of hot pixels was selected than of cold pixels for our analysis.

To identify cold pixels, the pixels with top 5% NDVI were first isolated. From these pixels, those with the lowest 20% of T_s were further selected. This provided a sample of 1% of the coldest pixels having the highest NDVI and lowest temperature. From this 1% set, the cold pixels were then chosen to be within ± 0.2 K of the average temperature and within $\pm$ 0.02 of the albedo threshold calculated using the equation provided in Allen et al. [19], which was parameterized for Southern Idaho, as follows:

$$\alpha = 0.001343\,\beta + 0.3281\exp(-0.0188\,\beta) \tag{5}$$

where α is the albedo threshold and β is the sun angle above the horizon.

Computing the albedo threshold using Equation (5) resulted in most of the cold pixel albedo values falling between 0.18 and 0.25, which is within the suggested range of albedo for a representative reference alfalfa (0.23). Also, since the study was performed during growing seasons with higher β values, we did not change the parameters of Equation (5).

Like NDVI, the Soil Adjusted Vegetation Index (SAVI) [43] is another common vegetation index that provides useful information on vegetation growth. However, since NDVI has a more linear means of specifying the reference ET fraction compared to SAVI [19], we only used NDVI during the calibration pixel selection process. To verify this assumption, SAVI values were calculated for both cold and hot candidate pixels. The distributions of these values are shown in the Supplementary Materials section of this paper (Figure S1). The horizontal line where SAVI equals 0.69 indicates when Leaf Area Index (LAI) becomes "saturated" (LAI $\geq$ 6.0). Above this, the leaf area does not vary significantly. In Figure S1, we note that SAVI values can exceed the full LAI conditions. However, ET_rF and crop coefficients (K_c) tend to reach maximum values near full-cover conditions, same as the NDVI [44]. Thus, in this case, NDVI was the correct choice for the cold pixel selection process.

To identify hot pixels, the lowest 10% NDVI pixels isolated from the candidate pixels were first selected. From these, the hottest 20% of T_s pixels were then selected. This provided a sample of 2% of the hottest pixels with lowest NDVI from the candidate pixels. Hot pixels within ± 0.2 K of the average temperature were then chosen from this 2% set. The reference ET_rF for this pixel was then computed using ET_rF_{bare} from the water balance model provided in FAO-56 [10]. Hot pixels were selected from regions of homogenous agricultural land use having bare soil conditions and, therefore, low values of NDVI. Bare soil pixels in some cases have vegetation residue but still have low NDVI values [45]. These pixels with organic cover need to be removed from the calibration set because the insulation provided by this cover can reduce soil heat flux (G) which may not be accurately captured by the relationship developed to estimate G in our study. Some of these low NDVI pixels with vegetative cover have higher albedo values than dry bare soil [46]. So, to remove pixels with organic cover within our hot candidate calibration pixel set, we removed hot pixels which had albedo values higher than 0.30~0.35 [19]. The sets of hot and cold pixels were then used to determine the dT vs. T_s relation for calibration of the model.

2.4.2. Determination of the dT vs. T_s Relationship from Selected Candidate Pixels

Values of dT computed from each hot and cold pixel selected from the step above were used to plot dT vs. T_s and identify the variations possible in calibration relations. Five sets of calibration pixel pairs were then selected from these possible calibration relations (Figure 3). These were selected so that the intercept and slope of the calibration line ("a" and "b" from Equation (4)) would encompass every possible relation within these available sets of calibration pixels.

The first set of calibration pixel pairs consists of a cold pixel with the lowest dT value and a hot pixel with the lowest dT value (Figure 3). This pair of pixels is referred to as the "*Min Min*" set hereafter (minimum cold pixel dT and minimum hot pixel dT). Similarly, the second pair of calibration pixels consists of a cold pixel which produced the highest dT value and a hot pixel with the highest *dT* value—the "*Max Max*" set (maximum cold pixel *dT* and maximum hot pixel *dT*). The third pair of calibration pixels consists of the cold pixel from the *Min Min* set and the hot pixel from the *Max Max* set—the "*Min Cold Max Hot*" set (minimum cold *dT* and maximum hot *dT*). Similarly, the fourth set of pair of calibration pixels consists of the cold pixel from the *Max Max* set and the hot pixel from the *Min Min* set—the "*Max Cold Min Hot*" set (maximum cold *dT* and minimum hot *dT*). The fifth pair of calibration pixels consists of the closest cold and hot pixels from the weather station—the "*Closest*" set.

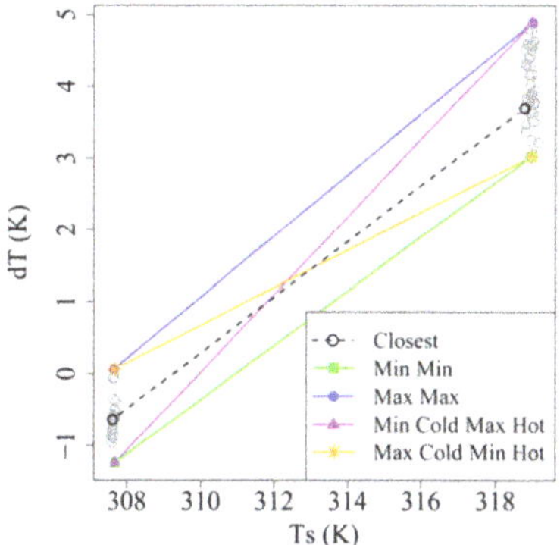

Figure 3. *dT* vs. T_s relationship for each candidate hot and cold pixel. Five selected sets are drawn as lines to encompass every possible combination for calibration of the model. (The locations of these points are shown in Figure 6).

2.5. Determination of Daily ET

These five sets of anchor pixels were used individually to calibrate the model and determine sensible heat flux (*H*). The reference ET fraction (*ETrF*, assumed constant within one day) was then computed as the ratio of instantaneous ET to reference ET. We did not use sloping terrain corrections since the study area was mostly level. Daily ET (ET_{24}) was then computed as

$$ET_{24} = (ETrF)(ET_{24}).\tag{6}$$

Five sets of ET_{24} were computed by calibrating the model with each anchor pixel pair.

An evaluation of the effect of calibration pixel selection on the model was done based on the model's ability to accurately predict daily ET values at the flux tower; common model evaluation criteria such as Pearson's correlation coefficient squared (r^2), Nash–Sutcliffe efficiency (NSE) [47], Root-Mean-Squared Error (RMSE), and mean percent bias (PBias) [48] were used to evaluate model performance.

Daily ET values from the flux tower were computed using the 30 min data provided through US-AmeriFlux (http://ameriflux.lbl.gov/). Data filtering for spikes or instrument malfunctions, cross wind and humidity correction for the sonic anemometer, and removal of fluxes for low friction velocity were already performed before data distribution [22].

To correct daily flux data, which had an energy balance closure of ~15%, an energy balance correction factor was computed using 30 min data having all energy balance components available. This method of forcing was used in Twine et al. [49]. Since this method assumes that the Bowen ratio is correct, we first filtered the flux data to remove data close to sunrise and sunset. When net radiation was less than 50 W/m^2, flux values were assumed to be close to sunset and sunrise and were not used to find the correction factor. The correction factor (*CF*) is computed for every half hour as

$$CF = \frac{(R_n\_FT - G\_FT)}{(H\_FT + LE\_FT)}\tag{7}$$

where R_n_FT is the net radiation measured at the flux tower, G_FT is the soil heat flux measured at the flux tower, H_FT is the sensible heat flux measured at the flux tower, and LE_FT is latent heat flux measured at the flux tower. Any CF values outside 1.5 times the interquartile range were filtered and a sliding window of ± 15 days was used to compute the median value of CF. These values were first used to compute the correct LE_FT, which was then used to compute daily ET values. These ET values were in turn used as data points to evaluate METRIC ET results.

3. Results

3.1. Properties of Candidate Calibration Pixels

Given that METRIC calibration requires two anchor pixels which are usually selected based on the NDVI and T_s of the image [50], NDVI and T_s values of the candidate hot and cold pixels are expected to have low variability. Results reflect this, with the coefficient of variation (standard deviation $\times$ 100/mean) of NDVI and T_s of candidate pixels (Figure 4) having low variability across the images used. However, the variation of available energy within these candidate pixels is significantly greater than the variation of NDVI and T_s, showing that the calibration process also depends on available energy ($R_n - G$). Scaling the relative variability of the three properties between 0 and 1 clearly highlights this difference in the coefficients of variation (scaled_NDVI, scaled_T_s, and scaled_Available_Energy; Figure 4).

Individual values in the boxplot (Figure 4) show the coefficient of variation in the candidate pixel set for each image. An example from 24 July 2014 of what a value in this boxplot would denote is presented in Figure 5. For this date, the coefficient of variation of scaled temperature is 0.3% for the cold pixel set and 0.4% for the hot pixel set (the least variation compared to NDVI and available energy). For the same date, the coefficient of variation of scaled NDVI is 1.3% for the cold pixel set and 9.4% for the hot pixel set. Finally, the coefficients of variation of scaled available energy are 20% for the cold pixel set and 66% for the hot pixel set, both notably higher than for NDVI and T_s. Since the calibration pixel selection process exclusively uses NDVI and T_s, these variables necessarily have low variability across all the images, as shown in Figure 4. The figure also clearly illustrates that variability within hot candidate pixels is generally higher than within cold candidate pixels.

Figure 3 shows that the cold pixel dT ranged from $-1.2\,^\circ$C to $0.1\,^\circ$C and hot pixel dT ranged from $3.0\,^\circ$C to $4.9\,^\circ$C for 24 July 2014. The median value of dT was $0.5\,^\circ$C for cold pixels and $4.8\,^\circ$C for hot pixels across all images.

Figure 5 shows the available energy for pixels which produced maximum and minimum dT values in the calibration process, as well as the available energy for the closest pixel. Contrary to what might be expected based on the nature of hot and cold pixels, the pixels with maximum and minimum dT values among the set of calibration pixels did not precisely correspond respectively to the pixels with highest and lowest values of available energy; the cold pixel with minimum dT is about 15 W/m^2 greater than the minimum available energy in this case. However, this is only 16% of the total variation in available energy.

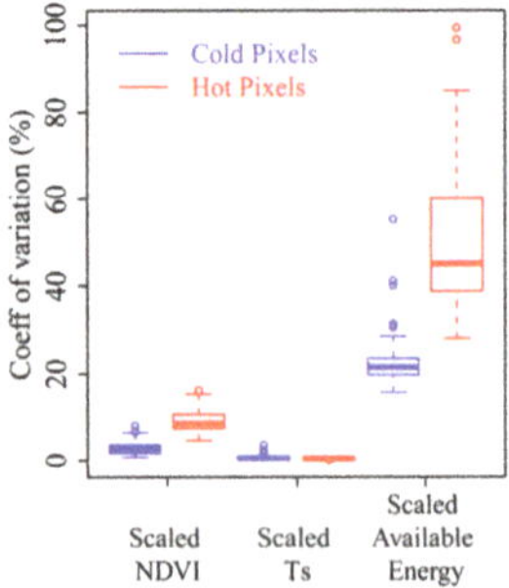

Figure 4. Coefficient of variation of scaled values of normalized difference vegetation index (NDVI), T_s, and Available Energy for hot (red) and cold (blue) pixels for all images.

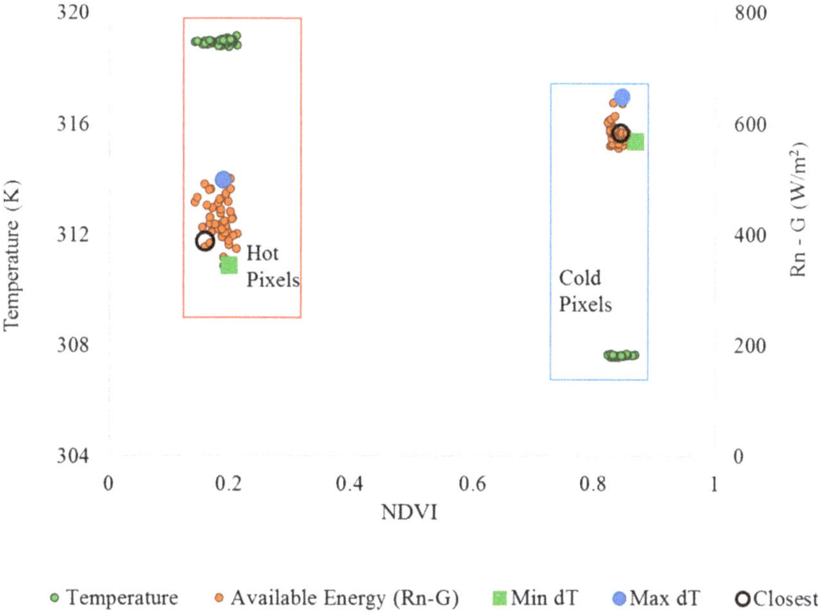

Figure 5. Variability in temperature, NDVI, and available energy within candidate anchor pixels for 24-07-2014.

The location of these sets of calibration pixels from 24 July 2014 is presented in Figure 6, along with the spatial distribution of NDVI, T_s, and available energy ($R_n - G$). From this figure, it can be observed that the image has two distinct regions—a region with less vegetation, higher surface temperature, and lower available energy (eastern side) and another region with more vegetation, lower surface temperature, and higher available energy (western side). However, rather than being concentrated in only these two regions, hot and cold pixels are scattered across the study area. In some instances, where clouds were not filtered correctly, cold candidate pixels were sometimes concentrated in areas with clouds and shadows due to lower temperatures around the area. So, each candidate pixel selection result needs to be manually checked for such errors.

With the automated pixel selection process, there were a total of 47 hot and 38 cold candidate pixels selected for this day. Manual inspection of these pixels in NDVI, RGB, and T_s images as suggested by Allen et al. [19] and Bhattarai et al. [18] indicated that any of them could feasibly be considered an anchor end-member pixel for calibration. That selection would depend subjectively on the modeler.

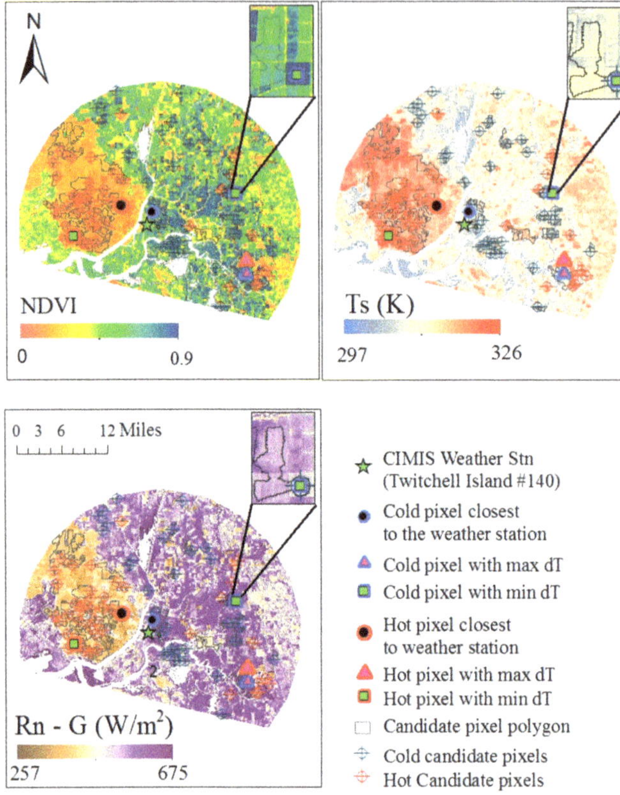

Figure 6. Location of hot and cold calibration pixel sets as polygons (24-07-2014). Inset shows a polygon from which the cold calibration pixel was selected and has minimum *dT*.

3.2. Effect of Calibration Pixel Selection on ET

The effect of selecting various calibration pixel sets on ET within the study area was evident as shown by the variability in ET in Figures 7 and 8. Figure 7 illustrates the spatial variability in daily ET computed using the sets of calibration pixels for a specific date, while Figure 8 illustrates the frequency distribution of these ET values. Together, these figures show that, in the study area, highest ET resulted from the *Min Min* set (5.8 mm) and lowest ET resulted from the *Max Max* set (4.5 mm). This was an expected result since the *Min Min* set calibrates using a low *dT* value, which would provide lower *H* values and, thus, higher *LE* and ET values. Similarly, the *Max Max* set calibrates using higher *dT*, which would provide higher *H* values and, thus, lower *LE* and ET values. The highest variability was in the *Max Cold Min Hot* set with a standard deviation of 2.5 mm, while the least variability was in the *Min Cold Max Hot* set with standard deviation of 1.8 mm.

The frequency distribution of these ET maps, along with their means and standard deviations, is presented in Figure 8. The least variability is observed when the model was calibrated using the *Min Cold Max Hot* set with a standard deviation of 1.8 mm, while the highest variability is observed when the calibration was done using the *Max Cold Min Hot* set with a standard deviation of 2.5 mm. The peaks for the *Max Max* set and the *Min Cold Max Hot* set occur at approximately 6.5 mm while the peaks for the *Min Min* set and the *Max Cold Min Hot* set occur at approximately 7.5 mm. ET distributions do not have a discernable scaling or skewing factor according to the calibration sets.

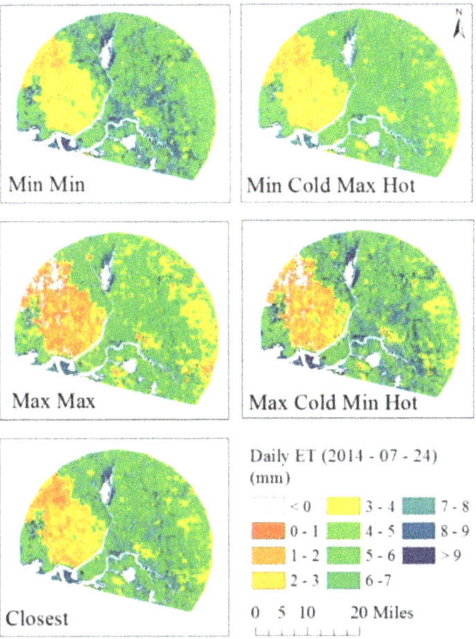

Figure 7. Spatial variation in daily ET based on various calibration sets.

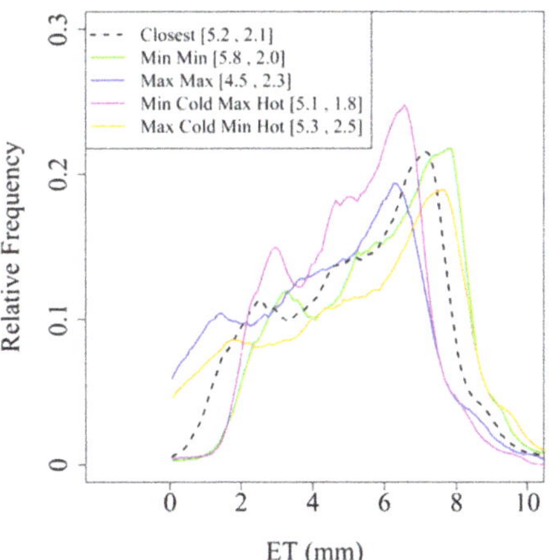

Figure 8. Frequency distribution of ET for different calibration sets for 2014-07-24. The values in the legends are, respectively, [Mean, standard deviation].

The range of variability in ET due to calibration pixel selection can be seen through the coefficient of variation of mean ET values (4% to 20%) computed using different calibration sets, as illustrated in Figure 9. This figure also shows that variation in ET due to calibration pixel selection is not affected by antecedent moisture conditions due to rainfall, but variability in ET due to calibration does have an increasing trend with time within the growing season. This increase in the coefficient of variation could possibly be attributed to higher variability in lower ET values later in the growing season (Figures 9 and 10).

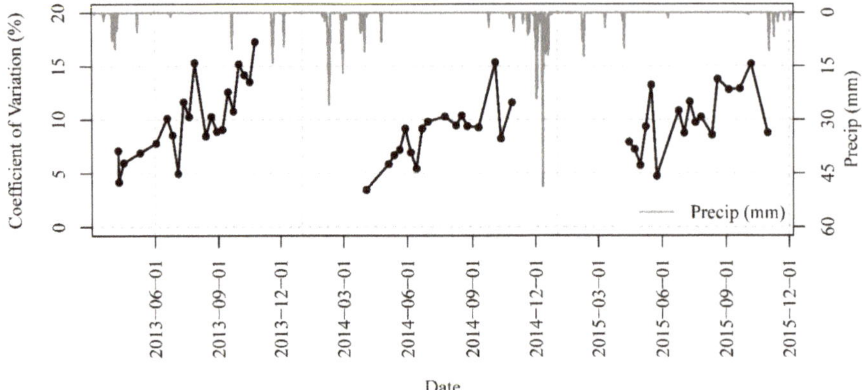

Figure 9. Coefficient of variation of mean ET calibrated by four sets of calibration pixels (*Min Min*, *Max Max*, *Min Max*, and *Max Min*). Precipitation added to the plot shows that antecedent moisture conditions did not affect the variability.

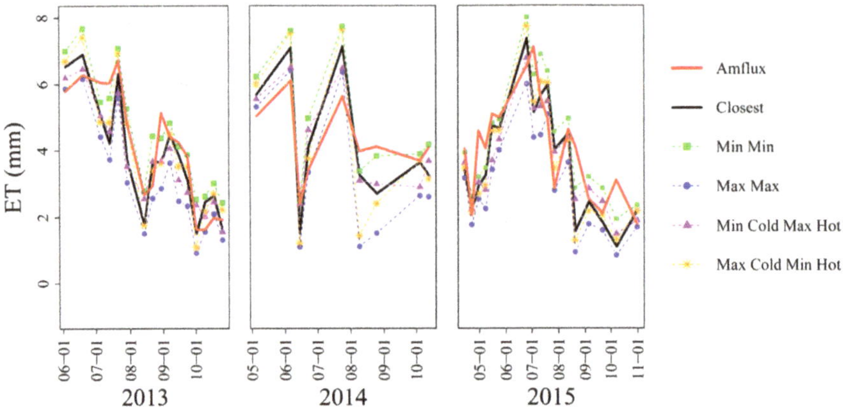

Figure 10. Comparison of 24-h ET computed using the selected five calibration schemes with eddy covariance flux tower data for 2014, 2015, and 2016. With any of these calibration sets, the temporal pattern of ET over the year is well captured. The line corresponding to "closest" presents results with the traditional method of anchor pixel selection (closest to the weather station).

3.3. Evaluation of Daily ET Estimates from Different Calibration Sets

Comparisons of modeled results with ET estimated at the flux tower are presented in Table 3 and Figure 10. We selected daily ET for validation since daily values are considered to be more useful than instantaneous ET, especially for agricultural applications [16]. Additionally, diurnal storage of heat averages out over daily time scales, which causes better energy balance closure and thus provides better estimates of ET [51]. The *closest* set results reflect what the values would be if the anchor pixel were selected using the traditional method, closest to the weather station, as suggested by Allen et al. [19] and Bhattarai et al. [18]. Additional results from Table 3 indicate that although r^2 is similar for all the sets for each year, ET computed using the cold pixel that provided the minimum dT and the hot pixel that provided the maximum dT (the *Min Cold Max Hot* set) performed the best, with higher NSE, lower RMSE, and lower Pbias. For the results of 2015, the *Min Min* set performed as well as the *Min Cold Max Hot* set. The worst set for calibration was the *Max Max* set, which had lower

r^2, lower NSE, higher RMSE, and higher Pbias values. These results also indicate that METRIC overall underestimated ET, resulting in negative biases for most cases.

At any date in the time series, the difference in ET between the flux tower value and ET from any of the calibration sets was no more than ~70% (Figure 10), showing that the temporal pattern of ET is well captured when using any of the calibration sets for computation of ET. The temporal pattern was better captured for 2013 and 2015 with higher NSE values than 2014.

Table 3. Model evaluation results comparing 24-h ET estimates computed using five different calibration schemes with measured ET at the AmeriFlux eddy covariance (EC) station. (NSE: Nash–Sutcliffe Efficiency, RMSE: Root Mean Squared Error).

	r^2	NSE	RMSE (mm)	Pbias (%)
		2013		
Closest	0.78	0.75	0.87	−4.0
Min Min	0.85	0.79	0.80	18.0
Max Max	0.80	0.47	1.28	−25.0
Min Cold Max Hot	0.80	0.75	0.88	−4.0
Max Cold Min Hot	0.82	0.79	0.79	−1.0
		2014		
Closest	0.83	0.18	0.95	−7.0
Min Min	0.81	0.04	1.03	11.0
Max Max	0.72	−1.13	1.54	−29.0
Min Cold Max Hot	0.74	0.46	0.77	−4.0
Max Cold Min Hot	0.77	−0.82	1.42	−14.0
		2015		
Closest	0.62	0.44	1.13	−9.0
Min Min	0.68	0.57	0.99	8.0
Max Max	0.63	0.01	1.50	−28.0
Min Cold Max Hot	0.67	0.56	1.00	−9.0
Max Cold Min Hot	0.65	0.41	1.16	−10.0

4. Discussion

4.1. Candidate Calibration Pixel Properties

The variability in ET computed using different sets of anchor pixel pairs shows that there are properties of candidate calibration pixels that cannot be completely quantified using only NDVI and T_s. As the candidate pixels have very low variability in T_s for a consistent ET result across the study area regardless of which candidate pixel pair is chosen, it was expected that the iterative calibration process would converge to the same dT. Based on suggestions in previous studies, H could be expected to absorb the biases introduced at earlier steps of the model run. However, our results indicate that not all candidate pixel pairs converge to the same dT, and the differences in dT at these pixel locations are large enough to have a significant effect on final ET results. Furthermore, calibration of the model using a specific dT vs. T_s (*Min Cold Max Hot*) relation provided consistently better results across multiple years despite little variation in T_s, and dT values coincide closely with available energy. Therefore, it can be presumed that the large differences in dT among the candidate pixels were not random noise, and that the inclusion of available energy into the calibration pixel selection process could potentially improve the calibration process and provide better estimates of ET.

When comparing properties of candidate pixels, we see that cold pixel sets have lower variability compared to the hot pixel set, which can be explained by the properties of the study area and the nature of these extreme pixels. Cold pixels are more readily available than hot pixels because it is easier to find well-watered agricultural pixels in a region having large areas of irrigated agriculture, since these have a relatively consistent set of surface conditions. Hot pixels, on the other hand, lie within dry, bare agricultural areas with higher T_s, and can have more varied surface conditions [21]. The higher variability in surface properties of hot candidate pixels leads to higher variability in dT.

The automated calibration processes in previous studies [18,19,21] have used a specific property of pixels (surface temperature or T_s) to evaluate the performance of the candidate pixel selection process. However, our study shows that evaluating the automated calibration process using temperature might not be useful because of the small variation in T_s between candidate pixels. While the evaluation performed by Bhattarai et al. [18] did compare daily ET to EC flux tower data, the final anchor pixel selection in their work was based on a temperature window of 0.25 K with an NDVI window of 0.01 from which the closest pixel to weather station was chosen. Based on our analysis, these temperature and NDVI windows are comparable to the variability within candidate pixels, meaning that ET variability would not be reduced. As an example, in Figure 5, NDVI variability ranges between 0.05 and 0.07 while T_s variability ranges between 0.1 K and 0.4 K for cold and hot pixels, respectively. This shows that ET computed with any pixels in the windows specified by Bhattarai et al. [18] should have similar variability in ET as that shown in our work.

A surface property that affects the calibration process is the actual weather station surface roughness. Any application of METRIC using past weather data requires us to estimate the surface roughness of the weather station using remote sensing data. Given the wind speed at a specific known height and an assumed logarithmic wind profile, this surface roughness is used to approximate the wind speed at a height 200 m above the ground surface. This predicted wind speed is considered to be constant throughout the image. Any error in this extrapolation can lead to error in the calibration process and final ET results.

4.2. Model Performance

Other studies have attempted to investigate the variation in the end-member pixel's impact on model accuracy [21,50]. These studies explain that by selecting pixels that have increased (or decreased) temperature of hot and cold extremes together, the values of evaporative fraction (EF) increase (or decrease) with similar impact. These studies also explain that varying end-members should not substantially affect the standard deviation of EF frequency distribution. Yet, as shown in our study, there were differences not only in total ET, but also in standard deviation when end-members were varied. Since METRIC's performance depends strongly on candidate pixel selection, one would expect differences in model performance with the five calibration pixel sets, and this is indeed what we observe.

Higher ET values are expected at the lower end of the distribution when calibrated using a cold pixel with minimum dT and lower ET values are expected at the higher end of the distribution when calibrated using a hot pixel with maximum dT. However, ET values can vary around the center of the distribution without discernable differences among results from various calibration sets.

As shown in the results, the model's performance varied not only with calibration pixel selection, but also with the analysis year. The poor performance of the model for 2014 could be due to the error involved with estimating net solar radiation in METRIC, which was inspected using flux tower data (Figure 11). Other models (e.g., SEBAL or Two Source Energy Balance) which use evaporative fraction (EF) as the ratio of instantaneous LE to available energy ($R_n - G$) to compute daily ET might perform better for conditions when the estimation of net radiation has larger biases. This effect can be seen in Figure 12, which shows a comparison between instantaneous and daily EF and $ETrF$ values computed using flux tower energy balance components. The higher r^2 value for ETrF than EF (between instantaneous and daily measurements) shows that instantaneous values of energy balance components transfer more consistently to daily scales when using $ETrF$. Consequently, the daily values of ET are more sensitive to errors in instantaneous values. Gonzalez-Dugo et al. [52] also showed that for models other than METRIC, use of EF improved the performance of the models as measured by reduction of the root-mean-squared error and increases in r^2 when compared to $ETrF$. However, we expect that METRIC, although it does not use EF, performs better for most other cases since it uses

ETrF, which can capture advection. Significantly higher precipitation in 2014 than in the other years (Figures 9 and 13) provides another possible explanation for the poor model performance that year. This might have caused the adjustment offset (T_{fac}, which is subtracted from T_s when selecting the hot pixel) to be erroneous since the relation used for T_{fac} in our work was developed for a drier location in Southern Idaho during 2000.

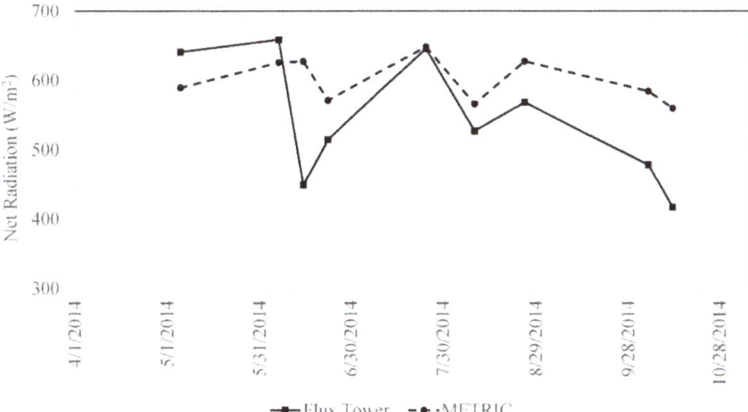

Figure 11. Validation of net radiation for 2014.

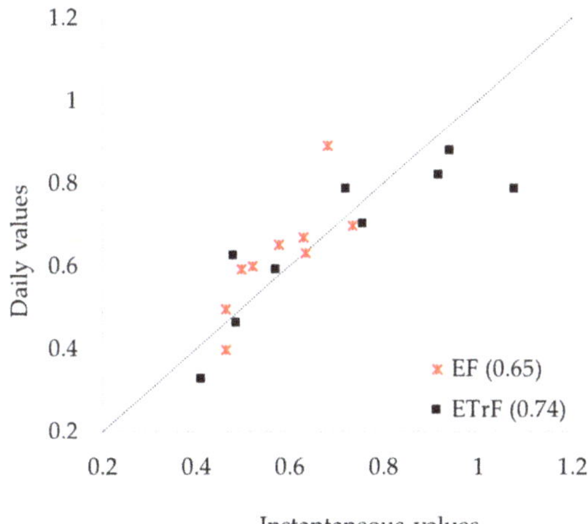

Figure 12. Comparison between instantaneous and daily Evaporative Fraction (*EF*) and Reference ET fraction (*ETrF*) computed using flux tower energy balance components. r^2 values are presented in parentheses.

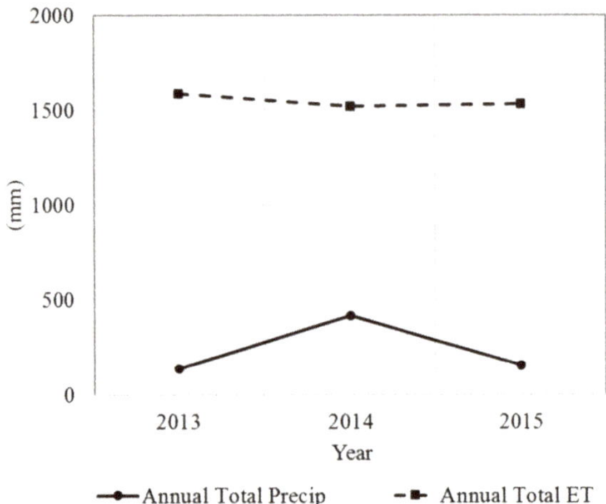

Figure 13. Total annual precipitation and evapotranspiration for the selected years.

4.3. Use and Limitations of This Approach to Estimate Variability in METRIC Due to Calibration Pixel Selection

The results obtained here suggest that there can be substantial variation in ET values within calibration pixels when following the techniques outlined in Allen et al. [19]. The process outlined in our work provides a simple method to estimate variability associated with model calibration. It can now be more readily used by less experienced users since it requires much less expert intervention and can allow for quicker analysis using an automated setting. These factors enable its use to compute ET on an operational basis. Although the computational requirement might be higher with multiple results of each individual imagery, using a cloud-based platform such as Google Earth Engine significantly reduces the computational time and storage required.

There are some issues inherent to METRIC that remain challenges for successful application of the approach. Naturally, the results can only be as good as the model inputs. Another major issue during the calibration process remains accurately identifying clouds, shadows, and agricultural lands. Cloud effects were particularly evident when selecting cold pixels, though filtering the clouds through a 7×7 filter helped avoid inaccurate identification of cold pixels close to the clouds. Overall, the F-mask algorithm [26] identified clouds in most images, but some images still needed visual inspection to identify and delineate the clouds.

Since each pixel property is important for calibration, the accuracy of the final ET results is highly dependent on accurate agricultural area selection. Although we used USGS National Land Cover Database (NLCD) 2011 agricultural land data for calibration pixel selection, various classification algorithms can be used to first identify land cover types then use the classified land cover for calibration of the model if no land use data is available.

Calibration of METRIC needs a range of pixels having a wide range of evaporative stresses and assumes that there are areas of dry and of wet pixels. Using the approach outlined, different sets of calibration pixels could be used for different climactic and ground conditions. As an example, if the study area received high precipitation, a calibration relationship with lower dT values might perform better owing to the lowered sensible heat flux.

An appropriate area selection for model calibration is important so that the calibration relation of dT vs. T_s remains linear. The area should have contrasting hydrologic regions (with highly irrigated and very dry areas), stable weather conditions, and relatively unchanging topography. For regions that do not have well-irrigated fields, researchers [50] have suggested using water bodies for selection of a cold calibration pixel. However, since the storage term in the energy balance formulation could have higher uncertainties, extra precaution and further study is needed prior to selecting a water body as the cold pixel.

Most major limitations of this study are the ones within METRIC. There are other sources of errors beyond calibration pixel selection, though Allen et al. [17] iterate that accurate calibration can reduce biases and errors in model. However, since there is no easy, specific method for identifying the accuracy of the calibration process, we assume that the calibration process reduced biases due to parameter selection in empirical relations used prior to calibration.

5. Conclusions

This study aimed to provide an approach for measuring the variability of watershed-scale ET estimates induced by human intervention in the calibration process. To do so, we examined the effect of anchor pixels on the performance of a satellite-based remote sensing ET model, METRIC. Differences resulting from anchor pixel set selection were shown to cause markedly different ET results, and we showed that including available energy when selecting a candidate pixel could improve the performance of the model. Based on this analysis, we demonstrate that the differences in ET, due to selection of a set of anchor pixels after the automatic selection process, of candidate pixels can range between 5% and 20%, based on the coefficient of variation. When compared to ET measurements from an eddy flux tower, the model's overall performance was best when the selected cold pixel had the minimum dT and the selected hot pixel had the maximum dT (from among all candidate pixels). Using these cold and hot pixels, ET results were more accurate than when using the traditional method of selecting a set of anchor pixels closest to the weather station (again from within a pool of calibration pixels). With an increasing volume of satellite image products, it can be expected that the use of models like METRIC will expand from roles as merely investigative research tools to serving as operational models for short-term and long-term decision-making. Our work aids in this transition by helping extend the domain of the model from well-trained modelers and experts to inexperienced and new users by providing a measure of uncertainty associated with the results. Models like METRIC must be calibrated for each specific image and being able to calibrate expediently is therefore crucial. With our study, a variability measure corresponding to selected calibration pixels around a specific weather station is easier to automate. Since our method of calibration does not strictly depend on closeness to a weather station, it can be more applicable for automating calibration of METRIC than the method outlined by Allen et al. [19] for cases when the reference weather station is not present within the image. It also provides a range of ET results that can be incorporated into uncertainty measures. Together, these will ultimately allow better water resources planning and management decisions.

Data Statement:

All raster data used in this study were obtained using from the GEE catalog and its functions. Use of GEE and its catalog requires creating an account with Google. The catalog provides a short description of the image, the bands available, and metadata or properties of the image. Images or raster data are provided as Image collections. For our work we used the following collections:

	Name	Image Collection ID	Number of Bands
1.	USGS Landsat 7 Collection 1 Tier 1 Raw Scenes	LANDSAT/LE07/C01/T1	10
2.	USGS Landsat 7 Collection 1 Tier 1 top-of-atmosphere (TOA) Reflectance	LANDSAT/LE07/C01/T1_TOA	10
3.	USGS Landsat 7 Surface Reflectance (SR) Tier 1	LANDSAT/LE07/C01/T1_SR	11
4.	USGS Landsat 8 Collection 1 Tier 1 Raw Scenes	LANDSAT/LC08/C01/T1	12
5.	USGS Landsat 8 Collection 1 Tier 1 top-of-atmosphere (TOA) Reflectance	LANDSAT/LC08/C01/T1_TOA	12
6.	USGS Landsat 8 Surface Reflectance Tier 1	LANDSAT/LC08/C01/T1_SR	12
7.	USGS National Elevation Dataset 1/3 arc-second	USGS/NED	1
8.	NLCD: USGS National Land Cover Database	USGS/NLCD	4

Data extraction and use:

We present an example (using JavaScript in Earth Engine Code Editor at https://code.earthengine.google.com) which shows how to obtain NDVI for a Landsat 8 image using the GEE catalog and functionalities.

```
// Load a Landsat 8 ImageCollection for a single path-row.
var collection = ee.ImageCollection('LANDSAT/LC08/C01/T1_TOA')
.filter(ee.Filter.eq('WRS_PATH', 44))
.filter(ee.Filter.eq('WRS_ROW', 34))
.filterDate('2015-06-25', '2015-07-01');
// Get the first image from the image collection
var img = ee.Image(collection.first());
// Define a function to compute NDVI.
var NDVI = function(image) {
return image.expression('float(b("B5") − b("B4")) / (b("B5") + b("B4"))');
};
// Use NDVI function
NDVI_img = NDVI(img);
// Display the NDVI image
Map.addLayer (NDVI_img, {min: 0, max: 1, palette: 'red, yellow, green'}, 'NDVI');
```

Supplementary Materials: Aupplementary materials can be found at http://www.mdpi.com/2072-4292/10/11/1695/s1, Figure S1: Distribution of SAVI values for hot and cold candidate pixels.

Author Contributions: Conceptualization: S.D. and M.E.B. Methodology: S.D. Software: S.D. Validation: S.D. Formal Analysis: S.D. Investigation: S.D. and M.E.B. Resources: M.E.B. Data Curation: S.D. Writing—Original Draft preparation: S.D. Writing—Review and Editing: M.E.B. Visualization: S.D. Supervision: M.E.B. Project Administration: M.E.B. Funding Acquisition: M.E.B.

Funding: This research was funded by Washington State Department of Ecology, Office of Columbia River.

Acknowledgments: This study is based on the work supported by Washington State Department of Ecology, Office of Columbia River. The authors would also like to thank US-AmeriFlux and Dennis Baldocchi at University of California Berkley for providing ET/Flux data measurements to the public. The authors would also like to express a sincere thank you to Mason Kreidler for providing constructive criticism of the manuscript.

Conflicts of Interest: The authors declare no conflict of interest.

References

1. Wang, K.; Dickinson, R.E. A review of global terrestrial evapotranspiration: Observation, modeling, climatology, and climatic variability. *Rev. Geophys.* **2012**, *50*, 1–54. [CrossRef]
2. Wallace, J.S. Increasing agricultural water use efficiency to meet future food production. *Agric. Ecosyst. Environ.* **2000**, *82*, 105–119. [CrossRef]
3. Farahani, H.J.; Howell, T.A.; Shuttleworth, W.J.; Bausch, W.C. Evapotranspiration: Progress in measurement and modeling in agriculture. *Trans. ASABE* **2007**, *50*, 1627–1638. [CrossRef]
4. Verstraeten, W.; Veroustraete, F.; Feyen, J. Assessment of evapotranspiration and soil moisture content across different scales of observation. *Sensors* **2008**, *8*, 70–117. [CrossRef] [PubMed]
5. Dugas, W.A.; Upchurch, D.R.; Ritchie, J.T. A weighing lysimeter for evapotranspiration and root measurements. *Agron. J.* **1985**, *77*, 821–825. [CrossRef]
6. Aubinet, M.; Vesala, T.; Papale, D. *Eddy Covariance a Practical Guide to Measurement and Data Analysis*; Springer: Dordrecht, The Netherlands; New York, NY, USA, 2012; pp. 59–82. ISBN 978-94-007-2351-1.
7. Ezzahar, J.; Chehbouni, A.; Hoedjes, J.C.B.; Er-Raki, S.; Chehbouni, A.; Boulet, G.; Bonnefond, J.M.; De Bruin, H.A.R. The use of the scintillation technique for monitoring seasonal water consumption of olive orchards in a semi-arid region. *Agric. Water Manag.* **2007**, *89*, 173–184. [CrossRef]

8.   Meiresonne, L.; Nadezhdin, N.; Cermak, J.; Van Slycken, J.; Ceulemans, R. Measured sap flow and simulated transpiration from a poplar stand in Flanders (Belgium). *Agric. For. Meteorol.* **1999**, *96*, 165–179. [CrossRef]

9.   Billah, M.M.; Goodall, J.L.; Narayan, U.; Reager, J.T.; Lakshmi, V.; Famiglietti, J.S. A methodology for evaluating evapotranspiration estimates at the watershed-scale using GRACE. *J. Hydrol.* **2015**, *523*, 574–586. [CrossRef]

10.   Allen, R.G.; Pereira, L.S.; Raes, D.; Smith, M. *Crop Evapotranspiration-Guidelines for Computing Crop Water Requirements*; FAO: Rome, Italy, 1998; pp. 87–101. ISBN 92-5-104219-5.

11.   Godfray, H.C.J.; Beddington, J.R.; Crute, I.R.; Haddad, L.; Lawrence, D.; Muir, J.F.; Pretty, J.; Robinson, S.; Thomas, S.M.; Toulmin, C. Food security: The challenge of feeding 9 billion people. *Science* **2010**, *327*, 812–818. [CrossRef] [PubMed]

12.   Bastiaanssen, W.G.M.; Noordman, E.J.M.; Pelgrum, H.; Davids, G.; Thoreson, B.P.; Allen, R. SEBAL model with remotely sensed data to improve water-resources management under actual field conditions. *J. Irrig. Drain. Eng.* **2005**, *131*, 85–93. [CrossRef]

13.   Bayat, B.; van der Tol, C.; Verhoef, W. Integrating satellite optical and thermal infrared observations for improving daily ecosystem functioning estimations during a drought episode. *Remote Sens. Environ.* **2018**, *209*, 375–394. [CrossRef]

14.   Chance, E.; Cobourn, K.; Thomas, V. Trend detection for the extent of irrigated agriculture in Idaho's Snake river plain, 1984–2016. *Remote Sens.* **2018**, *10*, 145. [CrossRef]

15.   Liou, Y.-A.; Kar, S. Evapotranspiration estimation with remote sensing and various surface energy balance algorithms: A review. *Energies* **2014**, *7*, 2821. [CrossRef]

16.   Allen, R.G.; Tasumi, M.; Trezza, R. Satellite-based energy balance for mapping evapotranspiration with internalized calibration (METRIC)—Model. *J. Irrig. Drain. Eng.* **2007**, *133*, 380–394. [CrossRef]

17.   Allen, R.; Irmak, A.; Trezza, R.; Hendrickx, J.M.H.; Bastiaanssen, W.; Kjaersgaard, J. Satellite-based ET estimation in agriculture using SEBAL and METRIC. *Hydrol. Process.* **2011**, *25*, 4011–4027. [CrossRef]

18.   Bhattarai, N.; Quackenbush, L.J.; Im, J.; Shaw, S.B. A new optimized algorithm for automating endmember pixel selection in the SEBAL and METRIC models. *Remote Sens. Environ.* **2017**, *196*, 178–192. [CrossRef]

19.   Allen, R.G.; Burnett, B.; Kramber, W.; Huntington, J.; Kjaersgaard, J.; Kilic, A.; Kelly, C.; Trezza, R. Automated calibration of the METRIC-Landsat evapotranspiration process. *JAWRA J. Am. Water Resour. Assoc.* **2013**, *49*, 563–576. [CrossRef]

20.   Bhattarai, N.; Shaw, S.B.; Quackenbush, L.J.; Im, J.; Niraula, R. Evaluating five remote sensing based single-source surface energy balance models for estimating daily evapotranspiration in a humid subtropical climate. *Int. J. Appl. Earth Obs. Geoinf.* **2016**, *49*, 75–86. [CrossRef]

21.   Morton, C.G.; Huntington, J.L.; Pohll, G.M.; Allen, R.G.; McGwire, K.C.; Bassett, S.D. Assessing calibration uncertainty and automation for estimating evapotranspiration from agricultural areas using METRIC. *JAWRA J. Am. Water Resour. Assoc.* **2013**, *49*, 549–562. [CrossRef]

22.   Eichelmann, E.; Hemes, K.S.; Knox, S.H.; Oikawa, P.Y.; Chamberlain, S.D.; Sturtevant, C.; Verfaillie, J.; Baldocchi, D.D. The effect of land cover type and structure on evapotranspiration from agricultural and wetland sites in the Sacramento–San Joaquin River Delta, California. *Agric. For. Meteorol.* **2018**, *256–257*, 179–195. [CrossRef]

23.   Oikawa, P.Y.; Sturtevant, C.; Knox, S.H.; Verfaillie, J.; Huang, Y.W.; Baldocchi, D.D. Revisiting the partitioning of net ecosystem exchange of CO_2 into photosynthesis and respiration with simultaneous flux measurements of $13CO_2$ and CO_2, soil respiration and a biophysical model, CANVEG. *Agric. For. Meteorol.* **2017**, *234–235*, 149–163. [CrossRef]

24.   Stannard, D.I. Comparison of Penman-Monteith, Shuttleworth-Wallace, and modified Priestley-Taylor evapotranspiration models for wildland vegetation in semiarid rangeland. *Water Resour. Res.* **1993**, *29*, 1379–1392. [CrossRef]

25.   Kilic, A.; Allen, R.; Trezza, R.; Ratcliffe, I.; Kamble, B.; Robison, C.; Ozturk, D. Sensitivity of evapotranspiration retrievals from the METRIC processing algorithm to improved radiometric resolution of Landsat 8 thermal data and to calibration bias in Landsat 7 and 8 surface temperature. *Remote Sens. Environ.* **2016**, *185*, 198–209. [CrossRef]

26.   Foga, S.; Scaramuzza, P.L.; Guo, S.; Zhu, Z.; Dilley, R.D.; Beckmann, T.; Schmidt, G.L.; Dwyer, J.L.; Joseph Hughes, M.; Laue, B. Cloud detection algorithm comparison and validation for operational Landsat data products. *Remote Sens. Environ.* **2017**, *194*, 379–390. [CrossRef]

27. Gorelick, N.; Hancher, M.; Dixon, M.; Ilyushchenko, S.; Thau, D.; Moore, R. Google Earth Engine: Planetary-scale geospatial analysis for everyone. *Remote Sens. Environ.* **2017**, *202*, 18–27. [CrossRef]

28. Ke, Y.; Im, J.; Park, S.; Gong, H. Downscaling of MODIS one kilometer evapotranspiration using Landsat-8 data and machine learning approaches. *Remote Sens.* **2016**, *8*, 215. [CrossRef]

29. Tasumi, M.; Allen, R.G.; Trezza, R. At-surface reflectance and albedo from satellite for operational calculation of land surface energy balance. *J. Hydrol. Eng.* **2008**, *13*, 51–63. [CrossRef]

30. Liang, S. Narrowband to broadband conversions of land surface albedo I: Algorithms. *Remote Sens. Environ.* **2001**, *76*, 213–238. [CrossRef]

31. Sobrino, J.A.; Jiménez-Muñoz, J.C.; Paolini, L. Land surface temperature retrieval from LANDSAT TM 5. *Remote Sens. Environ.* **2004**, *90*, 434–440. [CrossRef]

32. Jimenez-Munoz, J.C.; Cristobal, J.; Sobrino, J.A.; Soria, G.; Ninyerola, M.; Pons, X. Revision of the single-channel algorithm for land surface temperature retrieval from Landsat thermal-infrared data. *IEEE Trans. Geosci. Remote Sens.* **2009**, *47*, 339–349. [CrossRef]

33. Yu, X.; Guo, X.; Wu, Z. Land surface temperature retrieval from Landsat 8 TIRS—Comparison between radiative transfer equation-based method, split window algorithm and single channel method. *Remote Sens.* **2014**, *6*, 9829. [CrossRef]

34. Zanter, K. Landsat 8 (L8) Data Users Handbook. Available online: https://landsat.usgs.gov/sites/default/files/documents/LSDS-1574_L8_Data_Users_Handbook.pdf (accessed on 6 June 2018).

35. Bastiaanssen, W.G.M.; Menenti, M.; Feddes, R.A.; Holtslag, A.A.M. A remote sensing surface energy balance algorithm for land (SEBAL): Part 1. Formulation. *J. Hydrol.* **1998**, *212–213*, 198–212. [CrossRef]

36. Bastiaanssen, W.G.M.; Pelgrum, H.; Wang, J.; Ma, Y.; Moreno, J.F.; Roerink, G.J.; van der Wal, T. A remote sensing surface energy balance algorithm for land (SEBAL): Part 2: Validation. *J. Hydrol.* **1998**, *212–213*, 213–229. [CrossRef]

37. Businger, J.A.; Wyngaard, J.C.; Izumi, Y.; Bradley, E.F. Flux-profile relationships in the atmospheric surface layer. *J. Atmos. Sci.* **1971**, *28*, 181–189. [CrossRef]

38. Brutsaert, W. Evaporation. In *Hydrology: An Introduction*; Brutsaert, W., Ed.; Cambridge University Press: Cambridge, UK, 2005; pp. 117–158. ISBN 9780521824798.

39. Bastiaanssen, W.G.M. Regionalization of Surface Flux Densities and Moisture Indicators in Composite Terrain: A Remote Sensing Approach under Clear Skies in Mediterranean Climates. PhD Thesis, Wageningen Agricultural University, Wageningen, The Netherlands, 1995.

40. Allen, R.G.; Morton, C.; Kamble, B.; Kilic, A.; Huntington, J.; Thau, D.; Gorelick, N.; Erickson, T.; Moore, R.; Trezza, R. EEFlux: A Landsat-based evapotranspiration mapping tool on the Google Earth Engine. In Proceedings of the Joint ASABE/IA Irrigation Symposium 2015: Emerging Technologies for Sustainable Irrigation, Long Beach, CA, USA, 10–12 November 2015.

41. Allen, R.G.; Trezza, R.; Kilic, A.; Tasumi, M.; Li, H. Sensitivity of Landsat-scale energy balance to aerodynamic variability in mountains and complex terrain. *JAWRA J. Am. Water Resour. Assoc.* **2013**, *49*, 592–604. [CrossRef]

42. Irmak, A.; Ratcliffe, I.; Ranade, P.; Hubbard, K.G.; Singh, R.K.; Kamble, B.; Kjaersgaard, J. Estimation of land surface evapotranspiration with a satellite remote sensing procedure. *Great Plains Res.* **2011**, *21*, 73–88.

43. Huete, A.R. A soil-adjusted vegetation index (SAVI). *Remote Sens. Environ.* **1988**, *25*, 295–309. [CrossRef]

44. Burnett, B. A Procedure for Estimating Total Evapotranspiration Using Satellite-Based Vegetation Indices with Separate Estimates from Bare Soil. Master's Thesis, University of Idaho, Moscow, ID, USA, 2007.

45. Daughtry, C.S.T.; Hunt, E.R.; Doraiswamy, P.C.; McMurtrey, J.E. Remote Sensing the Spatial Distribution of Crop Residues. *Agron. J.* **2005**, *97*, 864–871. [CrossRef]

46. Horton, R.; Bristow, K.L.; Kluitenberg, G.J.; Sauer, T.J. Crop residue effects on surface radiation and energy balance—Review. *Theor. Appl. Clim.* **1996**, *54*, 27–37. [CrossRef]

47. Nash, J.E.; Sutcliffe, J.V. River flow forecasting through conceptual models part I—A discussion of principles. *J. Hydrol.* **1970**, *10*, 282–290. [CrossRef]

48. Yapo, P.O.; Gupta, H.V.; Sorooshian, S. Automatic calibration of conceptual rainfall-runoff models: Sensitivity to calibration data. *J. Hydrol.* **1996**, *181*, 23–48. [CrossRef]

49. Twine, T.E.; Kustas, W.P.; Norman, J.M.; Cook, D.R.; Houser, P.R.; Meyers, T.P.; Prueger, J.H.; Starks, P.J.; Wesely, M.L. Correcting eddy-covariance flux underestimates over a grassland. *Agric. For. Meteorol.* **2000**, *103*, 279–300. [CrossRef]

50. Long, D.; Singh, V.P. Assessing the impact of end-member selection on the accuracy of satellite-based spatial variability models for actual evapotranspiration estimation. *Water Resour. Res.* **2013**, *49*, 2601–2618. [CrossRef]
51. Foken, T. The energy balance closure problem: An overview. *Ecol. Appl.* **2008**, 1351–1367. [CrossRef]
52. Gonzalez-Dugo, M.P.; Neale, C.M.U.; Mateos, L.; Kustas, W.P.; Prueger, J.H.; Anderson, M.C.; Li, F. A comparison of operational remote sensing-based models for estimating crop evapotranspiration. *Agric. For. Meteorol.* **2009**, *149*, 1843–1853. [CrossRef]

Article

Evaluation of the SPARSE Dual-Source Model for Predicting Water Stress and Evapotranspiration from Thermal Infrared Data over Multiple Crops and Climates

Emilie Delogu [1,*], Gilles Boulet [1], Albert Olioso [2], Sébastien Garrigues [2], Aurore Brut [1], Tiphaine Tallec [1], Jérôme Demarty [3], Kamel Soudani and Jean-Pierre Lagouarde [5]

[1] Centre d'Etudes Spatiales de la Biosphère, Université de Toulouse, CNRS, CNES, IRD, UPS, INRA, 31401 Toulouse, France; gilles.boulet@cesbio.cnes.fr (G.B.); aurore.brut@cesbio.cnes.fr (A.B.); tiphaine.tallec@cesbio.cnes.fr (T.T.)

[2] EMMAH, INRA, Université d'Avignon et des Pays de Vaucluse, 84914 Avignon, France; albert.olioso@inra.fr (A.O.); sebastien.garrigues@inra.fr (S.G.)

[3] HSM, Univ Montpellier, CNRS, IRD, 34090 Montpellier, France; jerome.demarty@ird.fr

[4] Univ Paris-Sud, Laboratoire Ecologie Systématique et Evolution, AgroParisTech, CNRS, UMR 8079, Orsay, F-75231 Paris, France; kamel.soudani@u-psud.fr

[5] INRA, UMR 1391 ISPA, F-33140 Villenave d'Ornon, France; jean-pierre.lagouarde@inra.fr

*   Correspondence: emilie.delogu@cesbio.cnes.fr; Tel.: +33-561-558-513

Received: 10 October 2018; Accepted: 13 November 2018; Published: 15 November 2018

Abstract: Using surface temperature as a signature of the surface energy balance is a way to quantify the spatial distribution of evapotranspiration and water stress. In this work, we used the new dual-source model named Soil Plant Atmosphere and Remote Sensing Evapotranspiration (SPARSE) based on the Two Sources Energy Balance (TSEB) model rationale which solves the surface energy balance equations for the soil and the canopy. SPARSE can be used (i) to retrieve soil and vegetation stress levels from known surface temperature and (ii) to predict transpiration, soil evaporation, and surface temperature for given stress levels. The main innovative feature of SPARSE is that it allows to bound each retrieved individual flux component (evaporation and transpiration) by its corresponding potential level deduced from running the model in prescribed potential conditions, i.e., a maximum limit if the surface water availability is not limiting. The main objective of the paper is to assess the SPARSE model predictions of water stress and evapotranspiration components for its two proposed versions (the "patch" and "layer" resistances network) over 20 in situ data sets encompassing distinct vegetation and climate. Over a large range of leaf area index values and for contrasting vegetation stress levels, SPARSE showed good retrieval performances of evapotranspiration and sensible heat fluxes. For cereals, the layer version provided better latent heat flux estimates than the patch version while both models showed similar performances for sparse crops and forest ecosystems. The bounded layer version of SPARSE provided the best estimates of latent heat flux over different sites and climates. Broad tendencies of observed and retrieved stress intensities were well reproduced with a reasonable difference obtained for most of the points located within a confidence interval of 0.2. The synchronous dynamics of observed and retrieved estimates underlined that the SPARSE retrieved water stress estimates from Thermal Infra-Red data were relevant tools for stress detection.

Keywords: evapotranspiration; water stress; model; partition; remote-sensing

---

1. Introduction

Quantifying energy and water transfers throughout the soil-vegetation-atmosphere continuum is an essential issue to understand a wide range of processes involved in hydrological modeling, weather forecasting, and climate change impact assessment (Intergovernmental Panel on Climate Change, 2014).

Remote sensing in the thermal infrared (TIR) provides information on the surface energy balance, in particular, in relation to water stress level and on the partition of the available energy at the surface between sensible and latent heat fluxes. Available energy at the land surface, defined as the difference between net radiation (R_n) and soil heat flux (G), is mostly partitioned between sensible heat (H) and latent heat (LE) fluxes. As water needs energy to evaporate, evapotranspiration (ET) which combines evaporation from the soil and transpiration from the plants is a key component of the water and energy budgets ($ET = LE/L$ where L is the latent heat of vaporization).

Accurate estimates of soil latent heat flux (LE_s) and vegetation latent heat flux (LE_v) are needed for eco-agrohydrological applications such as drought monitoring and irrigation scheduling [1,2]. Since soil evaporation is considered as a water loss that does not contribute to biomass production, optimized management in agriculture consists in maximizing the crop transpiration:evaporation ratio [3]. Separate measurements of soil evaporation and vegetation transpiration are challenging. Accurate estimation of LE partitioning is also crucial to monitor and anticipate water stress which is quantified as the complementary part to unity of the ratio between actual and potential LE [4,5]. LE in potential conditions represents a theoretical value obtained if maximum LE is reached considering actual meteorological and plant development conditions.

Numerous soil-vegetation-atmosphere transfer (SVAT) modeling approaches were developed to estimate surface fluxes at leaf and canopy levels [6]. A SVAT model is based on the simultaneous solutions of water and energy budgets to compute the temporal dynamics of various prognostic variables, such as surface fluxes, temperatures, and water content profiles of the soil and the vegetation. SVAT models require many input data such as meteorological forcing, water supply chronicles (irrigation and rain) as well as information about thermal, hydraulic, optical, and biophysical surface properties. At a local-scale, when all surface properties are generally known, they are expected to perform well [7,8]. Nevertheless, SVAT models are more difficult to implement over large areas due to uncertainties in the spatial distribution of water inputs (irrigation or rain) and soil properties at a regional scale.

Estimates of ET can also be derived from the use of the remotely-sensed land surface temperature (T_{rad}) acquired from space in the TIR spectral domain. Conversely to SVAT models, most of these methods do not require information on water supply chronicles (irrigation and rain) and soil parameters. They rely on surface temperature, which is a relevant indicator of the water status of the surface and which can be used to retrieve the actual evapotranspiration and soil moisture status over large areas and various time scales [9]. Remotely-sensed ET products have reached a fairly satisfying level of accuracy from canopy to regional scales [2,8]. Most of the TIR-based methods employ instantaneous satellite data and compute an energy budget over the short integration period of the satellite overpass. They usually derive ET as the residual component of the surface energy balance. The most complex of these methods consider the soil and the vegetation as the two main sources of heat exchange in order to account for the partitioning of total LE into LE_s and LE_v [10]. A new two-source model named Soil Plant Atmosphere and Remote Sensing Evapotranspiration (SPARSE) based on the Two-Source Energy Balance (TSEB [10]) model rationale is proposed in [4]. The first innovative feature of SPARSE to the existing TSEB is similar to the post-processing step in the single-source Surface Energy Balance System (SEBS) model [11] but separately for the two components: soil and vegetation. It combines "retrieval" and "prescribed" modes by bounding each retrieved individual flux component (LE_s, LE_v) to its corresponding potential level deduced from running the model in prescribed potential conditions. This ensures that retrieved LE_s and LE_v values are below potential values considered as absolute maxima. In the usual "retrieval" mode, as in TSEB,

SPARSE solves the surface energy balance equations for the soil and the canopy, which means that two unknowns (LE_s and LE_v) can be solved simultaneously. In the "prescribed" mode, both energy balance equations are solved to compute transpiration and evaporation rates for given stress levels (for example minimal LE_s and LE_v in fully stressed conditions and maximal LE_s and LE_v in potential conditions). In the "prescribed" mode, the surface temperature is no longer an input of the model but an output. A second key feature of SPARSE is that it provides a "patch" and a "layer" approach to describe the soil–vegetation–atmosphere interactions [12], identical to TSEB. In the "layer" approach, also referred to as the series approach, the air is well mixed within the canopy, the air temperature at the aerodynamic level is homogeneous, and the vegetation layer uniformly covers the ground. Soil and vegetation heat sources are fully coupled through a resistance network organized in series. There is one aerodynamic resistance with the air above the canopy. In the "patch" approach, also referred to as the "parallel" approach, the vegetation layer is discontinuous so that the soil interacts directly with the air above the canopy, and soil and vegetation are modeled side by side. Soil and vegetation heat sources are thermally uncoupled and fluxes are computed with a parallel resistance scheme. In the "layer" or "series" approach, soil–vegetation radiative fluxes are taken into account whereas there are no such radiation exchanges between the soil and the vegetation patches in the patch approach.

The main objective of this paper is to assess the SPARSE model predictions classically for evapotranspiration components but also for water stress. It should be noted that good retrieval values for the total latent heat flux do not guarantee that total water stress is correctly simulated. As the detection of crop water stress is crucial for efficient irrigation water management, it appeared necessary to evaluate SPARSE ability to model accurately the water status of soil and plant. We addressed four major issues:

1. Patch vs. layer approach: Even though a "patch/parallel" approach was originally proposed for sparsely vegetated semi-arid regions, and the "layer/series" approach for denser vegetation [10,13,14], there is no consensus regarding which approach offers better results in semi-arid sparse vegetation. The TSEB layer version was more robust than the TSEB patch version even though layer and patch performances were close in [15,16]. This study will bring insights on the performances of the "patch" vs. the "layer" approaches to estimate evapotranspiration and its soil component over 20 irrigated and rainfed crops, including arable crops and orchards, and various climate conditions, from temperate or Mediterranean to semi-arid and tropical climates.

2. Benefit of bounding flux retrieval: The main improvement of SPARSE is the bounding of the output fluxes by their theoretical limit values. We will test whether this improves evapotranspiration retrieval performances by comparing bounded and unbounded retrieval methods.

3. Water stress retrieval: Estimates of potential surface evapotranspiration rates are fairly well constrained by soil and vegetation biophysical properties easily obtained from visible/near-infrared remote sensing data and can explain a large amount of the information contained in the actual ET. The added value of thermal infrared (TIR) data lies in the adequate amount of information introduced by the surface temperature itself. TIR data provides information on the difference between actual and potential evapotranspiration rates (i.e., water stress) and thus soil moisture-limited evaporation and transpiration rates. We assessed the capacities of the SPARSE model to monitor water stress by comparison to water stress index chronicles derived from field data.

4. Impact of the time of overpass of a TIR satellite: One of the auxiliary goals of this paper was to investigate the impact of the time of overpass of a TIR satellite on the performance of stress retrievals in order to take into account the operational constraints imposed by the existing or future satellite platforms. At present, specific studies are missing and no consensus has been reached on the best time of overpass for stress detection.

The paper is organized as follows: First, a brief description of the SPARSE model is presented, as well as the data set used to assess the performance of the model. Then, the performances themselves are described in a "result" section showing all criteria (total fluxes, water stress level, soil evaporation efficiency). The implication for future use of the model is discussed in the last section.

2. Materials and Methods

2.1. The SPARSE Model

2.1.1. Model Description

The SPARSE model ([4] see equations in Appendix A) computes the instantaneous equilibrium surface temperatures of soil (T_s) and vegetation (T_v) (for 30-min intervals in our study, corresponding to the measurement frequency of the meteorological forcing). These temperatures are used as separate signatures of the energy budget for the soil and the vegetation. SPARSE is a generalization of the TSEB model approach which consists in linearizing the full set of energy budget equations. It implements two approaches to describe the interactions between soil-vegetation-atmosphere, namely the 'patch' and 'layer' approach which correspond to fully uncoupled and fully coupled soil–vegetation–air exchanges, respectively, using a combination equation for potential fluxes (potential transpiration and potential evaporation) and expressions of the aerodynamic resistances scheme described in [17–19].

Five main equations are solved simultaneously, see Equation (1). The first two, express the continuity (layer version) or the summation (patch version) of the latent and the sensible heat fluxes from the soil and the canopy to the aerodynamic level and above. The third and fourth equations represent the energy budget of the soil and the vegetation. The fifth equation describes the link between the canopy radiative surface temperature T_{rad} that can be related to remote sensing measurements and the soil and the vegetation longwave radiative fluxes, which are related to the soil and the vegetation temperatures.

$$\begin{cases} H = H_s + H_v \\ LE = LE_s + LE_v \\ R_{ns} = G + H_s + LE_s \\ R_{nv} = H_v + LE_v \\ \sigma T_{rad}^4 = R_{atm} - R_{as} - R_{av} \end{cases} \tag{1}$$

R_{atm} is the atmospheric radiation (W m^{-2}), R_a is the net component longwave radiation (W m^{-2}), T_{rad} is the radiative surface temperature ($^\circ$K) as observed by the satellite (which is necessary to account for the measurement waveband and the directionality of the measurements), and σ is the Stefan–Boltzmann constant. LE is the latent heat flux, H is the soil sensible heat flux, R_n is the net radiation, and G is the heat flux in the soil; indexes "s" and "v" designate the soil and the vegetation, respectively.

For LE_s and LE_v retrieval, T_{rad} is known and derived from in situ thermal infrared observations. In order to compute the various fluxes of the energy balance, SPARSE follows the same approach employed by TSEB to estimate the soil evaporation and the plant transpiration from the knowledge of the sole surface temperature. As a first guess, the vegetation is supposed to transpire at a potential rate (i.e., unstressed conditions) and the system is solved to estimate LE_s. If a negative value is obtained for LE_s, the unstressed canopy assumption proves to be inconsistent. The vegetation is assumed to be affected by soil water stress. Then, LE_s is set to a minimum value of 30 W m^{-2} to take into account the contribution of vapor transfer from the superficial soil layer [20]. The system is then solved to estimate the vegetation latent heat flux (LE_v). If LE_v is negative, fully stressed conditions are imposed for both the soil and the vegetation independently of T_{rad}.

SPARSE can also be run in a forward mode from prescribed water stress conditions (from fully stressed to non-water-limited conditions). Actual conditions are expressed from potential conditions through the use of the efficiency coefficients which correspond to the ratio between actual and potential

LE rates, β_s and β_v, and are functionally equivalent to surface resistances ("s" for soil, "v" for vegetation). They range between 0 and 1. If $\beta_v = 1$, then the vegetation transpires at the potential rate. If $\beta_s = 1$, the soil evaporation rate is that of a saturated surface. $\beta_v = 0$ or $\beta_s = 0$ correspond to a non-transpiring or non-evaporating surface, respectively. This prescribed mode is implemented as a final step in the retrieval mode to provide theoretical limits corresponding to maximum reachable levels of sensible and latent heat for both the soil and the vegetation.

2.1.2. Model Implementation

In our study, no calibration was performed and the parameters were arbitrarily set to realistic levels: The minimum stomatal resistance was set to 100 s m^{-1} for herbaceous vegetation and crops [21] and 200 s m^{-1} for orchard and forest [22]. The $G{:}R_{ns}$ ratio is set to 40% for an arid climate and to 25% for others [23]. The displacement height and the roughness length for momentum exchange depend on the vegetation height. Soil albedo is set to a constant value for each site, see Table 1, depending on the soil texture in accordance with the characteristics of each site.

Table 1. Main characteristics of the data set, including maximum measured LAI (leaf area index) and number of identified stress periods. FR = France, TU = Tunisia, MO = Morrocco, NI = Niger.

Site Name (Country)	Ecosystem	Studied Year	Name Code	Maximal Observed LAI ($\text{m}^2 \text{ m}^{-2}$)	Number of Stress Periods	Soil Type (%Clay/%Sand)	Irrigation (mm)	Soil Albedo	Energy Balance Closure
Auradé (FR)	Wheat	2006	Aur W 2006	3.1	1	32/21	0	0.25	93%
Auradé (FR)	Sunflower	2007	Aur Su 2007	1.7	0	32/21	0	0.25	88%
Auradé (FR)	Wheat	2008	Aur W 2008	2.4	0	32/21	0	0.25	89%
Lamasquère (FR)	Wheat	2007	Lam W 2007	4.5	0	54/12	0	0.25	94%
Lamasquère (FR)	Wheat	2009	Lam W 2009	1.7	0	54/12	0	0.25	92%
Lamasquère (FR)	Wheat	2011	Lam W 2011	5.5	0	54/12	0	0.25	73%
Lamasquère (FR)	Wheat	2013	Lam W 2013	3.6	0	54/12	0	0.25	93%
Avignon (FR)	Peas	2005	Avi P 2005	2.8	0	33/14	100	0.25	95%
Avignon (FR)	Wheat	2006	Avi W 2006	5.5	0	33/14	20	0.25	94%
Avignon (FR)	Sorghum	2007	Avi So 2007	3.0	1	33/14	80	0.25	95%
Avignon (FR)	Wheat	2008	Avi W 2008	1.9	2	33/14	20	0.25	95%
Avignon (FR)	Wheat	2010	Avi W 2010	6.1	0	33/14	0	0.25	71%
Avignon (FR)	Wheat	2012	Avi W 2012	1.1	3	33/14	0	0.25	96%
Avignon (FR)	Sunflower	2013	Avi Su 2013	2.3	2	33/14	0	0.25	95%
Barbeau (FR)	Oak forest	2015	Bar Oa 2015	5.5	1	19/32	0	0.15	69%
Kairouan (TU)	Wheat	2012	Kai W 2012	2.1	0	31/40	0	0.25	60%
Kairouan (TU)	Olive	2012–2015	Kai Ol 2013	0.2	4	8/88	0	0.29	55%
Haouz (MO)	Wheat	2004	Hao W 2004	4.1	2	34/20	170	0.20	93%
Wankama-M (NI)	Millet	2009	Wan M 2009	0.4	1	13/85	0	0.30	91%
Wankama-F (NI)	Savannah	2009	Wan S 2009	0.3	0	13/85	0	0.30	91%

The SPARSE surface energy balance equations required a broadband brightness temperature. When sites are not equipped with a device allowing a broadband brightness temperature measurement (if not, specified in the experimental description below) but equipped with sensors measuring the brightness temperature in the 8 to 14 μm spectral band (see below), the equations were adapted to constrain the surface energy balance in the 8 to 14 μm spectral band. To do so, atmospheric radiation, atmospheric emissivity, and surface temperature were calculated in the 8 to 14 μm atmospheric window according to [24].

2.2. Experimental Data Sets Description

Twenty data sets collected over eight crop and forest sites, see Table 1, were used to assess the performance of SPARSE. Half-hourly observations of air temperature and humidity, wind speed, net radiation, and atmospheric pressure were continuously acquired above the ground or the canopy from micro-meteorological stations over the different sites. Soil heat flux (G) was also measured at each site. Sensible (*H*) and latent (*LE*) heat fluxes were computed every 30 min from eddy-covariance systems. For sites with an energy balance closure of less than 80%, the closure was forced with the residual method and *LE* was computed as R_n-*H*-*G*. For other sites with an energy balance closure over 80%, the half-hourly closure was achieved with the conservation of the Bowen ratio *H*/*LE* [25]; thus, *LE* was computed as $(R_n$-*G*$)/(1 + H/LE)$, see Table 1.

2.2.1. Auradé and Lamasquère Data Sets

The two cultivated plots Auradé and Lamasquère are located in the Occitanie region in France which exhibits a temperate climate. Data for Auradé were acquired in 2006 (wheat), 2007 (sunflower), and 2008 (wheat) while the data in Lamasquère were acquired in 2007, 2009, 2011, and 2013 (wheat). Surface radiative temperatures were measured with a precision infrared temperature sensor (IRTS-P, Campbell Scientific Inc, Logan, UT, USA) at 2.8 m above ground in the 6 to 14 μm spectral band in Auradé. Surface radiative temperatures were derived from longwave upwelling radiation measured by a 4-component net radiometer (CNR1 manufactured by Kipp and Zonen) at 3.65 m above ground in the 4.5 to 42 μm spectral band in Lamasquère. Leaf Area Index (LAI) was measured at key crop phenological stages (five to six measurements per crop cycle) using destructive methods and sampling schemes adapted to each crop. The leaf area was retrieved using a planimeter device. For a complete description of the site characteristics and more information on these data sets, see [26].

2.2.2. Avignon Arable Crop Data Sets

The "remote sensing and flux site" of INRA (National Institute of Agronomic Research) Avignon is located in South East France and characterized by a Mediterranean climate. Data were acquired in 2005 (peas), 2006 (wheat), 2007 (sorgho), 2008 (wheat), 2012 (wheat), and 2013 (sunflower). Surface radiative temperatures were derived from longwave upwelling radiation measured by a 4-component net radiometer (CNR1 manufactured by Kipp and Zonen) at 3 m above ground in the 4.5 to 42 μm spectral band. LAI was measured at key crop phenological stages (five to six measurements per crop cycle) using destructive methods and sampling schemes adapted to each crop. Leaf area was measured using a planimeter device. For a full description of the site characteristics and more information on these data sets, see [27].

2.2.3. Barbeau Forest Data Sets

Barbeau National Forest is a managed mature oak forest located 60 km southeast of Paris, France, in continental climatic conditions. Data covers the year 2015. Surface radiative temperatures were measured with an infrared temperature sensor (IR120, Campbell Scientific Inc., Logan, UT, USA) at 36 m above ground in the 8 to 14 μm spectral band. A complete description of the site characteristics is available in [28,29].

2.2.4. Tunisian Rainfed Wheat Data Set

The rainfed wheat was grown in 2012 in a semi-arid climate in central Tunisia, west of Kairouan. Surface temperature data were acquired with a nadir-looking Apogee thermoradiometer at 2.3 m above ground in the 8 to 14 μm spectral band. LAI was estimated with hemispherical photographs every 2 to 3 weeks depending on the phenological cycle. These data were evaluated using destructive measurements during key stages (growth and full cover). More information on that data set is available in [4].

2.2.5. Tunisian Olive Orchard Data Set

The olive orchard site is located in a semi-arid climate in central Tunisia, west of Kairouan. The site was equipped with infrared temperature sensors over the bare soil and the canopy (IR120, Campbell Scientific Inc, Logan, UT, USA) to measure the canopy and bare soil surface temperature at 9.8 m above ground in the 8 to 14 μm spectral band from March 2012. Data are available on the SEDOO OMP website with the assigned DOI: 10.6096/MISTRALS-SICMED.1479 [30].

2.2.6. Morocco Irrigated Wheat Data Set

Data for the irrigated wheat site were acquired during the 2004 growing season in the semi-arid Haouz plain in Morocco (B124 site, [31]). Surface temperature data were acquired with a nadir-looking

Apogee thermoradiometer at 2 m above ground in the 8 to 14 μm spectral band. LAI was estimated with hemispherical photography every 2 to 3 weeks depending on the phenological cycle, validated by destructive measurements during key stages (growth and full cover). For a complete description of the site characteristics and more information on the data sets, see [31].

2.2.7. Niger Crop and Fallow Data Set

The study area is located 60 km east of Niamey in the South West of the Republic of Niger, characterized by a tropical semi-arid climate. It consists of two plots of around 15 ha each in the AMMA-CATCH observatory [32,33]. The two data sets used in this study were collected in 2009 over a millet field and a fallow field. Surface temperature data were acquired with 10° incidence KT15 Heitronics at 2.9 m above ground in the 8 to 14 μm spectral band. LAI was derived from hemispherical photographs. For a recent description of both the site and data set, see [34].

2.3. Assessment of Simulated Surface Water Stress

We evaluated the instantaneous estimation of water stress levels computed from SPARSE over identified stress periods. The water stress index was defined as the ratio of the difference between potential and actual evapotranspiration to potential evapotranspiration. It varies theoretically from 0 (unstressed surface) to 1 (fully stressed surface).

$$S = \frac{LE_{pot} - LE_x}{LE_{pot}} \tag{2}$$

where LE_x refers to (i) LE_{obs} which is the observed instantaneous latent heat flux to compute "observed" water stress values or (ii) LE_{SPARSE} which is the simulated latent heat fluxes in actual conditions to compute "retrieved" water stress values from SPARSE. LE_{pot} is the simulated latent heat flux in potential conditions. Potential evapotranspiration rates were generated from SPARSE in prescribed conditions using the Penman–Monteith formulation.

The number of stress periods was determined for each data set, see Table 1. A water stress period was identified on the following basis: Stress starts when a large deviation (>40%) between the potential evapotranspiration LE_{pot} and the measured actual evapotranspiration rates is observed away from any rain event or any other income of water (i.e., irrigation) and ends with the next income of water. When this deviation is observed for more than 4 days in a row, we arbitrarily defined the period as stressed. As the ultimate goal of SPARSE is to retrieve ET from remotely-sensed TIR data, S was evaluated at the two nominal acquisition times by MODIS on board of TERRA and AQUA, 10:30 and 13:30.

2.4. Soil Evaporation Estimates

Individual estimates of soil evaporation and plant transpiration were not available in any site to evaluate the SPARSE simulations. Nevertheless, superficial soil moisture is a good proxy for soil evaporation, although topsoil moisture does not always react to small rainfall events and is also influenced by topsoil roots and capillarity rise of water from deeper layers. We evaluated the capacity of the SPARSE model to retrieve soil evaporation efficiency defined as $\beta_{s-SPARSE}$ (LEs_{SPARSE}/LEs_{pot}, where LEs_{SPARSE} and LEs_{pot} are the soil evaporation and the potential soil evaporation rate derived from running SPARSE in retrieval and prescribed potential conditions) by comparison to an independent estimation derived from the observed time series of superficial soil moisture θ_{surf} (top 5 cm for Auradé, Lamasquère, Avignon, Tunisia, and Morocco; top 10 cm for Niger). We used the efficiency model proposed in [35] to derive the empirical soil evaporation efficiency (β_{s-e}):

$$\beta_{s-e} = \left\{ 0.5 \times \left[1 - cos\left(\pi \frac{\theta_{surf}}{\theta_{sat}} \right) \right] \right\}^{p} \tag{3}$$

where θ_{sat} is the in situ water content at saturation and p is fixed according to soil texture to 0.5 for clay and 1 for loamy and sandy soils.

$\beta_{s-SPARSE}$ relates LE_s to LE_{pot} and ranges from 0 for a non-evaporative surface to 1 when the soil evaporation rate is equivalent to the flux from a saturated surface (see Appendix A).

$$LE_s = \beta_{s-SPARSE} \frac{\rho c_p}{\gamma} \frac{e_{sat}(T_s) - e_0}{r_{as}} \tag{4}$$

where ρc_p is the product of air density and specific heat, γ the psychrometric constant, r_{as} the soil to aerodynamic level, $e_{sat}(T_s)$ is the saturated vapor pressure at temperature T_s, and e_0 is the partial pressure of vapor at the aerodynamic level.

2.5. Experiment Design

Two sets of SPARSE simulations were derived for the layer and the patch versions of the model. In the first set, outputs were not limited by potential flux values (unbounded set) and in the second (bounded set), outputs were bounded by the potential (and fully stressed) flux rates LE_{pot} and LE_{pot}, considered as the absolute maximum reachable values for evaporation and transpiration, respectively. Soil sensible and plant sensible heat fluxes were also bounded by the maximum rates in fully stressed conditions. For both versions and both sets, the performances were calculated over the entire data set for each site (from sowing to harvest for the crops; during a calendar year for others).

Total water stress estimates were also evaluated to assess the amount of information introduced by the surface temperature. The inverse mode of SPARSE was used to assess the information on moisture-limited evaporation and transpiration rates which was introduced by the surface temperature. Retrieved and "observed" water stress indexes were generated over the 20 identified stress periods.

We also compared the retrieved soil evaporation efficiency from the layer version of SPARSE to the independent empirical model (β_{s-e}). As meteorological forcing can vary very quickly, LE_{pot} and T_{rad} can fluctuate significantly from one day to another and $\beta_{s-SPARSE}$ retrievals were highly variable. In order to smooth out the quick fluctuations of β_{s-e} and the scale differences between the information provided by the integrated soil moisture measurement (top 5 cm for Auradé, Lamasquère, Avignon, Tunisia, and Morocco; top 10 cm for Niger) and the information provided by the surface temperature (relative only to the first mm of the soil), 5-day moving averages were compared. This is consistent with the potential data assimilation method of β or LE estimated from TIR data that could be used in a SVAT model for example: A smoother is more likely to outperform a sequential assimilation algorithm for short observation windows since the former will naturally smooth out the high-order fluctuations due to high-order fluctuations of T_{rad}.

2.6. Performance Metrics

The simulations were quantitatively evaluated, comparing measured and simulated LE, H, R_n, and G from sowing to harvest for the seasonal crops and during the calendar year for others.

The simulation performance scores were quantified using the root mean square error (RMSE), the bias (Bias), and the Nash-Sutcliff Index (NI).

3. Results

3.1. Energy Balance Component Estimates

The global retrieval performances for the four energy balance components are reported in Table 2. For all data sets, the best LE estimates were obtained with the layer bounded version (global RMSE of 58 W m^{-2}). The performances of H estimations were almost similar in the four experiments (similar RMSE and NI, but a smaller negative bias for the layer version).

RMSE, Bias, and Nash Index (NI) from the unbounded layer and patch versions for each site (from sowing to harvest for crops; during a calendar year for others) are compared in Figure 1. RMSE presented a large dispersion for LE from 30 W m^{-2} to 100 W m^{-2} depending on sites whatever the version used. Half of the sites showed negative bias for LE, H, and R_n whereas both versions of the model were systematically overestimating G. Moreover, NI for G were almost always negative. The layer unbounded version was not able to reproduce LE dynamics on the Barbeau forest site (cross symbol) whereas performances of the patch version were relatively good. Conversely, for the two wheat cultures of Lamasquère in 2007 and of Avignon in 2006, the layer version showed much better performances than the patch version. Apart from over these particular sites, the two versions showed very similar performances.

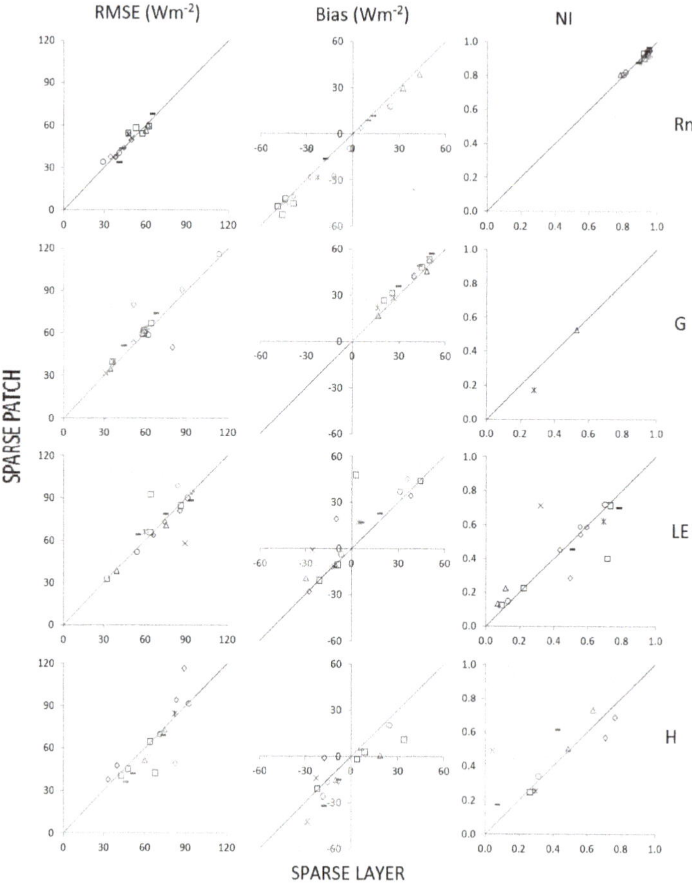

Figure 1. Performances of instantaneous retrievals from each unbounded model versions (patch vs. layer) at 13:30 for net radiation (R_n), soil heat flux (G), total latent heat flux (LE), and total sensible heat flux (H). (RMSE: root mean square error in W m^{-2}; Bias: in W m^{-2}; NI: Nash-Sutcliff efficiency Index). Grey zone matches area containing most of the points. Square symbols = Lamasquère, Circle symbols = Niger, Diamond symbols = Avignon, Triangle symbols = Tunisia, Cross symbol = Barbeau, String symbol = Auradé, Star symbol = Morocco.

The bounded experiment provided slightly better LE estimates, with RMSE reduced by 6 W m^{-2}, compared to the unbounded sets for the layer versions, as shown in Table 2. RMSE, Bias, and NI of the four energy balance components from the bounded layer and the unbounded layer versions were compared in Figure 2. LE retrievals were significantly better for seven sites with the bounded version compared to the non-bounded approach. LE dynamics of the Barbeau forest site were better reproduced with the bounded layer version (cross symbol). RMSE for R_n were often greater with the bounded version but biases were generally reduced. Actually, the bounded version constrains R_n with a prescribed value different of the observed one. Conversely, the unbounded version led to a simulated R_n very close in value to the measured one as R_n only depends on the incoming measured shortwave and longwave radiations and a linear expression of the surface temperature. Table 3 shows the performances of the bounded layer version for *LE* during each season. The RMSE and bias calculated over the fall and winter periods were the lowest over the year. These accurate performances of the model can somehow be explained by the low *LE* fluxes measured during these two seasons. Higher RMSE and bias were calculated over spring and, specifically, for the winter wheat cultures of Auradé, Lamasquère, and Avignon. For those crops, there was a strong overestimation by the model. In Table 3, NI were ranked into five classes to be interpreted as a non-parametric statistical test. Over the 60 seasons studied (whatever the site), 5 provided NI under 0 and attested for very poor retrievals, 14 reported between 0.25 and 0.5 (testifying for average performances), 20 reported between 0.5 and 0.75, and 15 reported over 0.75 (attesting for good to very good performances). NI showed that for two-thirds of the seasons studied, the model predicted fairly well the dynamic of *LE*.

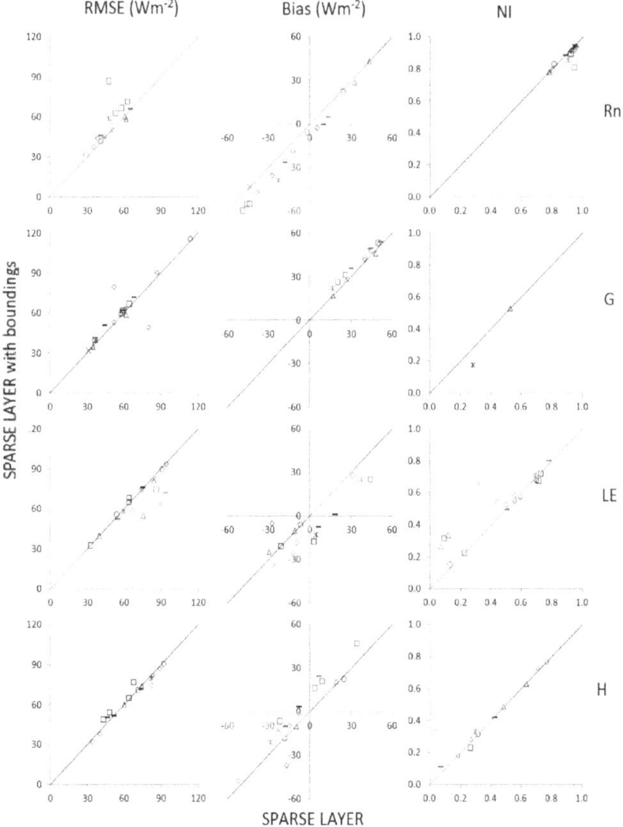

Figure 2. Performances of instantaneous retrievals from SPARSE bounded layer version vs. SPARSE simple layer version at 13:30 for net radiation (R_n), soil heat flux (*G*), total latent heat flux (*LE*) and total sensible heat flux (*H*). (RMSE: root mean square error in W m^{-2}; Bias: in W m^{-2}; NI: Nash-Sutcliff efficiency Index). Square symbols = Lamasquère, Circle symbols = Niger, Diamond symbols = Avignon, Triangle symbols = Tunisia, Cross symbol = Barbeau, String symbol = Auradé, Star symbol = Morocco.

Table 2. Global performances (among all sites) of instantaneous retrievals from each model version (patch vs. layer) at 13:30 for net radiation (R_n), soil heat flux (G), latent heat flux (LE), and sensible heat flux (H) (RMSE: root mean square error in W m^{-2}; NI: Nash-Sutcliff efficiency index).

	Boundings	no			yes		
LE	Performances	RMSE (W m^{-2})	NI	Bias (W m^{-2})	RMSE (W m^{-2})	NI	Bias (W m^{-2})
	SPARSE PATCH	61	0.62	10	59	0.65	2
	SPARSE LAYER	64	0.58	8	58	0.66	−4
	Boundings	no			Yes		
H	Performances	RMSE (W m^{-2})	NI	Bias (W m^{-2})	RMSE (W m^{-2})	NI	Bias (W m^{-2})
	SPARSE PATCH	68	0.52	−15	68	0.51	−10
	SPARSE LAYER	70	0.49	−11	70	0.48	−4
	Boundings	no			Yes		
Rn	Performances	RMSE (W m^{-2})	NI	Bias (W m^{-2})	RMSE (W m^{-2})	NI	Bias (W m^{-2})
	SPARSE PATCH	50	0.91	−6	56	0.88	−5
	SPARSE LAYER	50	0.91	−3	52	0.89	−4
	Boundings	no			Yes		
G	Performances	RMSE (W m^{-2})	NI	Bias (W m^{-2})	RMSE (W m^{-2})	NI	Bias (W m^{-2})
	SPARSE PATCH	56	−0.19	39	56	−0.19	38
	SPARSE LAYER	54	−0.09	35	54	−0.09	35

Table 3. Performances of instantaneous retrievals of total latent heat flux from the bounded layer versions at 13:30 for each season: summer (DOY 172 to 264), fall (DOY 264 to 355), winter (DOY 355 to 80), and spring (DOY 80 to 172). RMSE: root mean square error in W m^{-2}; NI: Nash-Sutcliff efficiency index. NI > 0.75: underlined green; 0.50 < NI < 0.75: green; 0.25 < NI < 0.50: orange; 0.00 < NI < 0.25: black; NI < 0.00: red.

	Whole Year			Summer			Fall			Winter			Spring		
	RMSE	NI	Bias	RMSE	NI	Bias	RMSE	NI	Bias	RMSE	NI	Bias	RMSE	NI	Bias
Aur W 2006	63	0.65	−5				41	0.85	13	40	0.76	−3	80	0.56	6
Aur Su 2007	87	0.12	17	76	0.25	12							98	−0.01	22
Aur W 2008	59	0.34	16				48	0.42	16	48	0.42	17	65	0.26	18
Lam W 2007	60	0.65	7				42	0.77	2	34	0.81	−3	68	0.74	11
Lam W 2009	74	0.18	32				30	0.74	−15	47	0.46	26	73	0.53	28
Lam W 2011	32	0.22	−20				32	0.23	−20	24	0.00	−24	63	0.61	12
Lam W 2013	70	0.81	−41				55	0.83	−38	35	0.92	−9	85	0.83	59
Avi P 2005	72	0.62	17							46	0.74	3	87	0.57	3
Avi W 2006	56	0.70	2				60	0.42	−49	30	0.83	−1	84	0.70	51
Avi So 2007	93	0.84	−2	86	0.87	5	53	−0.49	21				104	0.81	−36
Avi W 2008	79	0.44	−29				62	0.54	−51	62	0.52	−31	86	0.54	38
Avi W 2012	52	0.86	7				34	0.79	−21	38	0.73	14	72	0.85	22
Avi Su 2013	86	0.12	−3	76	0.25	−18							98	−0.01	−24
Bar Oa 2015	64	0.67	−25	76	0.61	−28				37	0.88	−4	68	0.58	−37
Kai W 2012	44	0.76	−14							40	0.73	1	47	0.77	−26
Kai Ol 2013	36	0.49	−11	35	0.58	−13	37	0.39	−16	22	0.41	−1	43	0.39	−13
Kai Ol 2014	43	0.23	−25	44	0.26	−32	41	0.22	−21	43	−0.04	−6	49	0.21	−32
Kai Ol 2015	49	0.18	−14	57	0.14	−13				39	0.15	−15	41	0.10	−12
Hao W 2004	57	0.63	−19							49	0.74	−16	69	0.42	−23
Wan F 2009	55	0.68	−10	79	0.61	−55	32	0.76	8				45	0.70	15
Wan M 2009	65	0.45	9	62	0.71	12	51	0.36	−37				83	0.02	66

3.2. Water Stress Index Estimation

During dry-downs, water stress was better retrieved from 13:30 fluxes estimations: at 10:30 only 5 RMSE values were lower than 0.2 while there were 12 values lower than 0.2 at 13:30; RMSE values at 13:30 were lower than those obtained at 10:30 for 12 periods within 16, see Table 4. The quality of water stress retrieval was highly dependent on the site and year. In general, water stress was fairly well retrieved for wheat crops. For the semi-arid sites (Tunisia, Niger, Morocco), water stress level retrievals showed RMSE values lower than 0.26 at 13:30. Figure 3 shows observed water stress estimates at 13:30 in grey and retrieved water stress estimates at 13:30 in black on four dry-downs, well-identified for the Avignon wheat in 2012 and for the Tunisian olive orchard in 2013. Missing points are related to missing data (observed data or/and inputs data). The overall magnitude of water stress is higher for the semi-arid Tunisian olive orchard than for Avignon. For the wheat crop, the observation indicates an abrupt change of stress between DOY 105 and 107, see Figure 3c whereas the SPARSE estimates displayed a smoother evolution. SPARSE properly captures the timing of water stress but could underestimate its magnitude compared to observations. For the Tunisian olive orchard, onsets of water stress were synchronous between observed and retrieved estimates. However, stress magnitude was lower for the simulated estimates, as shown in Figure 3d.

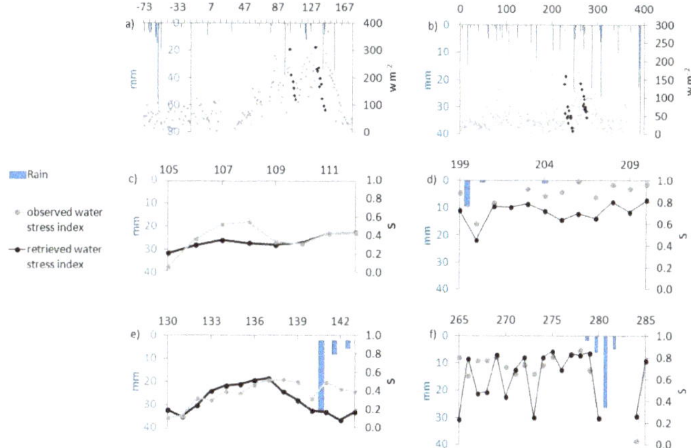

Figure 3. Evolution of the observed *LE* (W m^{-2}) for (**a**) the wheat culture on Avignon in 2012 and (**b**) the Tunisian olive orchard in 2013. Drydown periods are indicated in black. Parts (**c,e**) represent the observed water stress estimate at 13:30 in grey and retrieved water stress estimates at 13:30 in black for each of the two dry-downs observed for Avignon wheat 2012 and (**d,f**) the same for the Tunisian olive orchard 2013. The thick line represents the daily accumulated rain. The x-axis referred to the day of year.

Table 4. Performances of water stress estimates retrieval over each stress period at 10:30 and 13:30. RMSE: root mean square error. Statistics are calculated between observed water stress estimate and retrieved water stress estimates.

Acquisition Time			10:30	13:30
			RMSE	RMSE
Auradé (FR)	Wheat	2006	0.19	0.27
Avignon (FR)	Sorghum	2007	0.28	0.22
Avignon (FR)	Wheat	2008	0.15	0.10
Avignon (FR)	Wheat	2008	0.21	0.14
Avignon (FR)	Wheat	2012	0.35	0.11
Avignon (FR)	Wheat	2012	0.25	0.19
Avignon (FR)	Wheat	2012	0.25	0.24
Avignon (FR)	Sunflower	2013	0.25	0.14
Barbeau (FR)	Oak forest	2015	0.27	0.09
Sidi Rahal (MO)	Wheat	2004	0.24	0.26
Sidi Rahal (MO)	Wheat	2004	0.25	0.18
Niger C. (NI)	Millet	2009	0.93	0.09
Kairouan (TU)	Olive	2013	0.08	0.16
Kairouan (TU)	Olive	2013	0.22	0.17
Kairouan (TU)	Olive	2014	0.17	0.13
Kairouan (TU)	Olive	2015	0.11	0.11

3.3. Soil Evaporation Efficiency

SPARSE showed reasonable performances for the total *LE* retrieval. Here, we analyze its ability to retrieve the partitioning of *LE* between soil and vegetation fluxes. We focused the analysis on the "layer" version which was proven to provide more accurate estimates of total *LE* flux. Figure 4 shows that soil evaporation efficiency amplitudes varied among crop sites (less than 0.4 for the Tunisian olive orchard and from 0.3 to 1 for the Tunisian wheat crop for example). Figure 4 also shows that SPARSE properly reproduced the seasonal magnitude, particularly the growing season dynamics (e.g., Niger site). Wetting and drying cycles are fairly retrieved. However, in several situations, there were large discrepancies in magnitude between observed and simulated efficiencies. Auradé 2008 showed significant underestimation. For Avignon in 2005, the time evolution shown by SPARSE predictions was opposite to the evolution displayed in the observations. The lower limit of SPARSE evaporation efficiency (0) resulted from a negative soil latent heat flux (LE_s) (obtained because the assumption of an unstressed canopy proved to be inconsistent). In that case, the vegetation was assumed to suffer from water stress, the soil surface was assumed to have already long dried, and $\beta_{s-SPARSE}$ was zero.

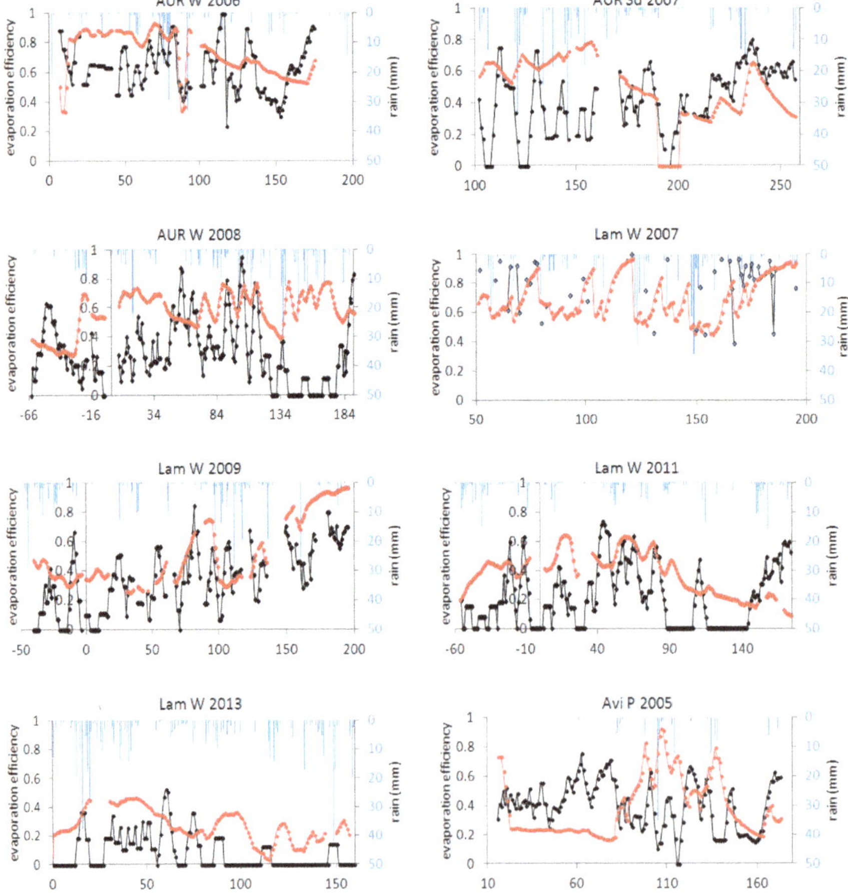

Figure 4. *Cont.*

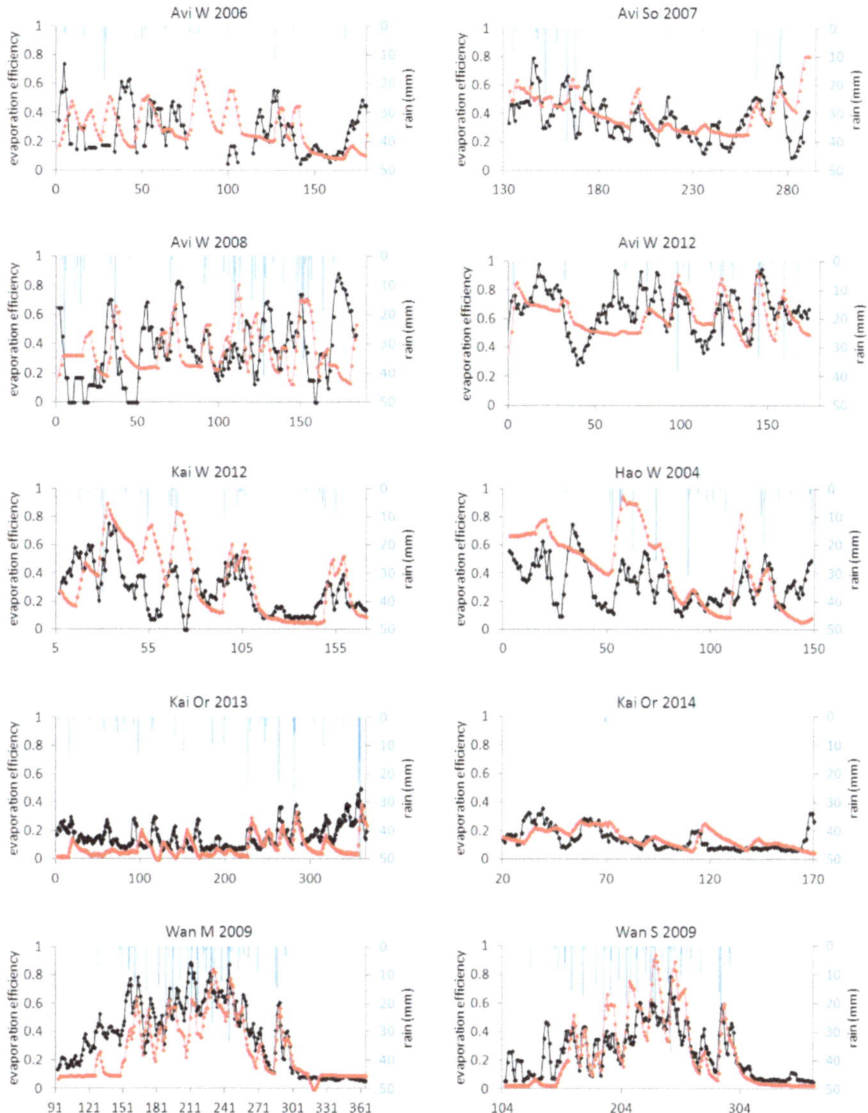

Figure 4. Evolution of the Soil Plant Atmosphere and Remote Sensing Evapotranspiration (SPARSE) layer retrieved evaporation efficiencies (black) compared to the empirical evaporation efficiency calculated using observed topsoil moisture β_{s-e} (red) for the different sites and years (Table 1) where the 5-cm top soil moisture was acquired. Vertical lines represent the daily accumulated rain (right y-axis).

4. Discussion

4.1. Overall Performance of the Dual-Source SPARSE Model

The performances of the SPARSE model for retrieving *LE* and its components from remotely-sensed surface temperatures were assessed over multiple crops, including orchards, and multiple climates. Over a large range of LAI values and for contrasting vegetation stress levels, the SPARSE model showed satisfactory retrieval performances of latent and sensible heat fluxes. For most of the data sets, RMSE ranged between 40 and 80 W m^{-2} for *LE* (bounded layer version) which is within the range of performances obtained by other models in the literature [15,36].

4.2. "Patch" vs. "Layer" Approach Performances

As expected for cereal covers, the layer version provided better estimates of *LE* than the patch version. Currently, homogeneity of cereal crops is usually well represented by such a layer approach. For sparse crops, such as orchards, or for the forest ecosystem, both patch and bounded layer schemes showed similar performances. The geometry of very sparse vegetation is better represented by the patch scheme [10] but interactions between soil and vegetation occur even for sparse vegetation [13]. A lot of usual land surface models such as CLM [37], ORCHIDEE [38], or SURFEX/MEB [39] are also based on a layer scheme. For now, using a layer or patch scheme is strongly linked to the study scale and needs to be specifically investigated to define the scale conditions corresponding to a patch approach. The patch approach could be justified at the scale where a boundary layer is fully developed over each patch and effects between patches are insignificant [40]. They also showed that the coupled model represented by the layer approach is a simplification of more complex and realistic models and that it is more widely applicable than the patch model.

4.3. Benefit of Bounding the Flux Estimates

Bounding fluxes by realistic limiting values based on potential conditions improved latent heat flux estimates in many cases, as it allowed to correct values of evaporation and transpiration which were sometimes retrieved beyond potential levels, especially in the layer approach. These inconsistencies appeared in two types of situations: senescent vegetation and oasis effect.

- For cereals in senescent situations at the end of the season, specific SPARSE parameter values in terms of canopy structure and stomatal conductance were not given. It is also essential to take into account the contribution of green stem and ear to the plant transpiration, especially for wheat [41,42]; this can explain the underestimation of transpiration for the Avignon and Auradé sites, see Figure 2.
- In semi-arid areas, transpiration and particularly evaporation were retrieved beyond potential levels during periods where the soil exhibits an important dry-down (in particular the Tunisian site; Figure 2). Sensible heat transfer to the crop from drier surrounding zones led to a great overestimation of transpiration by the model because of fully coupled soil–vegetation–air exchanges in the layer version. In these situations, bounding the outputs by realistic limiting flux values ensured model robustness and reduced *LE* RMSE values to 30 W m^{-2}, resulting in a reduction of global RMSE by 6 W m^{-2}. The bounded layer version of SPARSE provided the best estimates of LE over the different sites and climates.

4.4. Evaluation of the Capacity of SPARSE to Monitor Water Stress

The limit of stress retrieval from noisy TIR data was pointed out by previous studies [43,44]. The differences noticed between the observed and retrieved water stress intensities remained reasonable and the dynamics were well reproduced by SPARSE, with most points located within a confidence interval of 0.2. SPARSE could then be a relevant tool for stress detection, but the hypothesis used to define water stress here did not take into account differences in the crops ability to continue to grow in water stress conditions. Further studies would consist of assimilating the water stress efficiency simulated by SPARSE in a SVAT model in order to evaluate continuous evapotranspiration rates.

Soil evaporation efficiency was evaluated amongst sites and compared to a reconstructed time series relying on observed topsoil moisture. Overall, $\beta_{s-SPARSE}$ underestimated β_{s-e} which was a consequence of the two different approaches used to compute soil evaporation efficiencies. $\beta_{s-SPARSE}$

is not influenced by the same parameters as β_{s-e}. Indeed, β_{s-e} depends on the 5-cm topsoil moisture, a time-continuous variable, whose dynamics are strongly linked to large rain events while $\beta_{s-SPARSE}$ is mostly related to the porous network in direct contact with the atmosphere. Besides, $\beta_{s-SPARSE}$ and LE_{pot} are strongly impacted by a quick change in the meteorological forcing. T_{rad} is very sensitive to changes in atmospheric turbulence [45] whereas θ_{5cm}, and then β_{s-e}, are less reactive to these fluctuations. In our study, no calibration was performed and the parameters were arbitrarily set to realistic levels. This is consistent with the potential use of this model which aims at estimating LE and retrieving water stress estimations routinely from remote-sensing data with no additional calibration.

An additional issue is related to the uncertainty in the input variable, T_{rad}. Actually, many uncertainties and errors are known to affect the remotely-sensed surface temperature, including atmospheric correction and emissivity settings [24,46] in addition to the directional dependence of T_{rad} [47–49]. Particularly, angular emissivity effects can be important over heterogeneous cover where emission depends on many different individual emitters with contrasting emissivities. The TIR anisotropy comes from the heterogeneity of the observed object combined with a particular viewing angle. For sparse vegetations, directional effects can be important due to the soil contribution [48,49]. Moreover, T_{rad} retrieval can be affected by surface layer turbulence which can generate important temporal fluctuations [45]. At the end of almost every season (except for Niger, Lamasquère 2009 and 2013 and Tunisian wheat 2012), $\beta_{s-SPARSE}$ differed greatly from the β_{s-e} as it remained close to the potential rate. This could be related to the estimation of transpiration during the senescent period which is not properly simulated by SPARSE. Changes in soil-vegetation radiative exchanges and in canopy stomatal conductance occurring during senescence can lead to confusion over the transpiration:evaporation partition. Soil evaporation efficiencies derived from SPARSE were reasonably well retrieved for very sparse vegetation sites (Niger sites and olive orchard). In those sites, the coupling between surface temperature and evaporation reduction is properly simulated by SPARSE independent of the assumption on the water status of the vegetation.

4.5. Potential to be Driven by Earth Observation Data

One of the goals of this paper was to investigate the impact of the overpass time of a TIR satellite on the performance stress retrievals in order to take into account the operational constraints imposed by existing or future satellite platforms. The optimum time of overpass between the two tested in this work (10:30 and 13:30) was 13:30, which is in agreement with the theoretical study described in [43] based on an analytical estimation of peak latent heat flux as a response to a sinusoidal radiation forcing.

5. Conclusions

SPARSE showed satisfactory retrieval performances of latent and sensible heat fluxes, and the opportunity to bound fluxes by realistic limiting values based on potential conditions improved latent heat flux estimates in many cases, especially, in the layer approach in senescent situations at the end of the season and in semi-arid areas where transpiration and particularly evaporation were retrieved beyond potential levels.

The soil evaporation efficiencies estimated by SPARSE should be tested in order to be used to retrieve information on irrigation amount or precipitation inputs from TIR acquisitions.

Most models using information in the TIR domain like SPARSE rely on data acquired once a day within the constraints of the time of the satellite overpass, the revisit frequency, and the cloud cover. Consequently, the diurnal cycle of the energy budget is not accounted for and SPARSE will compute an instantaneous energy budget at the time of the satellite overpass and provide a single instantaneous latent heat flux. As a daily accumulation is usually required for hydrological applications for monitoring water stress, daily and seasonal LE need to be reconstructed from these retrieval instantaneous values. In order to encounter and evaluate the potential of SPARSE outputs with TIR acquisitions for the reconstruction of a continuous data set, future work will assess the impact of uncertainty on SPARSE model performances over operational methods to reconstruct *ET* at a daily and seasonal scale in order to fairly monitor water stress and irrigation.

Author Contributions: Data processing, data analysis, and results interpretation, E.D.; data analysis and results interpretation, G.B.; Avignon data processing, analysis, and discussions, S.G.; Auradé and Lamothe data processing and analysis, A.B. and T.T.; Barbeau data processing and analysis, K.S.; ideas and discussions, A.O., J.D. and J.-P.L.

Funding: This work was mostly supported by the French Space Agency (CNES) through TOSCA project PRECOSTRESS.

Conflicts of Interest: The authors declare no conflict of interest.

Appendix

Basic equations of the SPARSE "layer" model
Net solar radiation on the soil is:

$$R_{gs} = \frac{(1 - \alpha_s)(1 - f_c)}{1 - f_c \alpha_s \alpha_v} R_g$$

Net solar radiation on the vegetation is:

$$R_{gv} = (1 - \alpha_v) f_c \left[1 + \frac{\alpha_s (1 - f_c)}{1 - f_c \alpha_s \alpha_v} \right] R_g$$

Net longwave radiation for the soil is:

$$R_{as} = a_{ns} \sigma T_s^4 + b_{ns} \sigma T_v^4 + c_{ns}$$

Net longwave radiation for the vegetation is:

$$R_{av} = a_{nv} \sigma T_s^4 + b_{nv} \sigma T_v^4 + c_{nv}$$

where:

$$a_{ns} = -\frac{\varepsilon_s \left[(1 - f_c) + \varepsilon_v f_c \right]}{1 - f_c (1 - \varepsilon_s)(1 - \varepsilon_v)}$$

$$b_{ns} = a_{nv} = \frac{\varepsilon_v \varepsilon_s f_c}{1 - f_c (1 - \varepsilon_s)(1 - \varepsilon_v)}$$

$$c_{ns} = \frac{(1 - f_c) \varepsilon_s R_{atm}}{1 - f_c (1 - \varepsilon_s)(1 - \varepsilon_v)}$$

$$b_{nv} = -f_c \varepsilon_v \left[1 + \frac{\varepsilon_s + (1 - f_c)(1 - \varepsilon_s)}{1 - f_c (1 - \varepsilon_s)(1 - \varepsilon_v)} \right]$$

$$c_{nv} = f_c \varepsilon_v R_{atm} \left[1 + \frac{(1 - f_c)(1 - \varepsilon_s)}{1 - f_c (1 - \varepsilon_s)(1 - \varepsilon_v)} \right]$$

(α_s and ε_s are the albedo and the emissivity of the soil, α_v and ε_v are the albedo and the emissivity of the canopy, and R_g is the global incoming radiation; the vegetation cover fraction is $f_c = 1 - e^{-0.5 LAI / \cos \varphi}$ where φ is the view zenith angle; atmospheric radiation is $R_{atm} = 1.24 (e_a / T_a)^{1/7} \sigma T_a^4$ where T_a and e_a are the temperature and the vapor pressure of the air, respectively).

The various fluxes of the system of Equation (1) are expressed as:

$$R_{ns} = R_{gs} + R_{as}; \quad G = \zeta R_{ns}; \quad R_{nv} = R_{gv} + R_{av}$$

$$H_s = \rho c_p \frac{T_s - T_0}{r_{as}}, \quad H_v = \rho c_p \frac{T_v - T_0}{r_{av}} \quad et \quad H = \rho c_p \frac{T_0 - T_a}{r_a}$$

$$LE_s = \frac{\rho c_p}{\gamma} \beta_s \frac{e_{sat}(T_s) - e_0}{r_{as}}, \quad LE_v = \frac{\rho c_p}{\gamma} \beta_v \frac{e_{sat}(T_v) - e_0}{r_{av} + r_{stmin}} \quad et \quad LE = \frac{\rho c_p}{\gamma} \frac{e_0 - e_a}{r_a}$$

where r_a, r_{as}, and r_{av} are aerodynamic resistances between the aerodynamic level and the reference level, the soil and the aerodynamic level, and the vegetation and the aerodynamic level, respectively, while r_{stmin} is the minimum stomatal resistance; T_0 and e_0 are the temperature and the vapor pressure at the aerodynamic level, respectively; the two unknowns, T_s and T_v, as well as the soil evaporation efficiency β_s (or the transpiration β_v if the retrieved β_s is lower than a minimum value corresponding to $LE_s = 30 \text{ W/m}^2$), are solved simultaneously by inverting the system of Equation (1).

References

1. Er-Raki, S.; Chehbouni, A.; Boulet, G.; Williams, D.G. Using the dual approach of FAO-56 for partitioning ET into soil and plant components for olive orchards in a semi-arid region. *Agric. Water Manag.* **2010**, *97*, 1769–1778. [CrossRef]
2. Kustas, W.; Anderson, M. Advances in thermal infrared remote sensing for land surface modeling. *Agric. For. Meteorol.* **2009**, *149*, 2071–2081. [CrossRef]
3. Evett, S.R.; Tolk, J.A. Introduction: Can Water Use Efficiency Be Modeled Well Enough to Impact Crop Management? *Agron. J.* **2009**, *101*, 423–425. [CrossRef]
4. Boulet, G.; Mougenot, B.; Lhomme, J.-P.; Fanise, P.; Lili-Chabaane, Z.; Olioso, A.; Bahir, M.; Rivalland, V.; Jarlan, L.; Merlin, O.; et al. The SPARSE model for the prediction of water stress and evapotranspiration components from thermal infra-red data and its evaluation over irrigated and rainfed wheat. *Hydrol. Earth Syst. Sci.* **2015**, 4653–4672. [CrossRef]
5. Hain, C.R.; Mecikalski, J.R.; Anderson, M.C. Retrieval of an Available Water-Based Soil Moisture Proxy from Thermal Infrared Remote Sensing. Part I: Methodology and Validation. *J. Hydrometeorol.* **2009**, *10*, 665–683. [CrossRef]
6. Olioso, A.; Chauki, H.; Courault, D.; Wigneron, J.-P. Estimation of Evapotranspiration and Photosynthesis by Assimilation of Remote Sensing Data into SVAT Models. *Remote Sens. Environ.* **1999**, *68*, 341–356. [CrossRef]
7. Chirouze, J.; Boulet, G.; Jarlan, L.; Fieuzal, R.; Rodriguez, J.C.; Ezzahar, J.; Er-Raki, S.; Bigeard, G.; Merlin, O.; Garatuza-Payan, J.; et al. Intercomparison of four remote-sensing-based energy balance methods to retrieve surface evapotranspiration and water stress of irrigated fields in semi-arid climate. *Hydrol. Earth Syst. Sci.* **2014**, *18*, 1165–1188. [CrossRef]
8. Olioso, A.; Inoue, Y.; Ortega-FARIAS, S.; Demarty, J.; Wigneron, J.-P.; Braud, I.; Jacob, F.; Lecharpentier, P.; OttlÉ, C.; Calvet, J.-C.; et al. Future directions for advanced evapotranspiration modeling: Assimilation of remote sensing data into crop simulation models and SVAT models. *Irrig. Drain. Syst.* **2005**, *19*, 377–412. [CrossRef]
9. Boulet, G.; Chehbouni, A.; Gentine, P.; Duchemin, B.; Ezzahar, J.; Hadria, R. Monitoring water stress using time series of observed to unstressed surface temperature difference. *Agric. For. Meteorol.* **2007**, *146*, 159–172. [CrossRef]
10. Norman, J.M.; Kustas, W.P.; Humes, K.S. Source approach for estimating soil and vegetation energy fluxes in observations of directional radiometric surface temperature. *Agric. For. Meteorol.* **1995**, *77*, 263–293. [CrossRef]
11. Su, Z. The Surface Energy Balance System (SEBS) for estimation of turbulent heat fluxes. *Hydrol. Earth Syst. Sci.* **2002**, *6*, 85–100. [CrossRef]
12. Lhomme, J.P.; Montes, C.; Jacob, F.; Prévot, L. Evaporation from Heterogeneous and Sparse Canopies: On the Formulations Related to Multi-Source Representations. *Bound.-Layer Meteorol.* **2012**, *144*, 243–262. [CrossRef]
13. Blyth, E.M. Using a simple SVAT scheme to describe the effect of scale on aggregation. *Bound.-Layer Meteorol.* **1995**, *72*, 267–285. [CrossRef]
14. Verhoef, A.; De Bruin, H.A.R.; Van Den Hurk, B.J.J.M. Some Practical Notes on the parameter kB-1 for Sparse Vegetation. *Am. Meteorol. Soc.* **1997**, *36*, 560–572. [CrossRef]
15. Li, F.; Kustas, W.P.; Prueger, J.H.; Neale, C.M.U.; Jackson, T.J. Utility of Remote Sensing–Based Two-Source Energy Balance Model under Low- and High-Vegetation Cover Conditions. *J. Hydrometeorol.* **2005**, *6*, 878–891. [CrossRef]
16. Morillas, L.; García, M.; Nieto, H.; Villagarcia, L.; Sandholt, I.; Gonzalez-Dugo, M.P.; Zarco-Tejada, P.J.; Domingo, F. Using radiometric surface temperature for surface energy flux estimation in Mediterranean drylands from a two-source perspective. *Remote Sens. Environ.* **2013**, *136*, 234–246. [CrossRef]
17. Choudhury, B.J.; Monteith, J.L. A four-layer model for the heat budget of homogeneous land surfaces. *Q. J. R. Meteorol. Soc.* **1988**, *114*, 373–398. [CrossRef]
18. Shuttleworth, W.J.; Gurney, R.J. The theoretical relationship between foliage temperature and canopy resistance in sparse crops. *Q. J. R. Meteorol. Soc.* **1990**, *116*, 497–519. [CrossRef]

19. Shuttleworth, W.J.; Wallace, J.S. Evaporation from sparse crops-an energy combination theory. *Q. J. R. Meteorol. Soc.* **1985**, *111*, 839–855. [CrossRef]

20. Boulet, G.; Braud, I.; Vauclin, M. Study of the mechanisms of evaporation under arid conditions using a detailed model of the soil–atmosphere continuum. Application to the EFEDA I experiment. *J. Hydrol.* **1997**, *193*, 114–141. [CrossRef]

21. Gentine, P.; Entekhabi, D.; Chehbouni, A.; Boulet, G.; Duchemin, B. Analysis of evaporative fraction diurnal behaviour. *Agric. For. Meteorol.* **2007**, *143*, 13–29. [CrossRef]

22. Masson, V.; Champeaux, J.-L.; Chauvin, F.; Meriguet, C.; Lacaze, R. A Global Database of Land Surface Parameters at 1-km Resolution in Meteorological and Climate Models. *J. Clim.* **2003**, *16*, 1261–1282. [CrossRef]

23. Kustas, W.P.; Daughtry, C.S.T. Estimation of the soil heat flux/net radiation ratio from spectral data. *Agric. For. Meteorol.* **1990**, *49*, 205–223. [CrossRef]

24. Olioso, A. Estimating the difference between brightness and surface temperatures for a vegetal canopy. *Agric. For. Meteorol.* **1995**, *72*, 237–242. [CrossRef]

25. Twine, T.E.; Kustas, W.P.; Norman, J.M.; Cook, D.R.; Houser, P.R.; Meyers, T.P.; Prueger, J.H.; Starks, P.J.; Wesely, M.L. Correcting eddy-covariance flux underestimates over a grassland. *Agric. For. Meteorol.* **2000**, *103*, 279–300. [CrossRef]

26. Béziat, P.; Ceschia, E.; Dedieu, G. Carbon balance of a three crop succession over two cropland sites in South West France. *Agric. For. Meteorol.* **2009**, *149*, 1628–1645. [CrossRef]

27. Garrigues, S.; Olioso, A.; Calvet, J.C.; Martin, E.; Lafont, S.; Moulin, S.; Chanzy, A.; Marloie, O.; Buis, S.; Desfonds, V.; et al. Evaluation of land surface model simulations of evapotranspiration over a 12-year crop succession: Impact of soil hydraulic and vegetation properties. *Hydrol. Earth Syst. Sci.* **2015**, *19*, 3109–3131. [CrossRef]

28. Chemidlin Prévost-Bouré, N.; Soudani, K.; Damesin, C.; Berveiller, D.; Lata, J.-C.; Dufrêne, E. Increase in aboveground fresh litter quantity over-stimulates soil respiration in a temperate deciduous forest. *Appl. Soil Ecol.* **2010**, *46*, 26–34. [CrossRef]

29. Soudani, K.; Hmimina, G.; Delpierre, N.; Pontailler, J.-Y.; Aubinet, M.; Bonal, D.; Caquet, B.; de Grandcourt, A.; Burban, B.; Flechard, C.; et al. Ground-based Network of NDVI measurements for tracking temporal dynamics of canopy structure and vegetation phenology in different biomes. *Remote Sens. Environ.* **2012**, *123*, 234–245. [CrossRef]

30. Chebbi, W.; Boulet, G.; Le Dantec, V.; Lili Chabaane, Z.; Fanise, P.; Mougenot, B.; Ayari, H. Analysis of evapotranspiration components of a rainfed olive orchard during three contrasting years in a semi-arid climate. *Agric. For. Meteorol.* **2018**, *256–257*, 159–178. [CrossRef]

31. Boulet, G.; Olioso, A.; Ceschia, E.; Marloie, O.; Coudert, B.; Rivalland, V.; Chirouze, J.; Chehbouni, G. An empirical expression to relate aerodynamic and surface temperatures for use within single-source energy balance models. *Agric. For. Meteorol.* **2012**, *161*, 148–155. [CrossRef]

32. Cappelaere, B.; Descroix, L.; Lebel, T.; Boulain, N.; Ramier, D.; Laurent, J.-P.; Favreau, G.; Boubkraoui, S.; Boucher, M.; Bouzou Moussa, I.; et al. The AMMA-CATCH experiment in the cultivated Sahelian area of south-west Niger—Investigating water cycle response to a fluctuating climate and changing environment. *J. Hydrol.* **2009**, *375*, 34–51. [CrossRef]

33. Lebel, T.; Cappelaere, B.; Galle, S.; Hanan, N.; Kergoat, L.; Levis, S.; Vieux, B.; Descroix, L.; Gosset, M.; Mougin, E.; et al. AMMA-CATCH studies in the Sahelian region of West-Africa: An overview. *J. Hydrol.* **2009**, *375*, 3–13. [CrossRef]

34. Velluet, C.; Demarty, J.; Cappelaere, B.; Braud, I.; Issoufou, H.B.A.; Boulain, N.; Ramier, D.; Mainassara, I.; Charvet, G.; Boucher, M.; et al. Building a field- and model-based climatology of surface energy and water cycles for dominant land cover types in the cultivated Sahel. Annual budgets and seasonality. *Hydrol. Earth Syst. Sci.* **2014**, *18*, 5001–5024. [CrossRef]

35. Merlin, O.; Al Bitar, A.; Rivalland, V.; Béziat, P.; Ceschia, E.; Dedieu, G. An Analytical Model of Evaporation Efficiency for Unsaturated Soil Surfaces with an Arbitrary Thickness. *J. Appl. Meteorol. Climatol.* **2011**, *50*, 457–471. [CrossRef]

36. Kalma, J.D.; McVicar, T.R.; McCabe, M.F. Estimating Land Surface Evaporation: A Review of Methods Using Remotely Sensed Surface Temperature Data. *Surv. Geophys.* **2008**, *29*, 421–469. [CrossRef]

37. Jaeger, E.B.; Anders, I.; Lüthi, D.; Rockel, B.; Schär, C.; Seneviratne, S.I. Analysis of ERA40-driven CLM simulations for Europe. *Meteorol. Z.* **2008**, *17*, 349–367. [CrossRef]

38. Krinner, G.; Viovy, N.; de Noblet-Ducoudré, N.; Ogée, J.; Polcher, J.; Friedlingstein, P.; Ciais, P.; Sitch, S.; Prentice, I.C. A dynamic global vegetation model for studies of the coupled atmosphere-biosphere system. *Glob. Biogeochem. Cycles* **2005**, *19*, GB1015. [CrossRef]

39. Boone, A.; Samuelsson, P.; Gollvik, S.; Napoly, A.; Jarlan, L.; Brun, E.; Decharme, B. The interactions between soil-biosphere-atmosphere land surface model with a multi-energy balance (ISBA-MEB) option in SURFEXv8—Part 1: Model description. *Geosci. Model Dev.* **2017**, *10*, 843–872. [CrossRef]

40. Van den Hurk, B.J.J.M.; McNaughton, K.G. Implementation of near-field dispersion in a simple two-layer surface resistance model. *J. Hydrol.* **1995**, *166*, 293–311. [CrossRef]

41. Weiss, M.; Troufleau, D.; Barret, F.; Chauki, H.; Prévot, L.; Olioso, A.; Bruguier, N.; Brisson, N. Coupling canopy functioning and radiative transfer models for remote sensing data assimilation. *Agric. For. Meteorol.* **2001**, *108*, 113–128. [CrossRef]

42. Brisson, N.; Casals, M.L. Canopy senescence: A key factor in drought resistance of durum wheat. In Proceedings of the International Conference Land Water Resource Management in the Mediterranean Regions Tecnomack, Bari, Italy, 4–8 September 1994; Volume 5.

43. Gentine, P.; Entekhabi, D.; Polcher, J. Spectral Behaviour of a Coupled Land-Surface and Boundary-Layer System. *Bound.-Layer Meteorol.* **2010**, *134*, 157–180. [CrossRef]

44. Lagouarde, J.-P.; Bach, M.; Sobrino, J.A.; Boulet, G.; Briottet, X.; Cherchali, S.; Coudert, B.; Dadou, I.; Dedieu, G.; Gamet, P.; et al. The MISTIGRI thermal infrared project: Scientific objectives and mission specifications. *Int. J. Remote Sens.* **2013**, *34*, 3437–3466. [CrossRef]

45. Lagouarde, J.-P.; Irvine, M.; Dupont, S. Atmospheric turbulence induced errors on measurements of surface temperature from space. *Remote Sens. Environ.* **2015**, *168*, 40–53. [CrossRef]

46. Tardy, B.; Rivalland, V.; Huc, M.; Hagolle, O.; Marcq, S.; Boulet, G. A Software Tool for Atmospheric Correction and Surface Temperature Estimation of Landsat Infrared Thermal Data. *Remote Sens.* **2016**, *8*, 696. [CrossRef]

47. Duffour, C.; Olioso, A.; Demarty, J.; Van der Tol, C.; Lagouarde, J.-P. An evaluation of SCOPE: A tool to simulate the directional anisotropy of satellite-measured surface temperatures. *Remote Sens. Environ.* **2015**, *158*, 362–375. [CrossRef]

48. Lagouarde, J.P.; Dayau, S.; Moreau, P.; Guyon, D. Directional Anisotropy of Brightness Surface Temperature Over Vineyards: Case Study Over the Medoc Region (SW France). *IEEE Geosci. Remote Sens. Lett.* **2014**, *11*, 574–578. [CrossRef]

49. Luquet, D.; Vidal, A.; Dauzat, J.; Bégué, A.; Olioso, A.; Clouvel, P. Using directional TIR measurements and 3D simulations to assess the limitations and opportunities of water stress indices. *Remote Sens. Environ.* **2004**, *90*, 53–62. [CrossRef]

Article

Earth Observations-Based Evapotranspiration in Northeastern Thailand

Chaolei Zheng *, Li Jia, Guangcheng Hu and Jing Lu

State Key Laboratory of Remote Sensing Science, Institute of Remote Sensing and Digital Earth, Chinese Academy of Sciences, Beijing 100101, China; jiali@radi.ac.cn (L.J.); hugc@radi.ac.cn (G.H.); lujing@radi.ac.cn (J.L.)
* Correspondence: zhengcl@radi.ac.cn; Tel.: +86-10-6486-0422

Received: 5 December 2018; Accepted: 8 January 2019; Published: 12 January 2019

Abstract: Thailand is characterized by typical tropical monsoon climate, and is suffering serious water related problems, including seasonal drought and flooding. These issues are highly related to the hydrological processes, e.g., precipitation and evapotranspiration (ET), which are helpful to understand and cope with these problems. It is critical to study the spatiotemporal pattern of ET in Thailand to support the local water resource management. In the current study, daily ET was estimated over Thailand by ETMonitor, a process-based model, with mainly satellite earth observation datasets as input. One major advantage of the ETMonitor algorithm is that it introduces the impact of soil moisture on ET by assimilating the surface soil moisture from microwave remote sensing, and it reduces the dependence on land surface temperature, as the thermal remote sensing is highly sensitive to cloud, which limits the ability to achieve spatial and temporal continuity of daily ET. The ETMonitor algorithm was further improved in current study to take advantage of thermal remote sensing. In the improved scheme, the evaporation fraction was first obtained by land surface temperature—vegetation index triangle method, which was used to estimate ET in the clear days. The soil moisture stress index (SMSI) was defined to express the constrain of soil moisture on ET, and clear sky SMSI was retrieved according to the estimated clear sky ET. Clear sky SMSI was then interpolated to cloudy days to obtain the SMSI for all sky conditions. Finally, time-series ET at daily resolution was achieved using the interpolated spatio-temporal continuous SMSI. Good agreements were found between the estimated daily ET and flux tower observations with root mean square error ranging between 1.08 and 1.58 mm d^{-1}, which showed better accuracy than the ET product from MODerate resolution Imaging Spectroradiometer (MODIS), especially for the forest sites. Chi and Mun river basins, located in Northeast Thailand, were selected to analyze the spatial pattern of ET. The results indicate that the ET had large fluctuation in seasonal variation, which is predominantly impacted by the monsoon climate.

Keywords: ET; Thailand; ETMonitor; land surface temperature; Mun river basin; Chi river basin

---

1. Introduction

Under the context of climate change and rapid population development, water is increasingly becoming a scarce resource worldwide. Coping with water scarcity and growing competition for water among different sectors requires proper water management strategies and decision processes. Evapotranspiration (ET), the primary process of water transfer from the land surface to the atmosphere, is one of the most important components in hydrological cycle since it represents a loss of usable water from the hydrological supply for agriculture and natural resource [1,2]. Hence, ET plays a crucial role in understanding the response of water cycle to climate change and human activities ultimately emulating the water resource management to cope with the serious water resource shortages. Plants usually open their stomata under wet conditions, which are favorable for plant growth, while when the soil dries stomatal closure limits transpiration to prevent dehydration. And how to express the constrain of soil moisture is the key to ensure the accuracy of evapotranspiration algorithms.

Northeastern Thailand is characterized by tropical monsoon climate, with the wet season from May to October and the dry season from November to the next April. The annual mean air temperature is 20 °C and the annual precipitation is approximately 1300 mm that mostly falls during the rainy season (April to October). From the 2000s onwards, it is reported that the onset of the Asian monsoon and the start of the wet season is delayed in the whole of Indo China Peninsula, in association with El Nino Southern Oscillation (ENSO) events, and annual rainfall is likely to be reduced further [3]. This is followed by lowered normalized difference vegetation index (NDVI) values and higher surface temperatures in the widespread tropical forests, which will further exert significant influences on regional and global energy and water cycle [4]. Rain-fed paddy field, cassava and teak plantations are widely spread in this region, and they cover large portions of land use over Chi and Mun basins. The effects of climate change, including increasing temperatures, seasonal floods and droughts, severe storms and sea level rise, has threatened Thailand's agriculture for decades [5]. ET is one of the most important variables for crop water requirement estimation to rationalize the water consumption in the agricultural field under current and future climatic conditions. How to capture ET variation during the dry season is one of the most challenge issues, since ET during the dry season is regulated by soil moisture mostly rather than the meteorological variables. However, the spatiotemporal variation of ET and its trend in the Thailand is still not well studied, most likely due to the lack of high accurate ET dataset, which limit the local water resource management.

Several approaches have been developed to estimate land surface ET based on satellite earth observation data, which has proven to be able to obtain ET information at different spatial scales from regional to global coverage [6–15]. The widely used surface energy balance-based method is accurate and relies on the available land surface temperature (LST) data from satellite observations, and the basic theory is that LST could generally represent the land surface dry-wet condition. However, its application is limited in the clear sky, and large uncertainty exists when upscaling to cloudy sky conditions [16–18]. Thus, the process-based approach is more attractive to get the spatially and temporally continuous ET products with the increasing earth observation dataset availability [7–10]. For example, the process-based ETMonitor model driven by satellite observation has proven to be able to generate highly accurate ET estimation at the daily scale and 1 km spatial resolution in arid and semi-arid land by utilizing a variety of biophysical parameters derived from microwave and optical remote sensing observations [7]. One key process of ETMonitor is that it adopted soil moisture from microwave remote sensing to estimate root zone soil moisture empirically, which is utilized to estimate the canopy resistance with other climate variables, and it is crucial to ensure the estimation accuracy. Previously, the parameters in ETMonitor were assigned to locally arid and semi-arid basin based on related references, which should be carefully calibrated when applied to different regions. When applying ETMonitor to the monsoon climate in Thailand, an operational method should be developed to address the constrain of soil moisture on ET during the dry season. Considering the complementary advantages of the LST-based method and the process-based method, we hypothesize these two methods can be combined to take both advantages, and LST-based method can be utilized to parameterize the constraint of soil moisture on ET in ETMonitor, to improve the ET estimation scheme and obtain spatially and temporally continuous ET information.

In the current study, the ETMonitor algorithm was further improved to estimate ET in Thailand based on earth observation products. An operational method was proposed to parameterize the constrain of soil moisture on canopy resistance for ET estimation, mainly achieved based on the LST-based land surface energy balance method. The result was further validated based on eddy covariance flux tower observations to present the validity of current method. The spatial variation of ET was also analyzed to improve the understanding of the local scale ET information for Northeastern Thailand.

2. Materials and Methods

2.1. ET Estimation Scheme

The developed ET estimation scheme is a combination of process-based algorithm, ETMonitor, and LST-based energy balance algorithm. Different from the original ETMonitor, which adopt fixed parameters to estimate canopy resistance for ET estimation based on mainly soil moisture retrieved from microwave remote sensing, the developed scheme in current study also assimilates LST information retrieved by thermal remote sensing. For the clear sky, when both soil moisture and LST data are available, LST-VI (vegetation index) method is adopted to estimate EF and ET, which is taken to parameterize the regulation of soil moisture on the ET. This relationship between canopy resistance and soil moisture during the clear sky is built by introducing a new parameter named soil moisture stress index (SMSI), which is further interpolated and applied to cloudy days to estimate canopy resistance when only soil moisture data is available. Details on SMSI will be described in Sections 2.2 and 2.3. Hence, the canopy resistance time series is obtained and is finally applied to estimate ET.

The ETMonitor algorithm is designed to estimate the ET as the sum of different components, including plant transpiration (Ec), soil evaporation (Es), canopy rainfall interception loss (Ei), open water evaporation (Ew), and snow/ice sublimation (Ess). Due to the scarcity of snow and ice in the study region, sublimation process has been eliminated from current study. Detail on the parameterizations in ETMonitor can be found in [7]. Briefly, the canopy rainfall interception loss is estimated using a revised Gash analytical model, open water evaporation is estimated by the classical Penman equation. The soil evaporation and vegetation transpiration are calculated following the Shuttleworth–Wallace dual-source model [19]:

$$Ec = \frac{\Delta Rn_c + \rho c_p VPD_{0/r_a^c}}{\lambda\Delta + \lambda\gamma(1 + r_s^c/r_a^c)} \tag{1}$$

$$Es = \frac{\Delta(Rn_s - G) + \rho c_p VPD_{0/r_a^s}}{\lambda\Delta + \lambda\gamma(1 + r_s^s/r_a^s)} \tag{2}$$

where Rn_s and Rn_c represents the net radiation for soil and vegetation (W m^{-2}), respectively; G is the soil heat flux density (W m^{-2}); r_a^c and r_a^s are the bulk boundary layer resistance of vegetation and aerodynamic resistance between soil and canopy source height (s m^{-1}), estimated according to the canopy height; r_s^c is the bulk stomatal resistance of canopy (s m^{-1}), estimated by Jarvis-type model; r_s^s is the surface resistance of soil (s m^{-1}); Δ is the slope of saturation vapor pressure curve of air temperature (kPa K^{-1}); ϱ is the air density (kg m^{-3}); c_p is the specific heat of air at constant pressure (J kg^{-1} K^{-1}); λ is the latent heat of vaporization (MJ kg^{-1}); γ is the psychrometric constant (kPa K^{-1}); VPD_0 is the water vapor pressure deficit at canopy source height (kPa). Detail of these parameters retrieval can be found in [7].

The total Rn (Rn_{tot}) is partitioned vertically as Rn for canopy rainfall interception loss (Rn_i), Rn_c, and Rn_s:

$$Rn_{tot} = Rn_i + Rn_c + Rn_s \tag{3}$$

where the value of Rn_i stays zero on days with clear sky. For rainy days, Rn_i is estimated according to the interception loss by reversing the Penman-Monteith (PM) equation. Generally, Rn_{tot} is measured by remote sensing and Rn_i is determined from the Penman-Monteith equation, and these are subsequently used to determine Rn_c and Rn_s for Equations (1) and (2). The sum of Rn_c and Rn_s is estimated as the residual of $Rn_{tot} - Rn_i$, and is partitioned as:

$$Rn_c = Fc\,(Rn_{tot} - Rn_i) \tag{4}$$

$$Rn_s = (1 - Fc)(Rn_{tot} - Rn_i) \tag{5}$$

where Fc is the fraction of vegetation cover.

2.2. Canopy Resistance

Canopy resistance is estimated as the reciprocal of the canopy conductance, which is upscaled from the stomatal conductance of leaf. While the stomatal conductance was computed by the well-known Jarvis type model [20], which expresses stomatal conductance as the product of constraining functions of several environmental variables, including the solar radiation (Rd), vapor pressure deficit (VPD), air temperature (Ta), and root zone soil moisture (θ_{root}). These finally lead to the r_s^c estimation equation as follows:

$$r_s^c = r_{s,min} / \left(f(Rd)f(VPD)f(Ta)f(\theta_{root})LAI_{eff} \right) \tag{6}$$

where $r_{s,min}$ is the minimum stomatal resistance (s m^{-1}) associated with different plant functional types; LAI_{eff} represents the effective leaf area index (LAI), which has been applied for upscaling from leaf to canopy scale. LAI_{eff} was estimated following [21]:

$$LAI_{eff} = LAI / (0.3LAI + 1.2). \tag{7}$$

The constrain function of $f(Rd), f(VPD), f(Ta), f(\theta_{root})$, varying between 0 and 1, represent the constrains of solar radiation, VPD, air temperature, and root zone soil moisture to the stomatal conductance, expressed as:

$$f(Rd) = 1 - \exp(-Rd/k_1) \tag{8}$$

$$f(VPD) = 1 - k_2 VPD \tag{9}$$

$$f(Ta) = \left(\frac{Ta - T_{min}}{T_{opt} - T_{min}} \right) \left(\frac{T_{max} - T_{min}}{T_{max} - T_{opt}} \right)^{(T_{max} - T_{opt})/(T_{opt} - T_{min})} \tag{10}$$

$$f(\theta_{root}) = \begin{cases} 0 & for\ \theta_{root} < \theta_{min} \\ (\theta_{root} - \theta_{min})/(\theta_{max} - \theta_{min}) & for\ \theta_{min} < \theta_{root} < \theta_{max} \\ 1 & for\ \theta_{root} > \theta_{max} \end{cases} \tag{11}$$

where k_1 and k_2 are fitting parameters to describe the stomatal conductance sensitivity to radiation and VPD. θ_{max} and θ_{min} represents the saturating and wilting point soil moisture at which the plant stomata were totally open and close, respectively. T_{min}, T_{max}, and T_{opt} are the minimum, maximum and optimum air temperatures, respectively.

2.3. Parameterizing the Constrain of Soil Moisture

Since satellite soil moisture data can only provide surface soil moisture (θ_{surf}), previous study estimated θ_{root} according to an empirical equation ($\theta_{root} \propto \theta_{surf}$) and applied in Equation (11) for obtaining $f(\theta_{root})$ [7]. $f(\theta_{root})$ is directly proportional to θ_{surf}, hence SMSI is introduced in this study to express this relationship:

$$SMSI = f(\theta_{root})/\theta_{surf}. \tag{12}$$

Since θ_{surf} usually decreases rapidly compared to θ_{root} after rainfall events until it reaches a relatively stable level, SMSI is supposed to show decreasing trend with the decrease of θ_{surf} from wet to dry period. By introducing SMSI to expresses the sensitivity of $f(\theta_{root})$ to θ_{surf} compared with the original ETMonitor, the relationship between $f(\theta_{root})$ and θ_{surf} is simplified.

Generally, θ_{surf} can be obtained directly from microwave remote sensing soil moisture products, while θ_{root} can be estimated either from ground observation or soil water balance model, which could apply Equation (11) to estimate $f(\theta_{root})$ given that θ_{max} and θ_{min} are known. Unfortunately, θ_{max} and θ_{min} vary largely and depend on several factors including the soil characters, plant species, growing phases, and root development, a mathematical fitting was suggested to obtain θ_{min} based on field experiment [22]. Hence, we suggest to reverse Equation (6) to retrieve $f(\theta_{root})$, which is feasible for clear sky when ET is estimated by the LST-based method. And SMSI under cloud covered days can be interpolated linearly according to the clear sky SMSI. Accordingly, Equation (11) is rearranged as:

$$f(\theta_{root}) = \min\left\{1, \max\left(0, \theta_{surf} * SMSI\right)\right\}. \tag{13}$$

2.4. Clear Sky EF and ET Estimation

Different from the original ETMonitor, current study also utilize the LST-based energy balance to retrieve clear sky ET, which was further used to retrieve clear day SMSI. The clear sky ET and evaporative fraction (EF) were first estimated using LST-VI feature space method. The clear sky EF is based on reference [23]:

$$\text{EF} = \phi \frac{\Delta}{\Delta + \gamma} \tag{14}$$

where ϕ is a combined parameter for aerodynamic resistance, and is directly derived from remotely sensed data. The success of LST-VI triangle method for estimating EF and ET depends mainly on the choice of dry and wet edges in the LST-VI triangle space.

2.5. Data Collection

Table 1 lists the remote sensing dataset collected for ET estimation. For MOD11A1 and MYD11A1 daily LST data, only those under the clear sky were collected. The 8-day composites of LAI and albedo data were interpolated temporally to obtain daily 1 km LAI and albedo datasets. The 0.25° resolution precipitation and soil moisture products were spatially interpolated to a fine resolution of 1 km. The 500 m resolution MCD12Q1 land cover type data was resampled to 1 km.

Table 1. Main input remote sensing dataset for evapotranspiration (ET) estimation.

Input Variables	Products Name	Temporal Resolution	Spatial Resolution
Land Cover Types	MCD12Q1	Yearly	500 m
LST	MOD11A1 & MYD11A1	Daily	1 km
NDVI	MOD13A2 & MYD13A2	8 days	1 km
Albedo	GLASS	8 days	1 km
LAI	GLASS	8 days	1 km
Precipitation	CMORPH	Daily	0.25°
Soil Moisture	ESA CCI	Daily	0.25°

The gridded near-surface meteorological forcing data, including air temperature, air pressure, dew point temperature, wind speed, downward short-wave and long-wave radiation fluxes were retrieved from the ERA-Interim dataset (http://apps.ecmwf.int/datasets/). The meteorological forcing data with 0.25° resolution was downscaled to 1 km using statistical downscaling approaches [7,24,25].

To validate the estimated ET, flux observation data from 6 eddy covariance sites were collected (Table 2). The 30 min flux tower observed latent heat flux (LE) was summed to daily ET after careful quality check.

Table 2. Flux observation sites information.

Site ID	Description	Lat (°N)	Lon (°E)	Period	Reference
ctt007	Cassava field at Tak	16.90	99.43	2012–2015	[26]
dtt030	Diverse land surface at Tak	16.94	99.43	2003–2015	[26]
prt007	Paddy at Rachaburi	13.58	99.51	2011–2013	[26]
pst007	Paddy at Sukhothai	17.06	99.70	2004–2009	[26]
MKL	Forest at Sakaerat	14.59	98.84	2003–2004	[4]
SKR	Forest at Mae Klong	14.49	101.92	2001–2003	[4]

2.6. Application of ET Estimation in Thailand

The ET estimation flowchart in current study is shown in Figure 1, and ET from 2001 to 2015 are obtained for the study area.

For bare soil or water pixels, simple PM equation or classical Penman equation is applied to estimate bare soil evaporation and open water evaporation. For the vegetation-soil pixels, detail steps for ET estimation includes:

(1) The canopy rainfall interception loss during the rainy days estimated according a revised Gash model [27–29];

(2) The net radiation is partitioned according to Equations (3)–(5), while during the rainy days in the wet season Rn_i is estimated by reversing PM equation.

(3) Scatterplot of clear sky LST and NDVI is prepared, and triangle method is applied to retrieve clear sky EF during the Terra and Aqua Moderate Resolution Imaging Spectroradiometer (MODIS) pass time according to Equation (15) in the dry season. The daily EF is assumed to be equal to the EF during the satellite pass time, and daily ET is estimated according to the daily EF and daily available energy;

(4) Daily r_s^c and $f(\theta_{root})$ of clear days are obtained by reversing Equations (1) and (6), and SMSI is estimated using Equation (12) during these clear days;

(5) SMSI time series is reconstructed by linear interpolation of the SMSI during clear days, hence daily $f(\theta_{root})$ is obtained according to Equation (13) in the dry season; while in the wet season SMSI is taken as constant;

(6) $f(\theta_{root})$ time series is then applied in Equation (6) to estimate daily canopy stomatal resistance, and applied to Equations (1) and (2) to estimate plant transpiration and soil evaporation for all sky conditions.

Finally, these ET components (plant transpiration, soil evaporation, canopy rainfall interception loss, open water evaporation) are summed to total ET. And daily ET in Northeastern Thailand from 2001 to 2015 is achieved in current study.

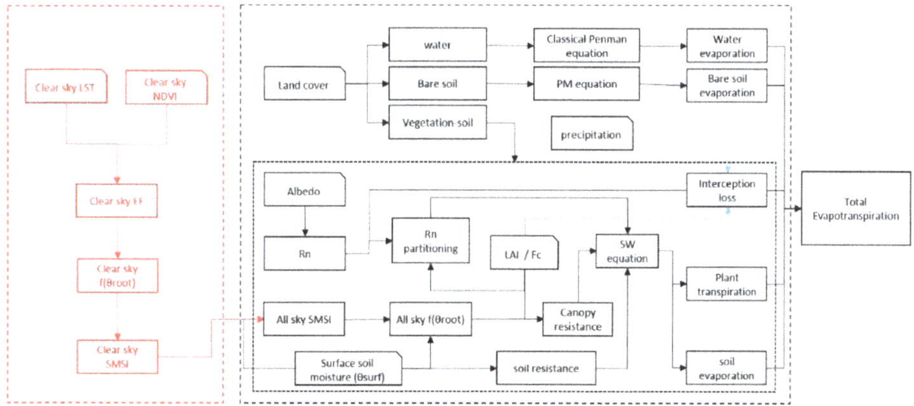

Figure 1. ET estimation flowchart in current study. The left block stands for the land surface temperature (LST)-based method to parameterize the regulation of soil moisture on the ET, while the right block illustrates the interpolation to the cloudy days and the estimation of daily ET.

3. Results

3.1. The Constriant of Soil Moisture on Canopy Resistance and ET

The interpolated daily SMSI for estimating daily canopy resistance was validated over MKL and SKR sites at the dry season. For each site, soil moisture at 10 cm (θ_{10cm}) and 50 cm (θ_{50cm}) depths were available, and surrogated for surface soil moisture and root zone soil moisture. θ_{50cm} is applied in Equation (11) to obtain the constrain of root zone soil moisture on the canopy resistance ($f(\theta_{50cm})$), which is taken as reference to further compare with $f(\theta_{root})$ by reversing Equations (1) and (6) further, while $f(\theta_{50cm})/\theta_{10cm}$ ($SMSI_{50cm}$) is considered as reference to compare with SMSI.

For the site scale application, when daily latent and sensible heat flux observation is available, mainly in the clear sky, $f(\theta_{root})$ was first retrieved by reversing Equations (1) and (6) ($f(\theta_{root})_{clear\,day}$), and SMSI was estimated ($SMSI_{clear\,day}$) according to Equation (12). $SMSI_{clear\,day}$ was interpolated linearly to obtain the daily SMSI ($SMSI_{daily}$), and applied in Equation (13) to obtain the estimated $f(\theta_{root})$ ($f(\theta_{root})_{daily}$). Note that $SMSI_{clear\,day}$ and $f(\theta_{root})_{clear\,day}$ are only for the days with valid latent and sensible heat flux observations, while $SMSI_{daily}$ and $f(\theta_{root})_{daily}$ are temporal continuous during the dry season (Figure 2).

Generally, in the beginning of wet-to-dry episode, the surface soil moisture drops down faster compared to the root zone soil moisture (Figure 2). Several days (e.g., roughly 20 days) from the beginning of dry season in 2003 for MKL site, θ_{root} drop to the level below θ_{max}, and $f(\theta_{root})$ start to decrease, and the decreasing rate is linearly related to the decreasing trend of surface soil moisture. This also lead the uneven decreasing rate of surface soil moisture and $f(\theta_{root})$. However, linear relationship was found between surface soil moisture and $f(\theta_{root})$ during a wet-dry episode, which was defined as the period between two rainy days (Figure 3). Hence, both $SMSI_{daily}$ and $f(\theta_{root})_{daily}$ could match the temporal variation during the dry season (Figure 2), and both showed low root mean square error (RMSE) and bias with the reference values (Figure 4), indicating the method to parameterize the constrain of soil moisture on canopy resistance based on EF in current study is reasonable.

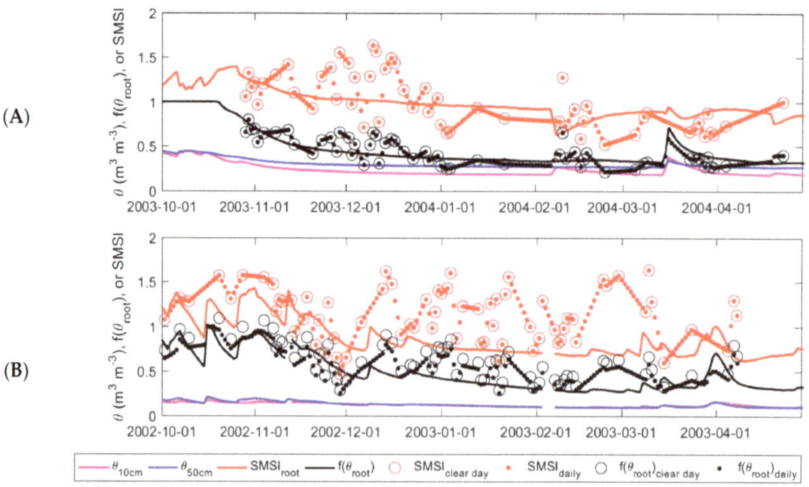

Figure 2. Time series of soil moisture stress index (SMSI) and $f(\theta_{root})$ at (**A**) MKL site and (**B**) SKR site. Soil moisture at 10 cm (θ_{10cm}) and 50 cm (θ_{50cm}) depths are also shown.

The uncertainty of SMSI illustrated in Figure 4A will contribute to $f(\theta_{root})$ in Figure 4B. As we can see in the Figure 4, RMSE of 0.31 in SMSI will result in RMSE of 0.09 in $f(\theta_{root})$, mostly because $f(\theta_{root})$ is also regulated by surface soil moisture. This is also shown in Figure 2 that $f(\theta_{root})_{daily}$ is very close to $f(\theta_{root})$, while $SMSI_{daily}$ bias relative large with $SMSI_{root}$. Hence, this interpolation method is considered valid and can be applied to estimate ET in the current study.

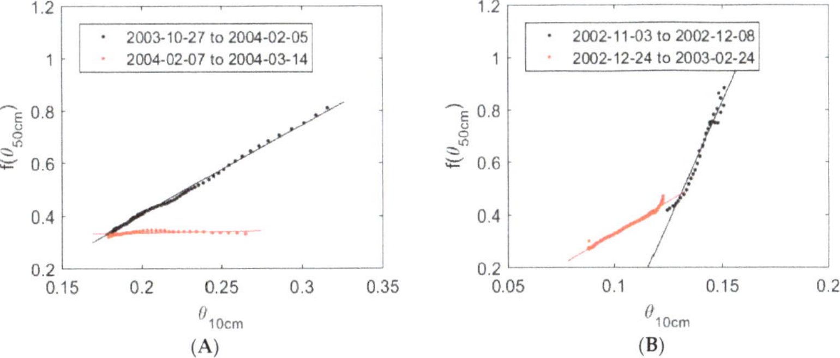

Figure 3. The relationship between soil moisture at 10 cm (θ_{10cm}) and $f(\theta_{50cm})$ at (**A**) MKL site and (**B**) SKR site at a wet-dry episode. Linear equations are used for fitting(y = 3.42 x − 0.28 with R^2 = 0.99 from 27 October 2003 to 5 February 2004 and y = 0.12 x − 0.31 with R^2 = 0.20 from 7 February 2004 to 14 March 2004 in MKL, y = 17.87 x − 1.87 with R^2 = 0.96 from 3 November 2002 to 8 December 2002 and y = 4.80 x − 0.15 with R^2 = 0.97 from 24 December 2002 to 24 February 2003 in SKR).

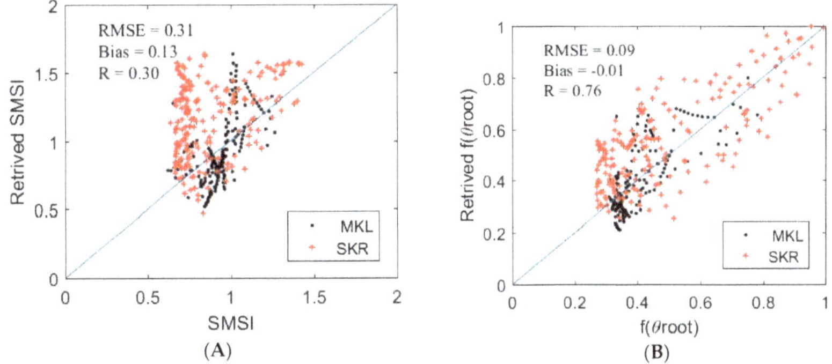

Figure 4. Scatterplot comparison of estimated (**A**) daily SMSI and (**B**) daily $f(\theta_{root})$ during dry season at MKL and SKR sites.

3.2. Validating Clear Sky EF by MODIS

The clear sky EF was first derived by LST-VI triangle method, and Figure 5 shows an example of the LST-VI scatter plot and the spatial variation of estimated EF for 20th January, of 2009. Generally, the dry edge and wet edge are well estimated both by Terra and Aqua MODIS. Furthermore, the Terra MODIS-based EF estimates is very close to that of the Aqua MODIS estimates (Figure 6), and both were applied for ET estimation.

To further validate the estimated EF, the estimated EF values over the 6 flux observation sites were extracted to compare with the ground observation, as shown in Figure 7. The observed EF at the satellite pass time is obtained by interpolating the ratio of observed half-hour LE to half-hour available energy to the Terra and Aqua MODIS pass time, while the daily EF is obtained as the ratio between daily LE and available energy. Generally the estimated clear sky EF agree well with observed half-hour and daily EF. Slight difference could be found in terms of RMSE and correlation coefficient, indicating the reasonability of constant EF assumption.

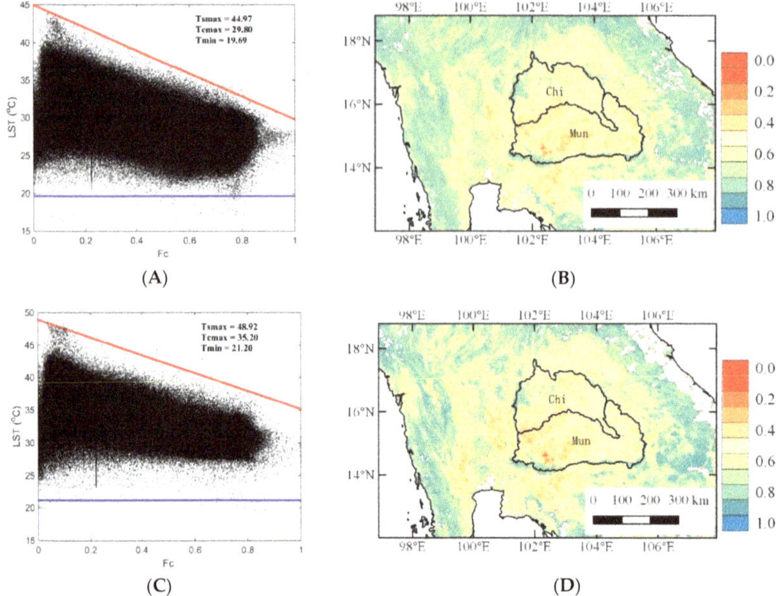

Figure 5. Example of LST-VI triangle scatterplot (**A,C**) and the spatial variation of estimated EF (**B,D**), by Terra (**A,B**) and Aqua Moderate Resolution Imaging Spectroradiometer (MODIS) (**C,D**) at January 20 of 2009. Red line in the LST-VI triangle scatter is the dry edge where ϕ =1.26 × *Fc*, while the blue line is the wet edge where ϕ = 1.26.

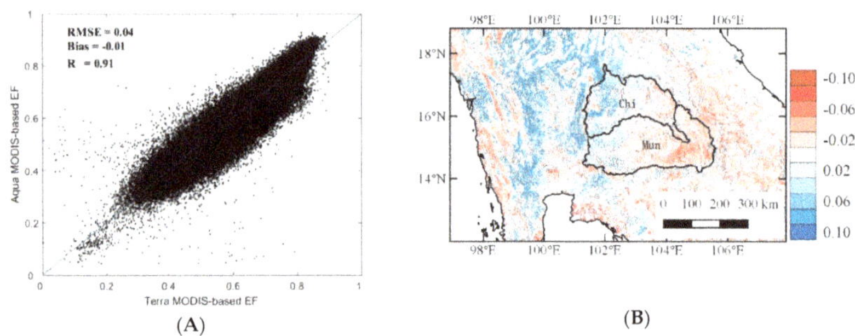

Figure 6. Comparison of EF based on Terra and Aqua MODIS on January 20 of 2009. (**A**) scatter plot of EF based on Terra and Aqua MODIS data; (**B**) spatial variation of EF difference (EF based on Terra minus that of the Aqua MODIS data).

3.3. Comparison of Estimated ET with Flux Tower Observations

The estimated daily ET agree well with the ground observations, with mean bias range from −0.44 to 0.89 mm d^{-1} and root mean square error (RMSE) range from 1.08 to 1.58 mm d^{-1} (Figure 8). The estimated ET could also catch the seasonal variations of ET (Figure 8). These indicate that the estimated ET has relatively good accuracy. Generally, better agreement could be found in the cropland sites than in the forest sites, most likely because significant seasonal variation of ET could be found in cropland while ET in the forest sites showed much less seasonal variation.

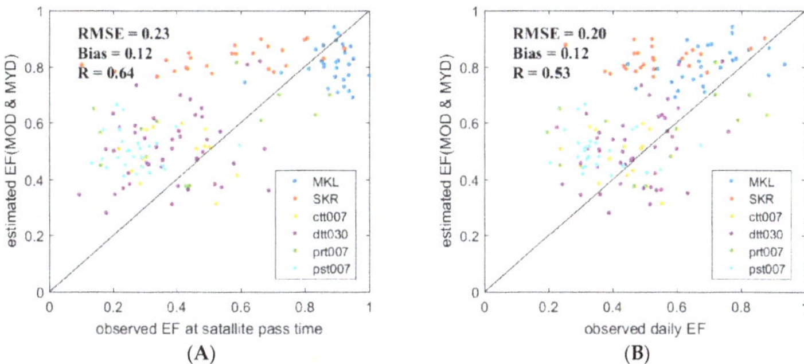

Figure 7. Comparison of estimated EF with (**A**) observed EF at satellite pass time and (**B**) daily EF obtained by ground observation.

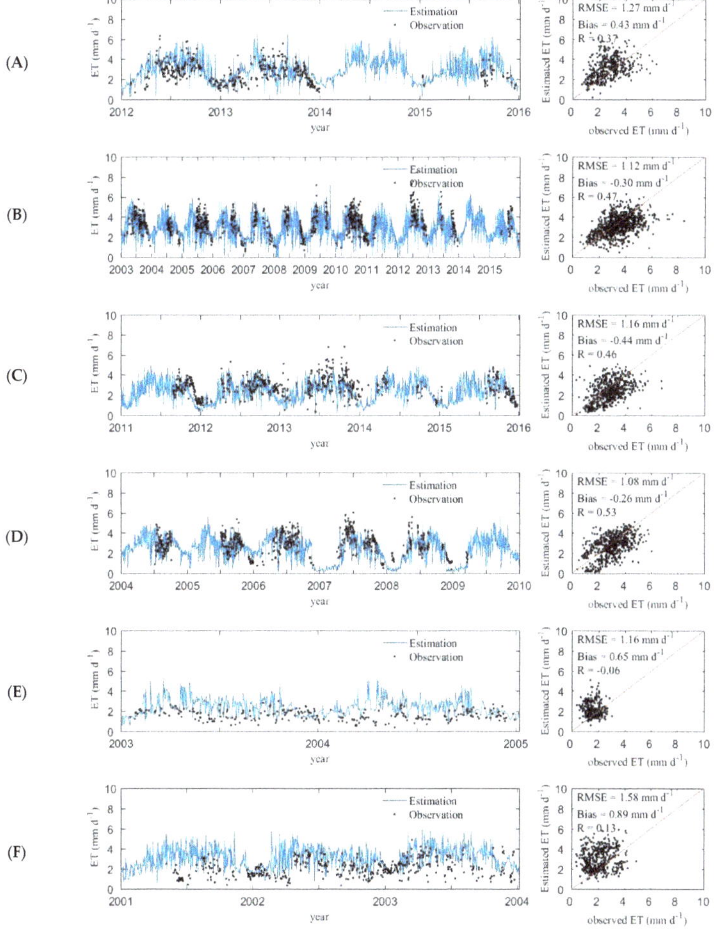

Figure 8. Comparison of ETMonitor estimated daily ET with flux tower based observations in (**A**) ctt007 site, (**B**) dtt030 site, (**C**) prt007 site, (**D**) pst007 site, (**E**) MKL site, and (**F**) SKR site.

Current estimation is compared with the results by original ETMonitor by Hu and Jia (2015) at daily time-step (Table 3). When original ETMonitor by Hu and Jia (2015) was applied, relatively large bias and RMSE could be found, especially for the forest sites. The accuracy is improved by parameter adapting for the forest area, resulting in reduction of RMSE from 1.92–2.12 mm d^{-1} to 1.16–1.39 mm d^{-1}. It also suggests the necessity to adapt model parameters regionally in the original ETMonitor when applying to different regions to achieve accurate ET estimation. The method developed in current study showed comparable accuracy with the regional adapted ETMonitor, indicating the developed method could be adopted to obtain ET with higher accuracy. This method is operational, and it could reduce the dependence of ground flux observation to calibrate the model parameters.

Table 3. Comparison of estimated daily ET by current improved ETMonitor and original ETMonitor by Hu and Jia (2015) with ground observed ET. Values in the baskets represent the results after adapting the parameters by Hu and Jia (2015) to the humid climate in Northeast Thailand.

Site ID	ETMonitor by Hu and Jia (2015)			Current Study Estimation		
	R	Bias (mm d^{-1})	RMSE (mm d^{-1})	R	Bias (mm d^{-1})	RMSE (mm d^{-1})
ctt007	0.48	0.02	1.06	0.37	0.43	1.27
dtt030	0.62 (0.45)	−1.70 (−0.77)	1.95 (1.25)	0.47	−0.3	1.12
prt007	0.45	−0.37	1.21	0.46	−0.44	1.16
pst007	0.51	−0.33	1.12	0.53	−0.26	1.08
MKL	−0.02 (0.00)	1.59 (0.68)	1.92 (1.16)	−0.06	0.65	1.16
SKR	0.14 (0.07)	1.67 (0.44)	2.12 (1.39)	0.13	0.89	1.58

Current estimation is also compared with the MOD16 ET product at 8-days resolution (Table 4 and Figure 9). Overall, the RMSE of estimated ET (RMSE = 0.97 mm d^{-1}) is much lower than MOD16 ET (RMSE = 1.54 mm d^{-1}) when compared with ground observation. Significant improvement is observed over forest sites (MKL and SKR) where the RMSE are reduced from 2.69 mm d^{-1} and 1.99 mm d^{-1} to 1.04 mm d^{-1} and 1.20 mm d^{-1}.

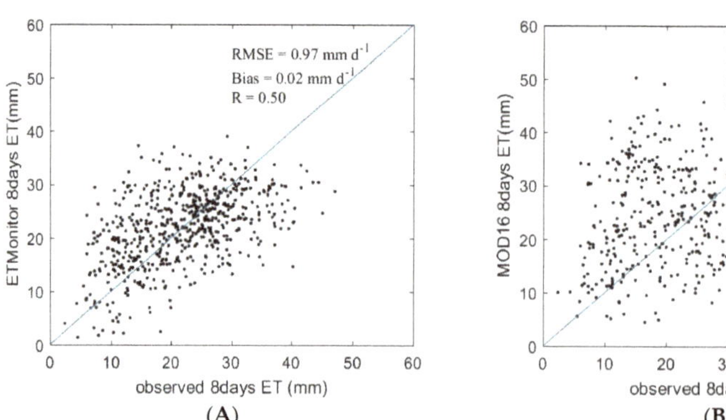

Figure 9. Comparison of (**A**) ETMonitor estimated 8 days ET in current study and (**B**) MOD16 ET with ground observed ET.

Table 4. Comparison of ETMonitor estimated 8 days ET and MOD16 8 days ET with ground observed ET.

Site ID	ETMonitor 8-Days ET			MOD16 8-Days ET		
	R	Bias (mm d^{-1})	RMSE (mm d^{-1})	R	Bias (mm d^{-1})	RMSE (mm d^{-1})
ctt007	0.59	0.44	0.94	0.30	0.36	1.19
dtt030	0.59	−0.31	0.92	0.50	−0.09	0.98
prt007	0.65	−0.37	0.82	0.23	−0.03	0.99
pst007	0.63	−0.45	0.97	0.59	−0.55	1.12
MKL	0.07	0.75	1.04	0.48	2.58	2.69
SKR	0.30	0.78	1.20	0.42	1.82	1.99

3.4. Spatial Variation of ET in Northeastern Thailand

Figure 10 shows the spatial variation of estimated annual mean ET, as well as the main ET components including plant transpiration, soil evaporation, and canopy rainfall interception in Northeastern Thailand from 2001 to 2015. The multi-annual mean ET in two largest river basins in this area, Chi and Mun river basins, are 938.8 mm yr^{-1} and 1023.7 mm yr^{-1}, respectively (Table 5). These two basins are mainly dominant by cropland, which account for 82.16% and 87.45% respectively according MCD12Q1 land cover classification data. Overall, annual precipitation and ET in the Mun river basin is higher than the Chi river basin (Table 5).

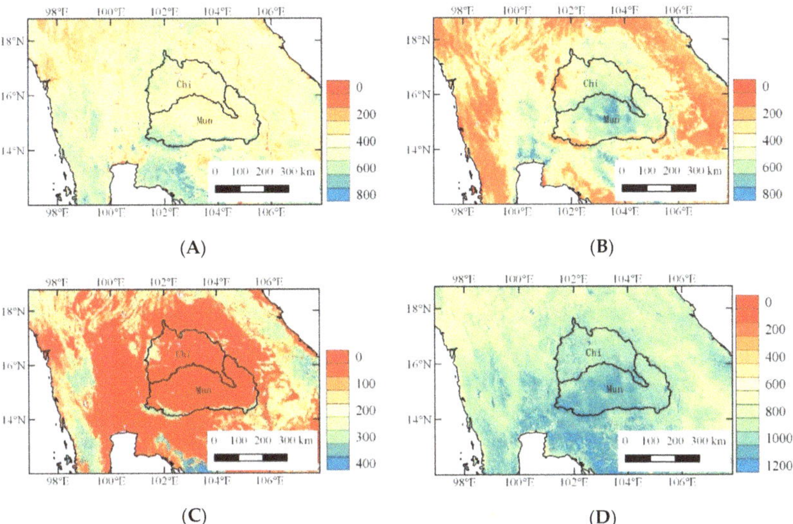

Figure 10. Spatial variation of (**A**) plant transpiration, (**B**) soil evaporation, (**C**) canopy rainfall interception, and (**D**) total ET (mm yr^{-1}), in Northeastern Thailand from 2001 to 2015. The boundaries of Chi river basin and Mun river basin are also plotted.

Table 5. Statistic information of Chi and Mun river basins.

	Basin Area (× 10^3 km^2)	Annual Precipitation (mm yr^{-1})	Annual ET (mm yr^{-1})	Cropland Coverage (%)	Forest Coverage (%)
Chi river basin	40.58	1269.52	938.80	82.16	5.56
Mun river basin	71.06	1374.50	1023.70	87.45	6.56

Plant transpiration and soil evaporation account for 43.32% and 51.48% of the total ET respectively. The canopy rainfall interception only account for 4.70% of total ET, since this region is mostly covered by low canopy crop, and the area of cropland account for over 80% of the total area, while 80% of the crop is paddy field. The forest only cover roughly 6% of the two basins, mostly located in the Northwest of Chi river basin and Southwest of Mun river basin, where relative high transpiration and interception could be found (Figure 10).

Figure 11 presents the spatial variation of wet season (April to October) and dry season (November to next March) ET in Northeastern Thailand from 2001 to 2015. The wet season ET in Chi and Mun river basins account for 62.30% and 61.27% of annual total ET, while dry season ET in Chi and Mun river basins account for the rest 37.70% and 38.73%.

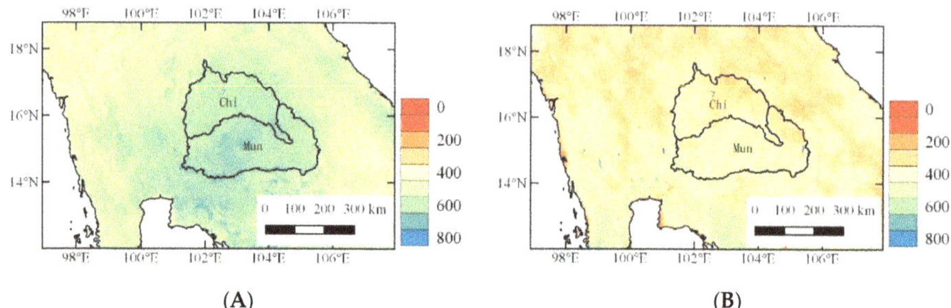

(A) **(B)**

Figure 11. Spatial variation of (**A**) wet and (**B**) dry season ET (mm) in Northeastern Thailand from 2001 to 2015.

4. Discussion

The spatially and temporally continuous ET information will advance our knowledge on the mean state and spatial and temporal variability of this significant component of the water cycle, and is highly needed for understanding the interactions between land surface and atmosphere and subsequently improving the water resource management [30]. The LST based surface energy balance method has been proved to provide accurate ET information, but inefficient to produce time series ET due to the cloud [31]. The process-based approach is better for obtaining the spatially and temporally continuous ET products, e.g., the MOD16 ET algorithm. The procedure provided by current study generally combines the advantage of LST-based energy balance method to improve the ET estimation accuracy and the advantage of process model to achieve the spatially and temporally continuous ET estimation, thus it can provide accurate ET estimation, as we showed that the RMSE of estimated ET is low (Figures 8 and 9, Table 3).

It is generally accepted that the RMSE of estimated daily ET range from 1–2 mm d^{-1} based on satellite forcing, and RMSE of about 50 W m^{-2} is common reported for latent heat flux (equal to 1.75 mm d^{-1} ET) [7,10,32]. The RMSE of estimated daily ET in current study range from 1.08 to 1.58 mm d^{-1}, with an averaged RMSE of 1.22 mm d^{-1}, which is less compared to original ETMonitor by Hu and Jia (2015) and MOD16 ET. Generally, MOD16 tend to overestimate forest ET and underestimate cropland ET, which is consistent with several past studies [33–35], though not to the same extent as found in this study. Several factors contribute to MOD16 ET uncertainties, e.g., parameterizing soil moisture constrain by meteorological factors (not soil moisture or LST), overestimating environmental stresses on canopy conductance, empirical parameter setting in complementary relationship hypothesis [35,36]. ETMonitor first presented by Hu and Jia (2015) is superior mostly because it is physical robustness, and it considers the impact of soil moisture on ET by introducing the microwave remote sensing-based surface soil moisture and has been demonstrated to be suitable to estimate ET at both regional and global scale [7,37,38]. Although the behavior of original ETMonitor does not meet the expectation in current study, it could be improved by simply adapting its regional parameter to local condition for achieving accurate ET estimation, and is clearly presented by the low RMSE of regional parameter adapted ETMonitor in Table 3. This also highlight the necessity to calibrate parameters to obtain accurate ET estimation when applying locally. Previously, empirical method or soil water balance model was adopted to estimate the root soil moisture based on surface soil moisture, which was further applied to estimate the constrain of soil moisture ($f(\theta_{root})$) based on fixed soil moisture sensitive parameter. Different from original ETMonitor presented by Hu and Jia (2015), who adopted empirical method to express the constrain of soil moisture on ET, the framework developed in current study provides an operational method to calibrate the algorithm regionally by parameterizing the constrain of soil moisture. The developed framework utilized the LST-VI triangle method to estimate the clear sky EF, which was further applied to estimate SMSI by reversing the canopy resistance equation. It is an operational approach to obtain the canopy resistance parameter for ETMonitor, and it could also be adopted by other ET algorithms.

It has been long recognized that the important variables to determine canopy resistance (or canopy stomatal conductance) include air temperature, humidity, solar radiation, and soil moisture [39,40]. Plants usually open their stomata under wet conditions, which are favorable for plant growth, while when the soil dries stomatal closure limits transpiration to prevent dehydration. Method obtained to express the constrain of soil moisture and climate factors on canopy resistance and evapotranspiration is the key to ensure the accuracy of evapotranspiration algorithms [8,41,42]. The accuracy of ETMonitor estimated ET is also sensitive to the canopy resistance parameter [7]. Different from the traditional calibration method, the developed approach utilized the clear sky EF maps obtained by LST-VI triangle method to retrieve the canopy resistance parameter. One advantage is that it can provide the pixel-to-pixel parameter. The traditional method usually relied on the heavy field work to retrieve the plot scale canopy resistance parameter, which was further applied to regional or global scale according to the land cover map [10]. Thus, traditionally a fixed parameter was usually adopted for a land cover type. However, there exist strong variability in drought tolerance across different plant species, sites, and environmental conditions, which limit the accuracy of global land surface model to simulate the ecosystem response to the decreasing soil moisture, hence much comprehensive calibration method should be addressed [42,43]. The SMSI retrieved by clear sky EF maps obtained by LST-VI triangle method is based on satellite remote sensing image, and can capture the spatial and temporal variation.

LST-VI method is chosen in this study mostly because of its simplicity and accuracy, making it acceptable in the current study (Figure 7). It is also noted that LST-VI method may suffer from domain dependence, and it may impact the accuracy of wet and dry boundary derivation [44]. This spatial-scale effect is common in other anchor-based ET algorithm, e.g., surface energy balance algorithm for land (SEBAL), and caution should be paid when applied to different images in extreme large regions [45,46].

For large regional application, e.g., continents scale, algorithm that is independent of domain size is suggested as alternative. The uncertainty in the input data also contribute to the error of estimated ET in current study. Former study already presented that remote sensing ET algorithms are very sensitive to input variables, e.g., LST, Rn, NDVI, and meteorological variables, depending on which algorithms are adopted [16,32,44,45]. The impact of land cover on ET estimation should also be addressed in ETMonitor, since some sensitive parameters like the minimum stomatal resistance are set according to land cover types. Hence, the uncertainty of land cover classification also contribute to the error in estimated ET.

The linear interpolation of SMSI works well during a wet-dry episode since SMSI generally presents monotonic deceasing trend. However caution should be paid when applying to the frequent rainfall period when the monotonic trend will be disturbed by rainfall. Meanwhile, it is not suitable for those regions where root-zone soil moisture and surface soil moisture are decoupled since the surface soil moisture by satellite remote sensing generally cannot represent the surface wetness condition.

5. Conclusions

ET in Northeastern Thailand was estimated by process-based ETMonitor algorithm based on mainly satellite earth observation datasets. Meanwhile, a new scheme was developed and applied in ETMonitor to take the advantage of LST-based energy balance method. In this scheme, the soil moisture stress index (SMSI) was defined to express the sensitivity of canopy resistance to surface soil moisture, and it was estimated by reversing the canopy resistance equation during the clear sky when EF could be achieved by LST-based energy balance method. The clear sky SMSI was further interpolated to the cloudy days to estimate canopy resistance based on temporal-continuous surface soil moisture data for continuous ET estimation. The estimated daily ET generally agreed well with the flux tower observation with RMSE ranging between 1.08 and 1.58 mm d^{-1}. The RMSE values over the forest sites are considerably lower compared to MOD16 products, indicating its better accuracy.

Author Contributions: C.Z. conceived and designed the study, and wrote the paper; L.J. contributed to the design and revision of the paper; G.H. and J.L. contributed to analyze the clear sky EF estimation and revision of the paper.

Funding: This study is supported by the National Key Research and Development Program of China (Grant no. 2016YFE0117300, no. 2017YFC0405802, and no. 2016YFA0600302), Strategic Priority Research Program of the Chinese Academy of Sciences (Grant no. XDA19080303, and no. XDA19030203), National Natural Science Foundation of China (NSFC) (Grant no. 41801346 and 41501405), National Key Basic Research Program of China (Grant no. 2015CB953702), and 13th Five-Year Plan of Civil Aerospace Technology Advanced Research Projects of China (Grant no. Y7D0070038).

Acknowledgments: We thank Peejush Pani for helping to revise the language. We also thank the reviews by three anonymous reviewers.

Conflicts of Interest: The authors declare no conflict of interest.

References

1. Schwalm, C.R.; Williams, C.A.; Schaefer, K. Carbon consequences of global hydrologic change. *J. Geophys. Res.-Biogeosci.* **2011**, *116*, 1948–2009. [CrossRef]
2. Oki, T.; Kanae, S. Global hydrological cycles and world water resources. *Science* **2006**, *313*, 1068–1072. [CrossRef] [PubMed]
3. Zhang, Y.S.; Li, T.; Wang, B.; Wu, G.X. Onset of the summer monsoon over the Indochina Peninsula: Climatology and interannual variations. *J. Clim.* **2002**, *15*, 3206–3221. [CrossRef]
4. Tanaka, N.; Kume, T.; Yoshifuji, N.; Tanaka, K.; Takizawa, H.; Shiraki, K.; Tantasirin, C.; Tangtham, N.; Suzuki, M. A review of evapotranspiration estimates from tropical forests in Thailand and adjacent regions. *Agric. For. Meteorol.* **2008**, *148*, 807–819. [CrossRef]
5. Prabnakorn, S.; Maskey, S.; Suryadi, F.X.; de Fraiture, C. Rice yield in response to climate trends and drought index in the Mun River Basin, Thailand. *Sci. Total Environ.* **2018**, *621*, 108–119. [CrossRef] [PubMed]
6. Bastiaanssen, W.G.M.; Menenti, M.; Feddes, R.A.; Holtslag, A.A.M. A remote sensing surface energy balance algorithm for land (SEBAL)—1. Formulation. *J. Hydrol.* **1998**, *212*, 198–212. [CrossRef]
7. Hu, G.C.; Jia, L. Monitoring of Evapotranspiration in a Semi-Arid Inland River Basin by Combining Microwave and Optical Remote Sensing Observations. *Remote Sens.* **2015**, *7*, 3056–3087. [CrossRef]
8. Bastiaanssen, W.G.M.; Cheema, M.J.M.; Immerzeel, W.W.; Miltenburg, I.J.; Pelgrum, H. Surface energy balance and actual evapotranspiration of the transboundary Indus Basin estimated from satellite measurements and the ETLook model. *Water Resour. Res.* **2012**, *48*. [CrossRef]
9. Miralles, D.G.; Holmes, T.R.H.; De Jeu, R.A.M.; Gash, J.H.; Meesters, A.; Dolman, A.J. Global land-surface evaporation estimated from satellite-based observations. *Hydrol. Earth Syst. Sci.* **2011**, *15*, 453–469. [CrossRef]
10. Mu, Q.Z.; Zhao, M.S.; Running, S.W. Improvements to a MODIS global terrestrial evapotranspiration algorithm. *Remote Sens. Environ.* **2011**, *115*, 1781–1800. [CrossRef]
11. Mu, Q.; Heinsch, F.A.; Zhao, M.; Running, S.W. Development of a global evapotranspiration algorithm based on MODIS and global meteorology data. *Remote Sens. Environ.* **2007**, *111*, 519–536. [CrossRef]

12. Nishida, K.; Nemani, R.R.; Glassy, J.M.; Running, S.W. Development of an evapotranspiration index from aqua/MODIS for monitoring surface moisture status. *IEEE Trans. Geosci. Remote Sens.* **2003**, *41*, 493–501. [CrossRef]

13. Norman, J.M.; Kustas, W.P.; Humes, K.S. Source approach for estimating soil and vegetation energy fluxes in observations of directional radiometric surface-temperature. *Agric. For. Meteorol.* **1995**, *77*, 263–293. [CrossRef]

14. Su, Z. The Surface Energy Balance System (SEBS) for estimation of turbulent heat fluxes. *Hydrol. Earth Syst. Sci.* **2002**, *6*, 85–99. [CrossRef]

15. Zhang, K.; Kimball, J.S.; Nemani, R.R.; Running, S.W. A continuous satellite-derived global record of land surface evapotranspiration from 1983 to 2006. *Water Resour. Res.* **2010**, *46*. [CrossRef]

16. Zheng, C.L.; Wang, Q.; Li, P.H. Coupling SEBAL with a new radiation module and MODIS products for better estimation of evapotranspiration. *Hydrol. Sci. J.-J. Des Sci. Hydrol.* **2016**, *61*, 1535–1547. [CrossRef]

17. Li, Z.L.; Tang, B.H.; Wu, H.; Ren, H.Z.; Yan, G.J.; Wan, Z.M.; Trigo, I.F.; Sobrino, J.A. Satellite-derived land surface temperature: Current status and perspectives. *Remote Sens. Environ.* **2013**, *131*, 14–37. [CrossRef]

18. Jia, L.; Xi, G.; Liu, S.; Huang, C.; Yan, Y.; Liu, G. Regional estimation of daily to annual regional evapotranspiration with MODIS data in the Yellow River Delta wetland. *Hydrol. Earth Syst. Sci.* **2009**, *13*, 1775–1787. [CrossRef]

19. Shuttleworth, W.J.; Wallace, J.S. Evaporation from sparse crops-an energy combination theory. *Q. J. R. Meteorol. Soc.* **1985**, *111*, 839–855. [CrossRef]

20. Jarvis, P.G. The interpretation of the variations in leaf water potential and stomatal conductance found in canopies in the field. *Philos. Trans. R. Soc. Lond. B Biol. Sci.* **1976**, *273*, 593–610. [CrossRef]

21. Allen, R.G.; Pruitt, W.O.; Wright, J.L.; Howell, T.A.; Ventura, F.; Snyder, R.; Itenfisu, D.; Steduto, P.; Berengena, J.; Yrisarry, J.B.; et al. A recommendation on standardized surface resistance for hourly calculation of reference ETO by the FAO56 Penman-Monteith method. *Agric. Water Manag.* **2006**, *81*, 1–22. [CrossRef]

22. Matsumoto, K.; Ohta, T.; Nakai, T.; Kuwada, T.; Daikoku, K.; Iida, S.; Yabuki, H.; Kononov, A.V.; van der Molen, M.K.; Kodama, Y.; et al. Responses of surface conductance to forest environments in the Far East. *Agric. For. Meteorol.* **2008**, *148*, 1926–1940. [CrossRef]

23. Tang, R.L.; Li, Z.L.; Tang, B.H. An application of the T-s-VI triangle method with enhanced edges determination for evapotranspiration estimation from MODIS data in and and semi-arid regions: Implementation and validation. *Remote Sens. Environ.* **2010**, *114*, 540–551. [CrossRef]

24. Stahl, K.; Moore, R.D.; Floyer, J.A.; Asplin, M.G.; McKendry, I.G. Comparison of approaches for spatial interpolation of daily air temperature in a large region with complex topography and highly variable station density. *Agric. For. Meteorol.* **2006**, *139*, 224–236. [CrossRef]

25. Gao, X.J.; Giorgi, F. Increased aridity in the Mediterranean region under greenhouse gas forcing estimated from high resolution simulations with a regional climate model. *Glob. Planet. Chang.* **2008**, *62*, 195–209. [CrossRef]

26. Kim, W.; Miyata, A.; Ashraf, A.; Maruyama, A.; Chidthaisong, A.; Jaikaeo, C.; Komori, D.; Ikoma, E.; Sakurai, G.; Seoh, H.H.; et al. FluxPro as a realtime monitoring and surveilling system for eddy covariance flux measurement. *J. Agric. Meteorol.* **2015**, *71*, 32–50. [CrossRef]

27. Cui, Y.K.; Jia, L. A Modified Gash Model for Estimating Rainfall Interception Loss of Forest Using Remote Sensing Observations at Regional Scale. *Water* **2014**, *6*, 993–1012. [CrossRef]

28. Cui, Y.K.; Jia, L.; Hu, G.C.; Zhou, J. Mapping of Interception Loss of Vegetation in the Heihe River Basin of China Using Remote Sensing Observations. *IEEE Geosci. Remote Sens. Lett.* **2015**, *12*, 23–27.

29. Zheng, C.L.; Jia, L. Global rainfall interception loss derived from multi-source observations. In Proceedings of the 2016 IEEE International Geoscience and Remote Sensing Symposium, Beijing, China, 10–15 July 2016; pp. 3532–3534.

30. Fisher, J.B.; Tu, K.P.; Baldocchi, D.D. Global estimates of the land-atmosphere water flux based on monthly AVHRR and ISLSCP-II data, validated at 16 FLUXNET sites. *Remote Sens. Environ.* **2008**, *112*, 901–919. [CrossRef]

31. Zhang, Y.Q.; Pena-Arancibia, J.L.; McVicar, T.R.; Chiew, F.H.S.; Vaze, J.; Liu, C.M.; Lu, X.J.; Zheng, H.X.; Wang, Y.P.; Liu, Y.Y.; et al. Multi-decadal trends in global terrestrial evapotranspiration and its components. *Sci. Rep.* **2016**, *6*, 19124. [CrossRef]

32. Michel, D.; Jimenez, C.; Miralles, D.G.; Jung, M.; Hirschi, M.; Ershadi, A.; Martens, B.; McCabe, M.F.; Fisher, J.B.; Mu, Q.; et al. The WACMOS-ET project—Part 1: Tower-scale evaluation of four remote-sensing-based evapotranspiration algorithms. *Hydrol. Earth Syst. Sci.* **2016**, *20*, 803–822. [CrossRef]
33. Hu, G.C.; Jia, L.; Menenti, M. Comparison of MOD16 and LSA-SAF MSG evapotranspiration products over Europe for 2011. *Remote Sens. Environ.* **2015**, *156*, 510–526. [CrossRef]
34. Kim, H.W.; Hwang, K.; Mu, Q.; Lee, S.O.; Choi, M. Validation of MODIS 16 Global Terrestrial Evapotranspiration Products in Various Climates and Land Cover Types in Asia. *KSCE J. Civ. Eng.* **2012**, *16*, 229–238. [CrossRef]
35. Bhattarai, N.; Mallick, K.; Brunsell, N.A.; Sun, G.; Jain, M. Regional evapotranspiration from an image-based implementation of the Surface Temperature Initiated Closure (STIC1.2) model and its validation across an aridity gradient in the conterminous US. *Hydrol. Earth Syst. Sci.* **2018**, *22*, 2311–2341. [CrossRef]
36. Yang, Y.T.; Long, D.; Guan, H.D.; Liang, W.; Simmons, C.; Batelaan, O. Comparison of three dual-source remote sensing evapotranspiration models during the MUSOEXE-12 campaign: Revisit of model physics. *Water Resour. Res.* **2015**, *51*, 3145–3165. [CrossRef]
37. Zheng, C.L.; Jia, L.; Hu, G.; Lu, J.; Wang, K.; Li, Z. Global Evapotranspiration derived By ETMonitor model based on Earth Observations. In Proceedings of the 2016 IEEE International Geoscience and Remote Sensing Symposium, Beijing, China, 10–15 July 2016; pp. 222–225.
38. Jia, L.; Zheng, C.; Hu, G.; Menenti, M. Evapotranspiration. In *Comprehensive Remote Sensing*, 2nd ed.; Liang, S., Ed.; Elsevier: Oxford, UK, 2018; Volume 4, pp. 25–50, ISBN 9780128032213.
39. Klingberg, J.; Engardt, M.; Uddling, J.; Karlsson, P.E.; Pleijel, H. Ozone risk for vegetation in the future climate of Europe based on stomatal ozone uptake calculations. *Tellus Ser. A-Dyn. Meteorol. Oceanogr.* **2011**, *63*, 174–187. [CrossRef]
40. Anav, A.; Proietti, C.; Menut, L.; Carnicelli, S.; De Marco, A.; Paoletti, E. Sensitivity of stomatal conductance to soil moisture: Implications for tropospheric ozone. *Atmos. Chem. Phys.* **2018**, *18*, 5747–5763. [CrossRef]
41. Ershadi, A.; McCabe, M.F.; Evans, J.P.; Wood, E.F. Impact of model structure and parameterization on Penman-Monteith type evaporation models. *J. Hydrol.* **2015**, *525*, 521–535. [CrossRef]
42. Marechaux, I.; Bartlett, M.K.; Sack, L.; Baraloto, C.; Engel, J.; Joetzjer, E.; Chave, J. Drought tolerance as predicted by leaf water potential at turgor loss point varies strongly across species within an Amazonian forest. *Funct. Ecol.* **2015**, *29*, 1268–1277. [CrossRef]
43. Sitch, S.; Huntingford, C.; Gedney, N.; Levy, P.E.; Lomas, M.; Piao, S.L.; Betts, R.; Ciais, P.; Cox, P.; Friedlingstein, P.; et al. Evaluation of the terrestrial carbon cycle, future plant geography and climate-carbon cycle feedbacks using five Dynamic Global Vegetation Models (DGVMs). *Glob. Chang. Biol.* **2008**, *14*, 2015–2039. [CrossRef]
44. Li, Z.S.; Jia, L.; Lu, J. On Uncertainties of the Priestley-Taylor/LST-Fc Feature Space Method to Estimate Evapotranspiration: Case Study in an Arid/Semiarid Region in Northwest China. *Remote Sens.* **2015**, *7*, 447–466. [CrossRef]
45. Long, D.; Singh, V.P.; Li, Z.L. How sensitive is SEBAL to changes in input variables, domain size and satellite sensor? *J. Geophys. Res.-Atmos.* **2011**, *116*. [CrossRef]
46. Tang, R.L.; Li, Z.L.; Chen, K.S.; Jia, Y.Y.; Li, C.R.; Sun, X.M. Spatial-scale effect on the SEBAL model for evapotranspiration estimation using remote sensing data. *Agric. For. Meteorol.* **2013**, *174*, 28–42. [CrossRef]

Article

A Modeling Framework for Deriving Daily Time Series of Evapotranspiration Maps Using a Surface Energy Balance Model

Kul Khand [1],*, Saleh Taghvaeian [1], Prasanna Gowda [2] and George Paul [3]

[1] Department of Biosystems & Agricultural Engineering, Oklahoma State University, Stillwater, OK 74078, USA; Saleh.Taghvaeian@okstate.edu
[2] USDA-ARS Grazinglands Research Laboratory, El Reno, OK 73036, USA; Prasanna.Gowda@ars.usda.gov
[3] Formation Environmental LLC, Sacramento, CA 95816, USA; GPaul@formationenv.com
* Correspondence: kul.khand@okstate.edu; Tel.: +01-405-744-8419

Received: 11 January 2019; Accepted: 26 February 2019; Published: 2 March 2019

Abstract: Surface energy balance models have been one of the most widely used approaches to estimate spatially distributed evapotranspiration (ET) at varying landscape scales. However, more research is required to develop and test an operational framework that can address all challenges related to processing and gap filling of non-continuous satellite data to generate time series of ET at regional scale. In this study, an automated modeling framework was developed to construct daily time series of ET maps using MODIS imagery and the Surface Energy Balance System model. The ET estimates generated from this modeling framework were validated against observations of three eddy-covariance towers in Oklahoma, United States during a two-year period at each site. The modeling framework overestimated ET but captured its spatial and temporal variability. The overall performance was good with mean bias errors less than 30 W m^{-2} and root mean square errors less than 50 W m^{-2}. The model was then applied for a 14-year period (2001–2014) to study ET variations across Oklahoma. The statewide annual ET varied from 841 to 1100 mm yr^{-1}, with an average of 994 mm yr^{-1}. The results were also analyzed to estimate the ratio of estimated ET to reference ET, which is an indicator of water scarcity. The potential applications and challenges of the ET modeling framework are discussed and the future direction for the improvement and development of similar automated approaches are highlighted.

Keywords: MODIS; Surface Energy Balance System; Oklahoma Mesonet; Eddy-covariance

1. Introduction

Time series of remotely sensed evapotranspiration (ET) maps have extensive applications in agricultural, hydrological and environmental studies as they capture the spatiotemporal variability of vegetation consumptive use from field to continental scales. For example, spatial ET data have been used in agriculture sector for water right regulation, planning and monitoring [1], assessing irrigation and drainage performance [2–4], closing water balance at irrigation scheme levels [5] and managing agricultural water resources [6–8]. Recent studies have shown that remotely sensed ET can be used effectively for monitoring agricultural droughts [9–11] with the future potential of improving the performance of ET-integrated agricultural drought indices [12]. ET maps have been also used in assessing crop water productivity [13–15] and crop yield analysis [16,17].

Numerous studies have demonstrated the use of time series ET maps for ecological applications, such as capturing the progress of vegetation and wetland restoration [18], assessing the vulnerability of forest to fire and drought [19] and accounting water use from riparian vegetation and invasive species [20–23]. Remote sensing-based ET products have also been applied in improving the performances of hydrological models [24–26] and for climate studies to capture water feedbacks associated with seasonal cycles and soil moisture deficit at regional scales [27].

Among different approaches developed for mapping ET, the surface energy balance (SEB) approach has been widely used to acquire distributed ET at varying geographical scales [28–31]. Numerous SEB models have been proposed, including but not limited to Surface Energy Balance Index (SEBI) [32], Two-Source Energy Balance (TSEB) [33,34], Surface Energy Balance Algorithm for Land (SEBAL) [35], Simplified Surface Energy Balance Index (S-SEBI) [36], Surface Energy Balance System (SEBS) [37], Mapping Evapotranspiration at high Resolution with Internalized Calibration (METRIC) [38], Atmosphere-Land Exchange Inverse (ALEXI) [39], Regional ET Estimation Model (REEM) [40], Remote Sensing Evapotranspiration model (ReSET) [41], Operational Simplified Surface Energy Balance (SSEBop) [42] and Hybrid Dual-Source Scheme and Surface Energy Framework-Based Evapotranspiration Model (HTEM) [43]. Some of these models such as SEBAL and METRIC use manual selection of extreme pixels to compute sensible heat flux, which could result in variations in estimated ET [44] and may add uncertainty and errors based on the user's experience [45]. Other models such as TSEB, SEBS and SSEBop do not require human intervention so that the associated uncertainties are minimized. The selection of the SEB model and the quality of input data are likely key factors to determine the accuracy of modeled ET [46].

Developing time series of ET maps requires complex, multi-step analyses to deal with issues associated with pre-processing of remote sensing data and post-processing of resulting ET products. The choice of the SEB model and satellite data could vary depending on intended applications of ET maps, availability and requirements of input data and availability of resources (time, money and expertise) to run the model. In general, the SEB-based ET estimation process can be divided into six steps: (i) collation of remotely sensed and ground-based input data, (ii) quality assessment of collected datasets and preparation of all necessary inputs for the selected SEB model, (iii) running the SEB model (including all modules and algorithms) to obtain the instantaneous ET at the time of satellite overpass, (iv) extrapolation of instantaneous ET to daily estimates, (v) filling the gaps due to cloud coverage over a portion of the map and (vi) interpolation of daily ET between image acquisition dates to obtain ET for longer time scales.

The first two steps are performed to ensure the quality of input data, a critical requirement for any remote sensing data analysis. A thorough QA/QC procedure for weather data as presented in [47,48] is necessary as the accuracy of final product depends on the quality of these datasets. The quality assurance of weather dataset is even more critical in case of SEB models as they are sensitive to weather parameters. For example, Webster et al. [49] found air temperature and wind speed as influential inputs for HTEM and SEBS models, whereas, S-SEBI was less sensitive to meteorological inputs.

For small-scale applications with similar climatic conditions, weather data from a single ground station are usually used as input in most SEB models. However, for regional applications with varying climatic conditions, distributed datasets are required. Several recent studies [50,51] have applied gridded weather datasets for mapping daily ET due to the ease of their application for regional studies. However, users need to confirm the integrity of the datasets before processing the SEB model. A study [52] found overestimation of reference ET due to biases in air temperature and wind speed in the widely used reanalysis data—North American Land Data Assimilation System when compared to reference ET estimates from the Texas High Plains ET Network [53]. The study recommended using weather station datasets within agricultural settings, whenever possible, for precise applications of time series ET information such as in irrigation scheduling. A few studies have explored the applicability of developing distributed weather data from the point measurements of a network of ground stations to account for the spatial variability of weather parameters [54].

The third step is to run the selected SEB model, which involves several sub-models to solve the SEB equation as shown in Equation (1).

$$LE = R_n - G - H \tag{1}$$

where LE is latent heat flux, R_n is net radiation, G is soil heat flux and H is sensible heat flux. All parameters are in units of W m^{-2}. Based on the sensible heat flux computation approach, SEB models can be categorized into single-source and two-source models. The sensible heat fluxes for soil and vegetation are computed separately in two-source models, while a single value for each pixel is computed in single-source models. Each approach has its own advantages and caveats. In theory, two-source models could provide more accurate ET over sparse vegetation as they close the energy balance separately for soil and vegetation. Timmermans et al. [44] found better accuracy from a TSEB model compared to SEBS across sparsely vegetated grasslands in the Southern Great Plains. Kustas et al. [55] reported that two-source performed better in sub-humid tallgrass prairie, whereas greater accuracy was found for a single-source model in semiarid rangeland.

As mentioned before, some single-source models require an additional step in running the model, which involves the manual selection of extreme hot and cold pixels by user. To remove the subjectivity in the selection of extreme pixels in SEBAL, Long et al. [56] introduced a trapezoidal approach to define boundary conditions for the selection of these pixels based on the relationship between vegetation fraction and surface temperature. Automated approaches have been proposed in [57–59] to replace human intervention. Alternative approaches are also applied by [60,61] to estimate ET from a cold pixel as a function of normalized difference vegetation index when an ideal cold pixel is difficult to find within a satellite image.

The fourth step is to extrapolate the instantaneous ET to daily values. Evaporative fraction (Λ) [35,37,62,63] and ETrF (fraction of reference ET) [38,64] are the common methods to obtain daily ET. Both of these methods assume the instantaneous Λ or ETrF is the same as the daily Λ or ETrF. However, a study [65] reported that this assumption was not satisfied when the fractional vegetation cover was close to a maximum. In the Texas Panhandle, Colaizzi et al. [66] found a better agreement of ETrF method for cropland and Λ method for bare soil when compared with lysimeter measurements. Chavez et al. [67] evaluated six extrapolation approaches on corn and soybean fields and found smaller error from Λ method when compared with eddy covariance measurements. Another study [68] found Λ method advantageous during several water stress events, whereas ETrF approach performed better under advective conditions [38,64], which could be significant in arid environments.

The fifth step is to fill the gaps caused by cloud coverage over a portion of the daily ET maps. One approach is to apply linear interpolation of nearest reliable values within an image [50]. This method is suitable when the nearest pixels are under the same land cover as that of missing pixels. However, it may not be appropriate when the area with data gap is large and encompasses heterogeneous terrain. Another approach includes the use of time-weighted interpolation of preceding and following images [69]. This method adjusts the vegetation development using normalized difference vegetation index (NDVI) across vegetated areas and residual soil moisture differences for the areas with bare soil surface. Anderson et al. [39] applied the available water for the root zone and soil surface layer to fill the gaps. The available water for the clear and cloudy days is used to estimate the daily water depletion due to ET from the root zone and soil surface layer and the fraction of available water is used to fill the gaps.

The final step is the interpolation of ET maps between consecutive satellite overpass dates to construct daily ET time series. Several interpolation and data-fusion approaches have been implemented for this purpose. A common approach is to apply linear interpolation of Λ or ETrF images between consecutive satellite overpass dates [70]. Another approach is to apply a curvilinear function using more than two Λ or ETrF images. For example, at least one cloud-free image for each month was used for spline interpolation within METRIC to obtain monthly and seasonal ET [38,71,72]. Singh et al. [70] evaluated the performance of several interpolation methods and found no significant difference in seasonal ET among cubic spline, fixed ETrF and linear interpolations. A backward-average iterative approach has been also proposed to estimate ET in between Landsat overpass dates [73].

While numerous studies have been conducted to address the issues related to specific steps involved in generating remotely sensed ET time series based on SEB models, only a few have focused on developing automated modeling frameworks, covering all hierarchical steps mentioned above. Such modeling frameworks, if validated, could have significant value in providing end-users with daily ET time series for practical applications in improving land and water management. Furthermore, a comprehensive and detailed documentation of the entire process of deriving daily ET maps at regional scales could be a useful resource to potential end-users who currently need to understand and select appropriate approaches for each of the six steps from many sources. Developing and documenting a comprehensive framework that generates complete ET time series from raw input data enables potential users outside the research community to utilize this framework for making more informed decisions and policies. The main goal of this study was to develop and document a modeling framework to construct daily time series of ET maps for the entire state of Oklahoma, USA. The performance of this framework was also evaluated by comparing its results with ET estimates of flux towers in Oklahoma. Finally, long-term variations in ET across Oklahoma were investigated.

2. Materials and Methods

2.1. Study Area

The study area covered the entire state of Oklahoma, USA, with an area of about 181,200 km^2 (Figure 1). Oklahoma Climate is classified as humid subtropical at most parts of the state and cold semi-arid at far west [74]. The state has nine climate divisions (CD) delineated based on precipitation and temperature gradients. The normal (1981–2010) annual precipitation is about 925 mm yr^{-1}, with significant spatial variation across CDs. While southeast (CD9) receives the largest amount of 1301 mm yr^{-1} on average, the Panhandle (CD1) holds the smallest record of 520 mm yr^{-1}. The normal annual mean air temperature is 15.6 °C, with July and January being the hottest and coldest months, respectively. The southcentral (CD8) has the maximum mean annual temperature of 16.7 °C, whereas the Panhandle region has the minimum value at 13.6 °C. The top two land cover categories in Oklahoma are grassland (36.4%) and pastureland (11.3%) [75]. The elevation varies between 88 m above mean sea level at the southeast border with Arkansas and 1516 m at far-west border with New Mexico.

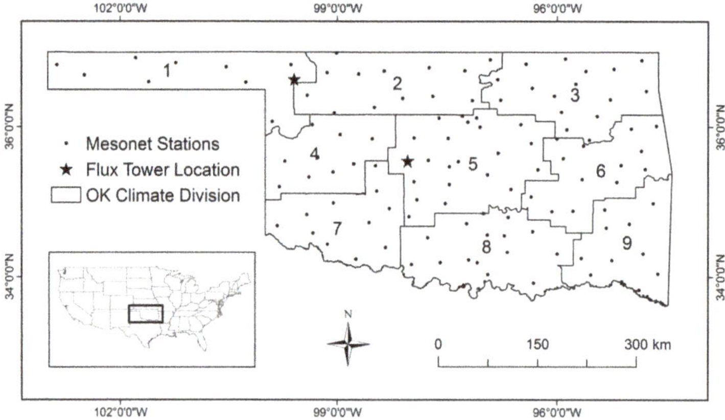

Figure 1. Map of Oklahoma and its nine climate divisions. The locations of Mesonet stations and flux towers are also specified.

2.2. Modeling Framework

The modeling framework was designed to use daily images from the MODIS Terra satellite as input data. The single-source SEBS model [37] was selected as the SEB model for estimating energy fluxes. The main reason for the selection of SEBS over other SEB models was its applicability over large areas with heterogeneous surfaces [31]. In addition, this model does not require intermittent human intervention, which facilitates the automation process. A graphical illustration of the proposed framework is shown in Figure 2, followed by detailed explanation of specific approaches selected for each of the six computational steps mentioned before.

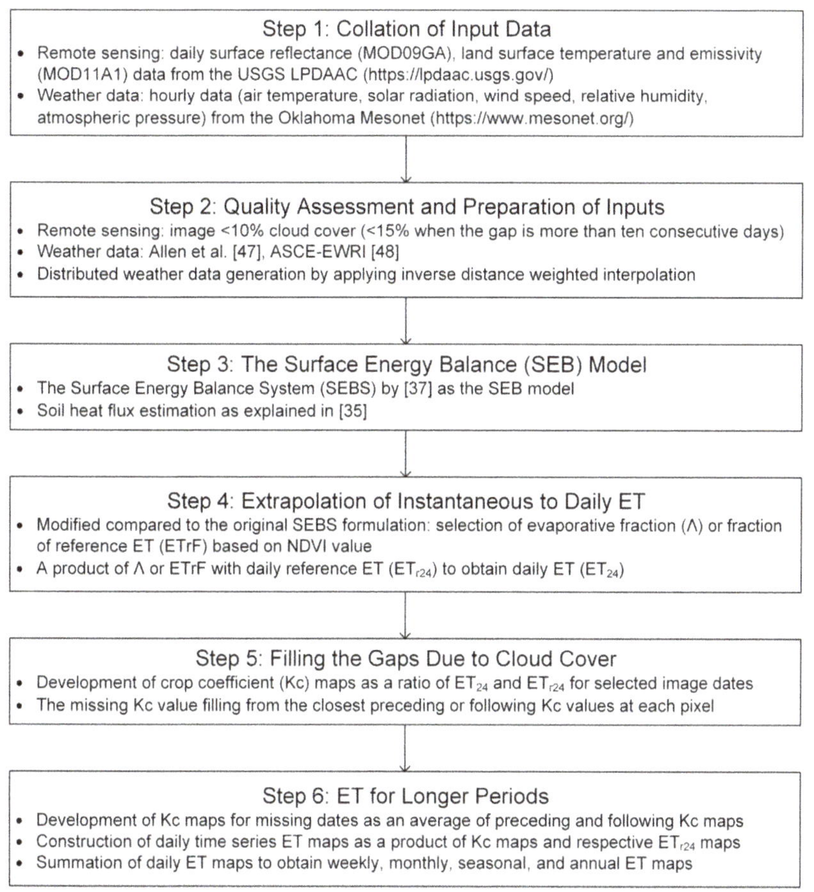

Figure 2. A descriptive flow diagram of the daily time series of evapotranspiration (ET) modeling framework.

2.2.1. Step 1: Collation of Input Data

The daily surface reflectance (MOD09GA, [76]), daily land surface temperature (LST) and emissivity (MDO11A1, [77]) data were downloaded from the US Geological Survey Land Processes Distributed Active Archive Center (https://lpdaac.usgs.gov/). Ground-based meteorological data included hourly air temperature, relative humidity, incoming shortwave solar radiation, wind speed and atmospheric pressure. These data were obtained from the Oklahoma Mesonet [78,79] weather stations installed across the state (Figure 1). The Oklahoma Mesonet is a world-class environmental monitoring network (https://www.mesonet.org/) consisting of 120 active stations with at least one station at each of the 77 counties in Oklahoma.

2.2.2. Step 2: Quality Assessment and Preparation of Inputs

The initial quality assessment of suitable MODIS images was based on cloud coverage. Images with less than 10% cloud cover were selected for further processing. Hence, any day when the cloud coverage was above 10% was assumed as a day with missing remotely sensed data. When the period of missing imagery was more than 10 consecutive days, images with less than 15% cloud cover were also included as acceptable quality. Then, cloud-covered pixels in each selected image were masked by applying a threshold of LST smaller than 250 K. These steps were repeated for all selected reflectance, LST and emissivity images. Since a single MODIS image tile was not sufficient to cover the entire state of Oklahoma, two image tiles (h09v05 and h10v05) were merged.

The quality assessment of each weather variables was performed as described in [47,48]. The solar radiation was checked against the upper limit under clear sky condition. Daily average temperature was compared against the average extreme temperatures to ensure the difference between them was within the acceptable range (2 °C) [48]. The quality of wind speed was maintained by considering gust factor threshold of more than 1. Relative humidity data were considered when the values were less than 100%. The missing weather data were filled by an average value of that parameter from four nearest Mesonet stations. Hourly alfalfa reference ET (ET_r) [48] was then computed at each station during the study period using the Bushland ET Calculator [80]. Daily ET_r estimates were obtained by summing 24-hour ET_r values. To incorporate the weather variability between the weather stations, spatial input data were generated by applying inverse distance weighted interpolation for all weather variables, including hourly and daily ET_r. As mentioned in the previous section, the Oklahoma Mesonet is a densely distributed weather station network, with about 1510 km^2 per station. This is a significantly finer spatial resolution than the 5000 km^2 per station value recommended by the World Meteorological Organization for evaporation stations on interior plains [81]. Hence, the adjustment of meteorological parameters with elevation was not considered during interpolation.

2.2.3. Step 3: The SEB Model

As mentioned before, the Surface Energy Balance System (SEBS) model of [37] was selected as the SEB model in the present study. However, other SEB models such as those reviewed in the Introduction section can be used in this step based on user resources, availability of input data and desired accuracy. Like other SEB models, SEBS estimates the latent heat flux (LE) as a residual of the land surface energy balance as shown in Equation (1). The R_n was calculated by applying the surface radiation balance equation:

$$R_n = (1 - \alpha)R_s + \varepsilon_s\,\varepsilon_a\,\sigma\,T_A^4 - \varepsilon_s\,\sigma\,T_S^4 \tag{2}$$

where R_S is incoming shortwave solar radiation, α is surface albedo (dimensionless) estimated following [82], ε_a and ε_s are emissivities (dimensionless) of atmosphere and surface, estimated following [83,84], respectively. σ is the Stefan-Boltzmann constant (5.67×10^{-8} W m^{-2} K^{-4}), T_A is air temperature (K) and T_s is the surface temperature (K), estimated as a ratio of brightness temperature to $\varepsilon_s^{-0.25}$. The G was estimated by applying the relationship developed by [35]:

$$\frac{G}{R_n} = \frac{(T_s - 273.15)}{100\alpha}\left(c_1\,\alpha + c_2\,\alpha^2\right)\left(1 - 0.98\text{NDVI}^4\right) \tag{3}$$

where c_1 and c_2 are calibration coefficients and were considered as 0.24 and 0.46, respectively.

SEBS uses similarity theories to estimate H: the bulk atmospheric similarity (BAS) theory for atmospheric boundary layer (ABL) scaling [85] and the Monin-Obukhov similarity (MOS) for atmospheric surface layer (ASL) scaling [86]. The ABL is a part of the atmosphere that is directly impacted by earth's surface and responds to surface forcing with a timescale of an hour or less, whereas ASL is usually the bottom 10% of ABL [37]. During unstable conditions, an appropriate atmospheric (BAS or MOS) scaling is determined as presented in [83]. For stable conditions, functions given by [83,87] are used for ABL and ASL scaling, respectively. In the ASL, the similarity relationships for mean wind speed (u) and the difference between potential temperature profiles are derived using the MOS theory as:

$$u = \frac{u_*}{k} \left[\ln\left(\frac{z - d_0}{z_{0m}}\right) - \psi_m\left(\frac{z - d_0}{L}\right) + \psi_m\left(\frac{z_{0m}}{L}\right) \right] \tag{4}$$

$$\theta_0 - \theta_a = \frac{H}{k\, u_*\, \rho_a\, C_p} \left[\ln\left(\frac{z - d_0}{z_{0h}}\right) - \psi_h\left(\frac{z - d_0}{L}\right) + \psi_h\left(\frac{z_{0h}}{L}\right) \right] \tag{5}$$

$$L = -\frac{\rho_a C_p u_*^3 \theta_v}{kgH} \tag{6}$$

where u_* is the friction velocity (m s^{-1}), k is the von Karman's constant (0.41), z is the height above the surface (m), d_0 is the zero plane displacement height (m), z_{0m} is the roughness height for momentum transfer (m) estimated using an empirical relationship with NDVI [88], z_{0h} is roughness height for heat transfer (m), θ_0 is the potential air temperature at surface (K), θ_a is the potential air temperature at z (K), θ_v is the potential virtual temperature near the surface (K), ρ_a is the air density (kg m^{-3}), C_p is the specific heat capacity of air (1013 J kg^{-1} K^{-1}) and g is the gravitational acceleration (9.8 m s^{-2}). ψ_m and ψ_h are the stability correction functions for momentum and sensible heat transfer, respectively and L is the Monin–Obukhov length (m).

The scalar roughness height for heat transfer, z_{0h}, is an important parameter to regulate the heat transfer between the land surface and the atmosphere and estimated as:

$$z_{0h} = \frac{z_{0m}}{\exp\left(kB^{-1}\right)} \tag{7}$$

where kB^{-1} is the Stanton number, a dimensionless heat transfer coefficient, estimated using a formulation from [89] as:

$$kB^{-1} = \frac{kC_d}{4C_t \frac{u_*}{u(h)}\left(1 - e^{\frac{-n_{ec}}{2}}\right)} f_c^2 + 2 f_c f_s \frac{k\left(\frac{u_*}{u(h)}\right)\left(\frac{z_{0m}}{h}\right)}{C_t^*} + kB_s^{-1} f_s^2 \tag{8}$$

The heat transfer coefficient in Equation (8) was formulated to account for three different land surface conditions. The first term follows the Choudhury and Monteith [90] model for full canopy, the second term accounts for the interaction between the vegetation and soil surface and the third term is for the bare soil surface given [83]. In this equation, f_c and f_s are canopy and soil fraction coverage, respectively, C_d is the drag coefficient for the foliage with a value of 0.2; C_t and C_t^* are the heat transfer coefficients of the leaf and soil, respectively. The value of C_t was taken as 0.03 and C_t^* was computed from Prandtl number and roughness Reynolds number (Re$_*$) [37]. The u(h) in Equation (8) is the horizontal wind speed at the canopy top (m s^{-1}) and h is canopy height (m) estimated as a ratio of z_{0m} to 0.136 [37]. The n_{ec} (within-canopy wind speed profile extinction coefficient) and Brutsaert term kB_s^{-1} (for bare soil surface) were calculated as:

$$n_{ec} = \frac{C_d\, LAI}{\frac{2u_*^2}{u(h)^2}} \tag{9}$$

$$kB_s^{-1} = 2.46(Re_*)^{0.25} - \ln(7.4) \tag{10}$$

where LAI is the leaf area index and estimated as a functional relation with NDVI [91].

SEBS requires estimation of H for dry (H_{dry}) and wet (H_{wet}) boundary conditions. Under dry conditions, the H_{dry} is equivalent to the available energy ($R_n - G$) as there is no evaporation due to the limitation of water availability and H_{wet} is calculated using the Penman-Monteith equation [92,93].

After computing H for boundary conditions, the relative evaporative fraction (Λ_r), the evaporative fraction (Λ) and ET are estimated. The steps and explanation are detailed in [37]

2.2.4. Step 4: Extrapolation of Instantaneous to Daily ET

The SEBS uses the Λ approach for scaling instantaneous ET to daily ET, assuming the Λ at the time of overpass is equal to the daily Λ. In this study, a modified approach was implemented where either Λ or ETrF is used for extrapolation of each pixel based on its NDVI value as shown in Equation (15).

$$\Lambda_r = 1 - \frac{H - H_{wet}}{H_{dry} - H_{wet}} \tag{11}$$

$$\Lambda = \frac{\Lambda_r (R_n - G - H_{wet})}{R_n - G} \tag{12}$$

$$ET_{inst} = \left(\frac{R_n - H - G}{\lambda} \right) \times 3600 \tag{13}$$

$$ETrF = \frac{ET_{inst}}{ET_r} \tag{14}$$

$$ET_{24} = [\Lambda \times ET_{r24} \text{ for NDVI} < 0.30] \text{ or } [ETrF \times ET_{r24} \text{ for NDVI} \geq 0.30] \tag{15}$$

where ET_{inst} and ET_r are the actual and reference ET at the hour of satellite overpass (mm h^{-1}), λ is the latent heat of vaporization (~2.45 MJ kg^{-1}). ET_{r24} is the daily reference ET and ET_{24} is the daily actual ET (mm d^{-1}). This modification was made to take the advantage of Λ and ETrF approach to better represent the water limited and energy limited conditions, respectively. The ETrF was estimated as a ratio of ET obtained from Step 3 to reference ET at the satellite overpass time (MODIS Terra satellite overpass local time around 10:30 AM).

2.2.5. Step 5: Filling the Gaps Due to Cloud Cover

Data-gaps due to cloud cover is a common issue in all space-borne satellites. In this study, crop coefficient (Kc) was used to fill the data-gaps. The Kc maps were created for all images as the ratio of ET_{24} and respective daily ET_r. To fill the Kc of a cloud covered (missing) pixel for a specific image date, the Kc value of the same pixel from the preceding image date was first used. If the same pixel was missing in the preceding image, the Kc value was obtained from the next Kc map. The latter step was repeated if the next day was missing until a date was found with a Kc value estimated for the same pixel. This interpolation method was suitable to fill the data gaps as most of the selected images were less than 10 days apart during the crop growing season (April to October).

2.2.6. Step 6: ET for Longer Periods

After filling the data gaps in daily ET maps due to clouds, the ET maps needed to be created for days when the cloud coverage was more than 10% (or 15%) and thus no input imagery was available. To fill these gaps, the average Kc of the preceding and following images closest to the image date of interest was used. The Kc images were then multiplied with respective daily ET_r to obtain complete time series of daily ET maps. Construction of weekly, monthly, seasonal and annual ET maps was accomplished by summation of daily ET maps over corresponding periods. The processing of all steps was executed in Python language within ArcGIS environment.

2.3. Comparison with Flux Tower Data

Daily ET time series from the modeling framework explained above were compared against observed ET from three flux towers: US-ARc (35.5464 N, 98.0400 W), US-ARb (35.5497 N, 98.0402 W) [94] and US-AR2 (36.6358 N, 99.5975 W) [95]. The US-ARc and US-ARb were located close to each other over native grassland in central Oklahoma. The US-AR2 was located over planted switchgrass in northwest Oklahoma. The 30-minute flux data from the towers were downloaded from the AmeriFlux data archive (http://ameriflux.lbl.gov/) for years 2005 and 2006 for US-ARc and US-ARb sites and years 2010 and 2011 for the US-AR2 site. The flux tower data usually have the issue with energy balance closure, therefore, the closure error was corrected by maintaining constant Bowen-ratio following [96]. The corrected 30-min data were averaged to obtain daily data. The daily observed ET was then compared with the average values of 3×3 pixels (~1390 m $\times$ 1390 m) from the SEBS ET at the flux tower locations. It should be noted that the three flux towers used for validating the performance of the modeling framework in this study represent only two land covers (native and managed grassland). Hence, the performance of the framework may be different from what is documented here over different types of land covers not included in the present analysis.

For statistical analysis, correlation coefficient (r), the coefficient of determination (R^2), mean absolute error (MAE), mean bias error (MBE) and root mean square error (RMSE) were used:

$$\text{MAE} = \frac{1}{n} \sum_{i=1}^{n} |\text{SEBS-ET} - \text{FT-ET}| \tag{16}$$

$$\text{MBE} = \frac{1}{n} \sum_{i=1}^{n} (\text{SEBS-ET} - \text{FT-ET}) \tag{17}$$

$$\text{RMSE} = \sqrt{\frac{1}{n} \sum_{i=1}^{n} (\text{SEBS-ET} - \text{FT-ET})^2} \tag{18}$$

where FT-ET is the observed flux tower daily ET and SEBS-ET is the estimated daily ET from the SEBS model.

2.4. Application of the Modeling Framework

After evaluating the accuracy of the modeling framework, it was used to estimate annual ET maps over the entire state of Oklahoma, as well as its nine climate divisions (CD), during the 2001–2014 period. The annual ET were also compared with publicly available MOD16 ET dataset [97,98] over the same period, which covers the most recent drought episode of 2011–2014. The degree of water availability for each pixel and CD within Oklahoma was assessed by estimating the ratio of annual ET from the modeling framework and the reference ET. This ratio is an indication of the portion of the atmospheric demand that is supplied at each pixel and CD. Areas with smaller ratios represent water scarcity since the actual ET from the model is far from the potential limits of ET. The information on annual ET variations and water availability across Oklahoma can assist state water managers with making critical decisions based on long-term objective data from the implemented framework. As mentioned before, the validation dataset only represented native and managed grassland. About half (47%) of all lands in Oklahoma are under rangeland and grassland. With winter wheat being the most dominant crop, the majority of croplands have similar canopy characteristics. Nevertheless, the lack of representation of other land covers (e.g., 21% of forest in Oklahoma) should be considered in applications and interpretations of the results of the modeling framework.

3. Results and Discussion

3.1. Comparison with Flux Tower Data

The comparison with flux tower data showed good agreement between daily SEBS-ET and FT-ET. The modeling approach captured the spatial and temporal variations in ET. However, the model overestimated ET at all sites and years (Figure 3), with average MBE of 20.1 W m^{-2}. The range of MBE was between 1.7 W m^{-2} at US-AR2 in 2011 and 29.3 W m^{-2} at US-ARb in 2006 (Table 1). The mean MAE and RMSE were 33.0 W m^{-2} and 42.7 W m^{-2}, respectively. The correlation coefficients varied from 0.61 to 0.81 and R^2 from 0.37 to 0.66.

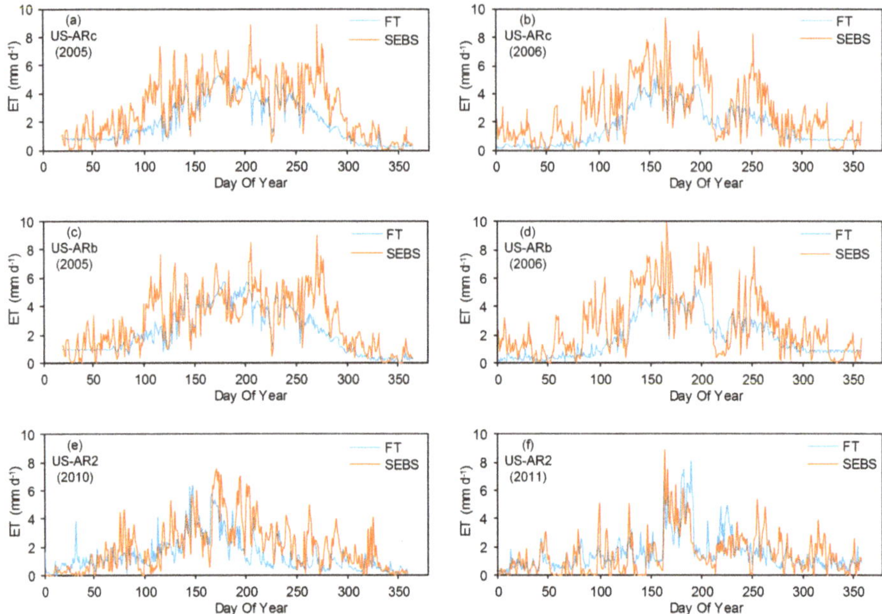

Figure 3. Comparison of daily ET from surface energy balances (SEBS) and flux tower (FT).

Table 1. Statistical indicators between SEBS and flux tower ET.

Site	Year	r	R^2	MAE (W m^{-2})	MBE (W m^{-2})	RMSE (W m^{-2})
US-ARc	2005	0.78	0.61	39.6	19.1	40.1
	2006	0.77	0.59	36.7	27.5	49.2
US-ARb	2005	0.81	0.66	31.9	26.6	43.3
	2006	0.78	0.61	35.9	29.3	47.7
US-AR2	2010	0.61	0.37	29.4	16.4	41.7
	2011	0.62	0.39	24.7	1.7	34.1

The errors in the ET estimates of the modeling framework are due to errors generated in each of the six steps outlined in previous section. A major step for error introduction is step three, that is, the surface energy balance model. Previous studies have reported uncertain characterization of kB^{-1} in water limited environments [99–101] and in low vegetation cover conditions [102]. Overestimating kB^{-1} under these conditions would lead to overestimating z_{0h}, underestimating H and consequently overestimating ET [99]. The overestimation errors observed in this study were within the range of errors in previous studies when using MODIS as the input imagery to SEBS model. For example, [103] reported ET overestimation with MBE of 6.1 W m^{-2}; [104] found MBE of 20.1 W m^{-2} and RMSE of 34.7 W m^{-2}; [105] reported MBE of 144.9 W m^{-2} when comparing SEBS-ET from cropland and grassland with flux tower estimates; [106] found overall MBE of 31 W m^{-2} and RMSE of 76 W m^{-2}; and, [107] reported MBE of 95.1 W m^{-2} and RMSE of 122.2 W m^{-2} across several land covers and climatic conditions. While several studies have reported overestimation error from SEBS, the mean absolute error from the current study was smaller than the threshold of 50 W m^{-2} suggested by [108].

Errors in other steps of the framework can contribute to biases in final ET estimates. A common source of error in estimating ET from satellite imagery is due to cloud contamination. A thin layer of cloud or a shaded area due to cloud presence over nearby pixels can result in underestimation of

LST and consequently, overestimation of ET. In practical applications, it is impossible to remove all these contaminated pixels from the entire image even after applying the LST thresholds during quality control. In this study, there were days with underestimated LST due to cloud presence. For example, the LST at the flux tower pixel area dropped by 10.6 K from Day of Year (DOY) 113 to 114, while both DOYs were identified as cloud-free and no precipitation was recorded. The instantaneous T_A increased by 3.2 K over the same period. The smaller LST on DOY 114 affected ET estimation for this day and the following days until another cloud-free image was obtained for DOY 117 (Figure 3a).

A sensitivity analysis study [109] on SEBS model reported LST as the most sensitive parameter, with up to 70% error in H from irrigated fields expected with 0.5 K bias in LST. Another study [110] found that error in H varied between −41% and 152% when LST bias ranged from −4 K to 10 K. These studies show that a small bias in LST can significantly impact H and ultimately ET. The magnitude of error may depend upon the sensitivity of SEBS to LST, including other parameters such as T_A, u, Δt [111] and could vary depending on whether the wet or dry limits have been reached [110,112].

In this study, the filtering criteria of less than 15% cloud cover limited the availability of cloud-free images. Applying this filter resulted in 125 and 154 cloud-free images for processing during 2005 and 2006, respectively. For the days with no cloud-free images, the ET estimate was dependent on the Kc approach explained before. However, the Kc approach may fail to account for the variability in pixel conditions, especially if land and weather conditions change dramatically during long periods of gaps in imagery. In this study, 10 to 15 cloud-free images each month were available for most months, which was assumed sufficient to capture general daily soil moisture and weather variations. In other periods, however, it was not possible to keep the length of gap periods short. For example, cloud-free images were not available for 17 consecutive days from DOY 270 to 286 in 2005, when larger differences between FT-ET and SEBS-ET were observed (Figure 3a,c).

The combined impact of LST bias due to cloud contamination and unavailability of cloud-free images significantly increase biases in ET estimation. The 15% cloud cover filter could be reduced to reduce cloud contamination issue but this would come at the cost of increasing the length of periods with no imagery at all. Increasing the filtering limit will have an opposite effect (more available imagery with larger cloud contamination within each image). Another solution is to manually inspect and select images. However, this increases the processing time and interrupts the automated nature of the ET modeling framework. Another factor that could play a significant role in increasing ET errors is the availability and quality of input weather data. Su et al. [113] reported about 40% increase in RMSE (from 73 W m^{-2} to 102 W m^{-2}) when using reanalysis dataset—Global Land Data Assimilation System within SEBS instead of ground-based weather data. In this study, the impact of this source of error is expected to be minimal since rigorous quality control was conducted on ground-based data and only less than 2% of data were missing during the study period.

As highlighted before, the daily ET results, uncertainty and potential biases of the proposed ET modeling framework were evaluated and discussed based on flux tower measurements over native and managed grassland at central and northwest Oklahoma. Flux tower data across other land covers were not available for comparison, thus the results from the framework may need further assessment to warrant the similar level of accuracy and uncertainty while applying the results to different land covers and climates across the state. In particular, the analysis and interpretation of results from current study may differ for vegetation with different canopy structure compared to grassland.

3.2. Application of the Modeling Framework

The automated operational ET modeling framework proposed in this study was used to create annual ET maps covering the entire state of Oklahoma for the period from 2001 to 2014. As expected, the annual ET followed the precipitation pattern and increased from southeast to Panhandle (Figure 4). When averaged over the entire 14 years, the southeast climate division (CD9) had the largest annual ET of 1272 mm yr^{-1} and the Panhandle climate division (CD1) had the smallest annual ET of 588 mm yr^{-1} (Table 2). The reference ET (ET$_r$) had an opposite pattern, with CD1 having the largest amount at 2140 mm yr^{-1} and CD9 the smallest (1360 mm yr^{-1}). This means that on average, about 94% of atmospheric demand was fulfilled at southeast, compared to only 27% in the Panhandle during the study period. In other words, water scarcity is a larger issue in CD1 compared to CD9 as available resources were not sufficient to keep up with atmospheric demand. The statewide average annual ET was 994 mm yr^{-1}, about 57% of the average annual ET$_r$ of 1755 mm yr^{-1}.

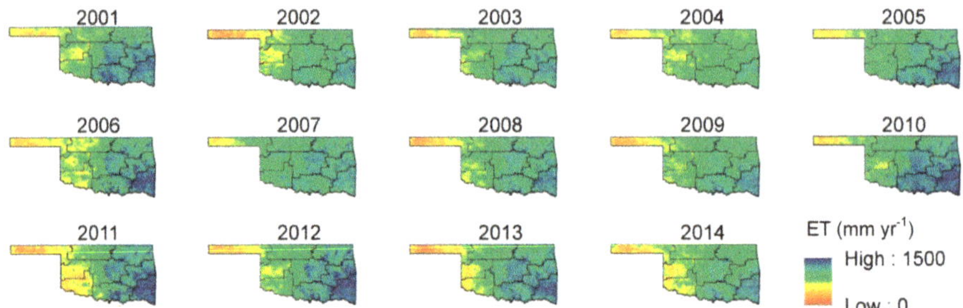

Figure 4. Annual ET maps (SEBS-ET) of Oklahoma from 2001 to 2014. The solid lines represent boundaries of the nine climate divisions. It should be noted that CDs 6 and 9 in southeast have a forested area of more than 29%. Hence, their ET estimates may not be accurate since the flux towers used in validation did not include forest land cover.

Table 2. Average annual SEBS-ET, MOD16-ET, ET$_r$ and the ratio of SEBS-ET to ET$_r$ for all Oklahoma climate divisions (CD) during the 2001–2014 period.

Climate Division	SEBS-ET (mm yr^{-1})	MOD16-ET (mm yr^{-1})	ET$_r$ (mm yr^{-1})	SEBS-ET ET$_r$$^{-1}$
CD1 (Panhandle)	588	259	2140	0.27
CD2 (North Central)	918	364	1871	0.49
CD3 (Northeast)	1098	657	1521	0.72
CD4 (West Central)	790	338	2018	0.39
CD5 (Central)	1095	531	1700	0.64
CD6 (East Central) *	1175	736	1492	0.79
CD7 (Southwest)	845	363	2009	0.42
CD8 (South Central)	1163	599	1683	0.69
CD9 (Southeast) *	1272	798	1360	0.94
Oklahoma	994	516	1755	0.57

* These CDs have a forested area of more than 29%. The results presented in this table may not be accurate for these CDs since the flux towers used in validation did not include forest land cover.

The average annual ET comparison between MOD16 and SEBS indicated large differences across all Oklahoma CDs (Table 2). The differences between MOD16-ET and SEBS-ET varied between 37% at CD9 to 60% at CD2, with an average of 48% lower ET rates from MOD16. Three eastern humid CDs (CD3, CD6, CD9) had smaller differences between MOD16-ET and SEBS-ET compared to three western CDs (CD1, CD4, CD7). The difference between SEBS-ET and MOD16-ET is possibly due to a combination of overestimations from SEBS and underestimation from MOD16. The underestimation of ET from MOD16 has been reported in previous studies, particularly in semi-arid and arid climates [114,115].

The ratio of SEBS-ET to ET$_r$ can be estimated on a pixel wise basis to provide information on water scarcity at a finer resolution for local water management and planning. This ratio is mapped in Figure 5. The general patterns are similar to those presented in Table 2, with western parts of the state under relatively larger water scarcity compared to the eastern parts. However, significant variability can be observed within some CDs. In CD1, for example, the western half of CD (Cimarron and Texas counties) had smaller ratios compared to the eastern half, suggesting a more severe water scarcity. CD2 was similar in terms of variations in the ET ratios across the CD. The surface water resources in western Oklahoma were visible in regions with a ratio value of more than 0.5. Examples include the riparian areas of Cimarron and North Canadian rivers in southwest of CD2, as well as Canton

Lake and Foss reservoir in CD4 and the five reservoirs in CD7 (Lugert-Altus, Tom Steed, Lawtonka, Ellsworth and Fort Cobb). Maps similar to the one in Figure 5 can be developed at varying temporal and spatial scales to monitor changes in water availability more closely. The ratio of actual ET to reference or potential ET has been used in the past in monitoring water stress and drought, such as in the Evaporative Stress Index [17].

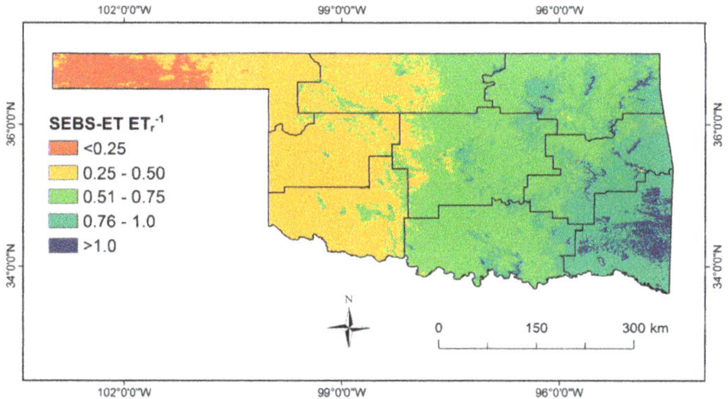

Figure 5. The ratio of average annual SEBS-ET to ET_r across Oklahoma during the period 2001–2014. It should be noted that CDs 6 and 9 in southeast have a forested area of more than 29%. Hence, their ET estimates may not be accurate since the flux towers used in validation did not include forest land cover.

The inter-annual variations in ET were also examined for each CD and for the entire state. Figure 6 demonstrates deviations in SEBS-ET as percentage of the average annual ET during the 2001–2014 period. The impact of the 2011–2014 drought in western Oklahoma can be observed in this graph, with the maximum reduction in ET occurring in 2011 for the three western CDs of CD1, CD4 and CD7. The percent deviations from average was −22%, −21% and −33% for the same CDs, respectively. According to the U.S. Drought Monitor (USDM) [116], more than 80% of the three CDs was under extreme drought (D3 category) from June. The drought condition worsened in July and remained under D4 category until December 2011. The three eastern CDs of CD3, CD6 and CD9 were above average in 2011, with percent deviations of 9%, 8% and 10%, respectively. The USDM indicated almost no drought at CD3 in 2011, whereas CD6 and CD9 had less than 40% of their area under extreme drought from August to November 2011. The middle three CDs registered close to long-term average ET. The largest positive deviations for the three western CDs occurred in 2007, a year that was characterized by above normal precipitation.

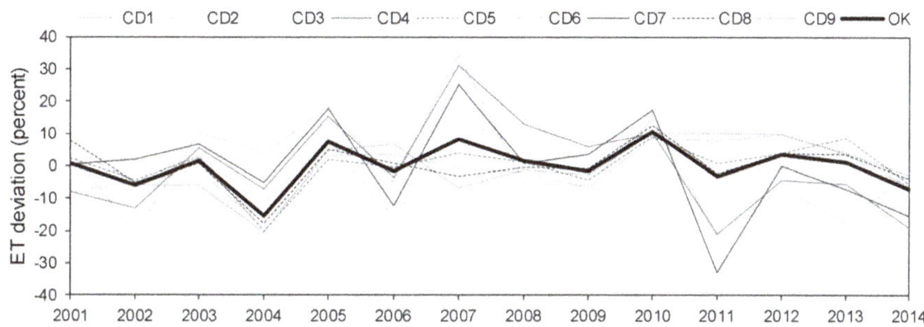

Figure 6. Annual ET deviation across climate divisions of Oklahoma from 2001 to 2014.

The ET modeling framework proposed in this study can automatically generate time series of daily ET maps on a continuous basis, with several applications beyond those mentioned in previous sections. For example, ET maps over agricultural areas can be analyzed in conjunction with yield data to evaluate the water use efficiency. However, this modeling framework has some limitations that must be considered and improved in future applications. One limitation is the size of MODIS pixels, which practically hinders the possibility of using the ET data at field scale. This limitation can be overcome by modifying the framework to use satellite imagery at finer resolution (e.g., Landsat). Another challenge is identifying and removing cloud contaminated pixels. The filters used in this study were not always effective in identifying pixels that were covered by thin layers of cloud or were in the shadow of a cloud. Thus, further investigation and application of robust methods to examine cloud contamination are needed. Finally, there were periods when no images were available for several days due to clouds covering the entire scene. This negatively affects the ability to capture ET fluctuations during those periods. Data-fusion approaches can be implemented in the modeling framework as a potential solution to improving ET interpolation for days with missing images.

4. Conclusions

An ET modeling framework was proposed to automatically construct daily time series of ET maps across Oklahoma by integrating MODIS imagery, ground-based weather data and surface energy balance model. The comparison of the results with daily observations at three flux towers (two years of data at each site) showed good performance of the modeling framework with mean bias errors less than 30 W m^{-2} and root mean squared errors less than 50 W m^{-2}. The results were then used to investigate spatial and temporal variations in ET across the state and its nine climate divisions (CD). The statewide annual ET varied between 841 and 1100 mm yr^{-1} during the period from 2001 to 2014, with an average of 994 mm yr^{-1}. A large difference in ET was observed among CDs, with Oklahoma Panhandle (CD1) having the smallest and southeast (CD9) the largest average annual ET of 588 and 1272 mm yr^{-1}, respectively. The ratio of estimated ET to reference ET was used as an indicator of water scarcity at pixel and CD levels. The deviations in annual ET from the 2001–2014 average ET were also studied and found to be in good agreement with temporal and spatial variations in drought. The proposed ET modeling framework provided a pathway to construct daily time series of ET maps with potential for a range of applications. However, further improvements are necessary to resolve the issues highlighted in the current study.

Author Contributions: Conceptualization, P.G. and G.P.; methodology, P.G. and G.P.; formal analysis, K.K., P.G. and G.P.; investigation, K.K., S.T. and P.G.; writing—original draft preparation, K.K. and S.T.; writing—review and editing, K.K., S.T. and P.G.; project administration, S.T. and P.G.; funding acquisition, S.T. and P.G.

Funding: This research was funded by a joint research and extension program funded by the Oklahoma Agricultural Experiment Station (Hatch funds) and Oklahoma Cooperative Extension (Smith-Lever funds) received from the National Institutes for Food and Agriculture, U.S. Department of Agriculture. Additional support was provided by the Oklahoma Water Resources Center through the U.S. Geological Survey 104(b) grants program.

Conflicts of Interest: The authors declare no conflict of interest.

References

1. Allen, R.G.; Tasumi, M.; Morse, A.; Trezza, R. A Landsat-based energy balance and evapotranspiration model in Western US water rights regulation and planning. *Irrig. Drain. Syst.* **2005**, *19*, 251–268. [CrossRef]
2. Droogers, P.; Bastiaanssen, W. Irrigation performance using hydrological and remote sensing modeling. *J. Irrig. Drain. Eng.* **2002**, *128*, 11–18. [CrossRef]
3. Santos, C.; Lorite, I.J.; Tasumi, M.; Allen, R.G.; Fereres, E. Performance assessment of an irrigation scheme using indicators determined with remote sensing techniques. *Irrig. Sci.* **2010**, *28*, 461–477. [CrossRef]
4. Taghvaeian, S.; Neale, C.M.; Osterberg, J.C.; Sritharan, S.I.; Watts, D.R. Remote Sensing and GIS Techniques for Assessing Irrigation Performance: Case Study in Southern California. *J. Irrig. Drain. Eng.* **2018**, *144*, 05018002. [CrossRef]

5. Taghvaeian, S.; Neale, C.M. Water balance of irrigated areas: A remote sensing approach. *Hydrol. Process.* **2011**, *25*, 4132–4141. [CrossRef]

6. Folhes, M.T.; Rennó, C.D.; Soares, J.V. Remote sensing for irrigation water management in the semi-arid Northeast of Brazil. *Agric. Water Manag.* **2009**, *96*, 1398–1408. [CrossRef]

7. Bastiaanssen, W.G.M.; Noordman, E.J.M.; Pelgrum, H.; Davids, G.; Thoreson, B.P.; Allen, R.G. SEBAL model with remotely sensed data to improve water-resources management under actual field conditions. *J. Irrig. Drain. Eng.* **2005**, *131*, 85–93. [CrossRef]

8. Anderson, M.C.; Allen, R.G.; Morse, A.; Kustas, W.P. Use of Landsat thermal imagery in monitoring evapotranspiration and managing water resources. *Remote Sens. Environ.* **2012**, *122*, 50–65. [CrossRef]

9. Yao, Y.; Liang, S.; Qin, Q.; Wang, K. Monitoring drought over the conterminous United States using MODIS and NCEP Reanalysis-2 data. *J. Appl. Meteorol. Climatol.* **2010**, *49*, 1665–1680. [CrossRef]

10. Anderson, M.C.; Hain, C.; Wardlow, B.; Pimstein, A.; Mecikalski, J.R.; Kustas, W.P. Evaluation of drought indices based on thermal remote sensing of evapotranspiration over the continental United States. *J. Clim.* **2011**, *24*, 2025–2044. [CrossRef]

11. Zhang, J.; Mu, Q.; Huang, J. Assessing the remotely sensed Drought Severity Index for agricultural drought monitoring and impact analysis in North China. *Ecol. Indic.* **2016**, *63*, 296–309. [CrossRef]

12. Moorhead, J.E.; Gowda, P.H.; Singh, V.P.; Porter, D.O.; Marek, T.H.; Howell, T.A.; Stewart, B.A. Identifying and evaluating a suitable index for agricultural drought monitoring in the Texas high plains. *J. Am. Water Resour. Assoc.* **2015**, *51*, 807–820. [CrossRef]

13. Li, H.; Zheng, L.; Lei, Y.; Li, C.; Liu, Z.; Zhang, S. Estimation of water consumption and crop water productivity of winter wheat in North China Plain using remote sensing technology. *Agric. Water Manag.* **2008**, *95*, 1271–1278. [CrossRef]

14. Ahmad, M.U.D.; Turral, H.; Nazeer, A. Diagnosing irrigation performance and water productivity through satellite remote sensing and secondary data in a large irrigation system of Pakistan. *Agric. Water Manag.* **2009**, *96*, 551–564. [CrossRef]

15. Teixeira, A.D.C.; Bastiaanssen, W.G.M.; Ahmad, M.U.D.; Bos, M.G. Reviewing SEBAL input parameters for assessing evapotranspiration and water productivity for the Low-Middle Sao Francisco River basin, Brazil: Part A: Calibration and validation. *Agric. For. Meteorol.* **2009**, *149*, 462–476. [CrossRef]

16. Cai, X.L.; Sharma, B.R. Integrating remote sensing, census and weather data for an assessment of rice yield, water consumption and water productivity in the Indo-Gangetic river basin. *Agric. Water Manag.* **2010**, *97*, 309–316. [CrossRef]

17. Anderson, M.C.; Zolin, C.A.; Sentelhas, P.C.; Hain, C.R.; Semmens, K.; Yilmaz, M.T.; Gao, F.; Otkin, J.A.; Tetrault, R. The Evaporative Stress Index as an indicator of agricultural drought in Brazil: An assessment based on crop yield impacts. *Remote Sens. Environ.* **2016**, *174*, 82–99. [CrossRef]

18. Oberg, J.W.; Melesss, A.M. Evapotranspiration dynamics at an ecohydrological restoration site: An energy balance and remote sensing approach. *J. Am. Water Resour. Assoc.* **2006**, *42*, 565–582. [CrossRef]

19. Nepstad, D.; Lefebvre, P.; Lopes da Silva, U.; Tomasella, J.; Schlesinger, P.; Solorzano, L.; Moutinho, P.; Ray, D.; Guerreira Benito, J. Amazon drought and its implications for forest flammability and tree growth: A basin-wide analysis. *Glob. Chang. Biol.* **2004**, *10*, 704–717. [CrossRef]

20. Bawazir, A.S.; Samani, Z.; Bleiweiss, M.; Skaggs, R.; Schmugge, T. Using ASTER satellite data to calculate riparian evapotranspiration in the Middle Rio Grande, New Mexico. *Int. J. Remote Sens.* **2009**, *30*, 5593–5603. [CrossRef]

21. Taghvaeian, S.; Neale, C.M.; Osterberg, J.; Sritharan, S.I.; Watts, D.R. Water use and stream-aquifer-phreatophyte interaction along a Tamarisk-dominated segment of the Lower Colorado River. In *Remote Sensing of the Terrestrial Water Cycle*; John & Sons, Inc.: Hoboken, NJ, USA, 2014; pp. 95–113.

22. Nagler, P.L.; Glenn, E.P.; Nguyen, U.; Scott, R.L.; Doody, T. Estimating riparian and agricultural actual evapotranspiration by reference evapotranspiration and MODIS enhanced vegetation index. *Remote Sens.* **2013**, *5*, 3849–3871. [CrossRef]

23. Khand, K.; Taghvaeian, S.; Hassan-Esfahani, L. Mapping Annual Riparian Water Use Based on the Single-Satellite-Scene Approach. *Remote Sens.* **2017**, *9*, 832. [CrossRef]

24. Chen, J.M.; Chen, X.; Ju, W.; Geng, X. Distributed hydrological model for mapping evapotranspiration using remote sensing inputs. *J. Hydrol.* **2005**, *305*, 15–39. [CrossRef]

25. Immerzeel, W.W.; Gaur, A.; Zwart, S.J. Integrating remote sensing and a process-based hydrological model to evaluate water use and productivity in a south Indian catchment. *Agric. Water Manag.* **2008**, *95*, 11–24. [CrossRef]

26. Herman, M.R.; Nejadhashemi, A.P.; Abouali, M.; Hernandez-Suarez, J.S.; Daneshvar, F.; Zhang, Z.; Anderson, M.C.; Sadeghi, A.M.; Hain, C.R.; Sharifi, A. Evaluating the role of evapotranspiration remote sensing data in improving hydrological modeling predictability. *J. Hydrol.* **2018**, *556*, 39–49. [CrossRef]

27. Vinukollu, R.K.; Wood, E.F.; Ferguson, C.R.; Fisher, J.B. Global estimates of evapotranspiration for climate studies using multi-sensor remote sensing data: Evaluation of three process-based approaches. *Remote Sens. Environ.* **2011**, *115*, 801–823. [CrossRef]

28. Gowda, P.H.; Chavez, J.L.; Colaizzi, P.D.; Evett, S.R.; Howell, T.A.; Tolk, J.A. Remote sensing based energy balance algorithms for mapping ET: Current status and future challenges. *Trans. ASABE* **2007**, *50*, 1639–1644. [CrossRef]

29. Kalma, J.D.; McVicar, T.R.; McCabe, M.F. Estimating land surface evaporation: A review of methods using remotely sensed surface temperature data. *Surv. Geophys.* **2008**, *29*, 421–469. [CrossRef]

30. Li, Z.L.; Tang, R.; Wan, Z.; Bi, Y.; Zhou, C.; Tang, B.; Yan, G.; Zhang, X. A review of current methodologies for regional evapotranspiration estimation from remotely sensed data. *Sensors* **2009**, *9*, 3801–3853. [CrossRef] [PubMed]

31. Liou, Y.A.; Kar, S.K. Evapotranspiration estimation with remote sensing and various surface energy balance algorithms—A review. *Energies* **2014**, *7*, 2821–2849. [CrossRef]

32. Menenti, M.; Choudhury, B. Parameterization of land surface evaporation by means of location dependent potential evaporation and surface temperature range. *Proc. IAHS Conf. Land Surf. Process.* **1993**, *212*, 561–568.

33. Norman, J.M.; Becker, F. Terminology in thermal infrared remote sensing of natural surfaces. *Agric. For. Meteorol.* **1995**, *77*, 153–166. [CrossRef]

34. Kustas, W.P.; Norman, J.M. Evaluation of soil and vegetation heat flux predictions using a simple two-source model with radiometric temperatures for partial canopy cover. *Agric. For. Meteorol.* **1999**, *94*, 13–29. [CrossRef]

35. Bastiaanssen, W.G.M.; Menenti, M.; Feddes, R.A.; Holtslag, A.A.M. A remote sensing surface energy balance algorithm for land (SEBAL): 1. Formulation. *J. Hydrol.* **1998**, *212*, 198–212. [CrossRef]

36. Roerink, G.; Su, Z.; Menenti, M. S-SEBI: A simple remote sensing algorithm to estimate the surface energy balance. *Phys. Chem. Earth Part B Hydrol. Oceans Atmos.* **2000**, *25*, 147–157. [CrossRef]

37. Su, Z. The surface energy balance system (SEBS) for estimation of turbulent heat fluxes. *Hydrol. Earth Syst. Sci.* **2002**, *6*, 85–99. [CrossRef]

38. Allen, R.G.; Tasumi, M.; Trezza, R. Satellite-based energy balance for mapping evapotranspiration with internalized calibration (METRIC)-model. *J. Irrig. Drain. Eng.* **2007**, *133*, 380–394. [CrossRef]

39. Anderson, M.C.; Norman, J.M.; Mecikalski, J.R.; Otkin, J.A.; Kustas, W.P. A climatological study of evapotranspiration and moisture stress across the continental United States based on thermal remote sensing: 1. Model formulation. *J. Geophys. Res. Atmos.* **2007**, *112*, D11112. [CrossRef]

40. Samani, Z.; Bawazir, A.S.; Skaggs, R.K.; Bleiweiss, M.P.; Piñon, A.; Tran, V. Water use by agricultural crops and riparian vegetation: An application of remote sensing technology. *J. Contemp. Water Res. Educ.* **2007**, *137*, 8–13. [CrossRef]

41. Elhaddad, A.; Garcia, L.A. Surface energy balance-based model for estimating evapotranspiration taking into account spatial variability in weather. *J. Irrig. Drain. Eng.* **2008**, *134*, 681–689. [CrossRef]

42. Senay, G.B.; Bohms, S.; Singh, R.K.; Gowda, P.H.; Velpuri, N.M.; Alemu, H.; Verdin, J.P. Operational evapotranspiration mapping using remote sensing and weather datasets: A new parameterization for the SSEB approach. *J. Am. Water Resour. Assoc.* **2013**, *49*, 577–591. [CrossRef]

43. Yang, Y.; Long, D.; Shang, S. Remote estimation of terrestrial evapotranspiration without using meteorological data. *Geophys. Res. Lett.* **2013**, *40*, 3026–3030. [CrossRef]

44. Timmermans, W.J.; Kustas, W.P.; Anderson, M.C.; French, A.N. An intercomparison of the surface energy balance algorithm for land (SEBAL) and the two-source energy balance (TSEB) modeling schemes. *Remote Sens. Environ.* **2007**, *108*, 369–384. [CrossRef]

45. Allen, R.G.; Pereira, L.S.; Howell, T.A.; Jensen, M.E. Evapotranspiration information reporting: I. Factors governing measurement accuracy. *Agric. Water Manag.* **2011**, *98*, 899–920. [CrossRef]

46. Fisher, J.B.; Whittaker, R.J.; Malhi, Y. ET come home: Potential evapotranspiration in geographical ecology. *Glob. Ecol. Biogeogr.* **2011**, *20*, 1–18. [CrossRef]
47. Allen, R.G. Assessing integrity of weather data for reference evapotranspiration estimation. *J. Irrig. Drain. Eng.* **1996**, *122*, 97–106. [CrossRef]
48. Environmental and Water Resources Institute for the American Society of Civil Engineers (ASCE-EWRI). *The ASCE Standardized Reference Evapotranspiration Equation*; Report of the ASCE-EWRI Task Committee on Standardization of Reference Evapotranspiration; ASCE: Reston, VA, USA, 2005.
49. Webster, E.; Ramp, D.; Kingsford, R.T. Spatial sensitivity of surface energy balance algorithms to meteorological data in a heterogeneous environment. *Remote Sens. Environ.* **2016**, *187*, 294–319. [CrossRef]
50. Senay, G.B.; Friedrichs, M.; Singh, R.K.; Velpuri, N.M. Evaluating Landsat 8 evapotranspiration for water use mapping in the Colorado River Basin. *Remote Sens. Environ.* **2016**, *185*, 171–185. [CrossRef]
51. Biggs, T.W.; Marshall, M.; Messina, A. Mapping daily and seasonal evapotranspiration from irrigated crops using global climate grids and satellite imagery: Automation and methods comparison. *Water Resour. Res.* **2016**, *52*, 7311–7326. [CrossRef]
52. Moorhead, J.; Gowda, P.; Hobbins, M.; Senay, G.; Paul, G.; Marek, T.; Porter, D. Accuracy assessment of NOAA gridded daily reference evapotranspiration for the Texas High Plains. *J. Am. Water Resour. Assoc.* **2015**, *51*, 1262–1271. [CrossRef]
53. Porter, D.; Gowda, P.; Marek, T.; Howell, T.; Moorhead, J.; Irmak, S. Sensitivity of grass-and alfalfa-reference evapotranspiration to weather station sensor accuracy. *Appl. Eng. Agric.* **2012**, *28*, 543–549. [CrossRef]
54. Elhaddad, A.; Garcia, L.A. ReSET-Raster: Surface energy balance model for calculating evapotranspiration using a raster approach. *J. Irrig. Drain. Eng.* **2011**, *137*, 203–210. [CrossRef]
55. Kustas, W.P.; Humes, K.S.; Norman, J.M.; Moran, M.S. Single-and dual-source modeling of surface energy fluxes with radiometric surface temperature. *J. Appl. Meteorol.* **1996**, *35*, 110–121. [CrossRef]
56. Long, D.; Singh, V.P. A modified surface energy balance algorithm for land (M-SEBAL) based on a trapezoidal framework. *Water Resour. Res.* **2012**, *48*, W02528. [CrossRef]
57. Kjaersgaard, J.H.; Allen, R.G.; Garcia, M.; Kramber, W.; Trezza, R. Automated selection of anchor pixels for Landsat based evapotranspiration estimation. In Proceedings of the World Environmental and Water Resources Congress, Kansas City, MO, USA, 17–21 May 2009.
58. Allen, R.G.; Burnett, B.; Kramber, W.; Huntington, J.; Kjaersgaard, J.; Kilic, A.; Kelly, C.; Trezza, R. Automated calibration of the METRIC-Landsat evapotranspiration process. *J. Am. Water Resour. Assoc.* **2013**, *49*, 563–576. [CrossRef]
59. Bhattarai, N.; Quackenbush, L.J.; Im, J.; Shaw, S.B. A new optimized algorithm for automating endmember pixel selection in the SEBAL and METRIC models. *Remote Sens. Environ.* **2017**, *196*, 178–192. [CrossRef]
60. Trezza, R.; Allen, R.G.; Tasumi, M. Estimation of actual evapotranspiration along the Middle Rio Grande of New Mexico using MODIS and Landsat imagery with the METRIC model. *Remote Sens.* **2013**, *5*, 5397–5423. [CrossRef]
61. Khand, K.; Numata, I.; Kjaersgaard, J.; Vourlitis, G.L. Dry Season Evapotranspiration Dynamics over Human-Impacted Landscapes in the Southern Amazon Using the Landsat-Based METRIC Model. *Remote Sens.* **2017**, *9*, 706. [CrossRef]
62. Brutsaert, W.; Sugita, M. Application of self-preservation in the diurnal evolution of the surface energy budget to determine daily evaporation. *J. Geophys. Res. Atmos.* **1992**, *97*, 18377–18382. [CrossRef]
63. Kustas, W.P.; Perry, E.M.; Doraiswamy, P.C.; Moran, M.S. Using satellite remote sensing to extrapolate evapotranspiration estimates in time and space over a semiarid rangeland basin. *Remote Sens. Environ.* **1994**, *49*, 275–286. [CrossRef]
64. Trezza, R. Evapotranspiration Using a Satellite-Based Surface Energy Balance with Standardized Ground Control. Ph.D. Thesis, Biological and Irrigation Engineering Department, Utah State University, Logan, UT, USA, 2002.
65. Gentine, P.; Entekhabi, D.; Chehbouni, A.; Boulet, G.; Duchemin, B. Analysis of evaporative fraction diurnal behaviour. *Agric. For. Meteorol.* **2007**, *143*, 13–29. [CrossRef]
66. Colaizzi, P.D.; Evett, S.R.; Howell, T.A.; Tolk, J.A. Comparison of five models to scale daily evapotranspiration from one-time-of-day measurements. *Trans. ASABE* **2006**, *49*, 1409–1417. [CrossRef]
67. Chávez, J.L.; Neale, C.M.; Prueger, J.H.; Kustas, W.P. Daily evapotranspiration estimates from extrapolating instantaneous airborne remote sensing ET values. *Irrig. Sci.* **2008**, *27*, 67–81. [CrossRef]

68. Delogu, E.; Boulet, G.; Olioso, A.; Coudert, B.; Chirouze, J.; Ceschia, E.; Le Dantec, V.; Marloie, O.; Chehbouni, G.; Lagouarde, J.P. Reconstruction of temporal variations of evapotranspiration using instantaneous estimates at the time of satellite overpass. *Hydrol. Earth Syst. Sci.* **2012**, *16*, 2995–3010. [CrossRef]

69. Kjaersgaard, J.; Allen, R.; Trezza, R.; Robison, C.; Oliveira, A.; Dhungel, R.; Kra, E. Filling satellite image cloud gaps to create complete images of evapotranspiration. In Proceedings of the Remote Sensing and Hydrology 2010 Symposium, Jackson Hole, WY, USA, 27–30 September 2010.

70. Singh, R.K.; Liu, S.; Tieszen, L.L.; Suyker, A.E.; Verma, S.B. Estimating seasonal evapotranspiration from temporal satellite images. *Irrig. Sci.* **2012**, *30*, 303–313. [CrossRef]

71. Kjaersgaard, J.; Allen, R.; Irmak, A. Improved methods for estimating monthly and growing season ET using METRIC applied to moderate resolution satellite imagery. *Hydrol. Process.* **2011**, *25*, 4028–4036. [CrossRef]

72. Khand, K.; Kjaersgaard, J.; Hay, C.; Jia, X. Estimating impacts of agricultural subsurface drainage on evapotranspiration using the Landsat imagery-based METRIC model. *Hydrology* **2017**, *4*, 49. [CrossRef]

73. Dhungel, R.; Allen, R.G.; Trezza, R.; Robison, C.W. Evapotranspiration between satellite overpasses: Methodology and case study in agricultural dominant semi-arid areas. *Meteorol. Appl.* **2016**, *23*, 714–730. [CrossRef]

74. Kottek, M.; Grieser, J.; Beck, C.; Rudolf, B.; Rubel, F. World map of the Köppen-Geiger climate classification updated. *Meteorol. Z.* **2006**, *15*, 259–263. [CrossRef]

75. Homer, C.; Dewitz, J.; Yang, L.; Jin, S.; Danielson, P.; Xian, G.; Coulston, J.; Herold, N.; Wickham, J.; Megown, K. Completion of the 2011 National Land Cover Database for the conterminous United States–representing a decade of land cover change information. *Photogramm. Eng. Remote Sens.* **2015**, *81*, 345–354.

76. Vermote, E.; Wolfe, R. *MOD09GA MODIS/Terra Surface Reflectance Daily L2G Global 1 km and 500 m SIN Grid V006*; NASA EOSDIS LP DAAC; NASA: Washington, DC, USA, 2015. [CrossRef]

77. Wan, Z.; Hook, S.; Hulley, G. *MOD11A1 MODIS/Terra Land Surface Temperature/Emissivity Daily L3 Global 1km SIN Grid V006*; NASA EOSDIS LP DAAC; NASA: Washington, DC, USA, 2015. [CrossRef]

78. Brock, F.V.; Crawford, K.C.; Elliott, R.L.; Cuperus, G.W.; Stadler, S.J.; Johnson, H.L.; Eilts, M.D. The Oklahoma Mesonet: A technical overview. *J. Atmos. Ocean. Technol.* **1995**, *12*, 5–19. [CrossRef]

79. McPherson, R.A.; Fiebrich, C.A.; Crawford, K.C.; Kilby, J.R.; Grimsley, D.L.; Martinez, J.E.; Basara, J.B.; Illston, B.G.; Morris, A.D.; Kloesel, K.A.; et al. Statewide monitoring of the mesoscale environment: A technical update on the Oklahoma Mesonet. *J. Atmos. Ocean. Technol.* **2007**, *24*, 301–321. [CrossRef]

80. Gowda, P.H.; Ennis, J.; Howell, T.A.; Marek, T.H.; Porter, D.O. The ASCE Standardized Equation-Based Bushland Reference ET Calculator. In Proceedings of the World Environmental and Water Resources Congress, Albuquerque, NM, USA, 20–24 May 2012.

81. WMO. Hydrology–From Measurement to Hydrological Information. In *Guide to Hydrological Practices*; WMO: Geneva, Switzerland, 2008; Volume I.

82. Liang, S. Narrowband to broadband conversions of land surface albedo I: Algorithms. *Remote Sens. Environ.* **2001**, *76*, 213–238. [CrossRef]

83. Brutsaert, W.H. *Evaporation into the Atmosphere*; D. Reidel: London, UK, 1982.

84. Liang, S. *Quantitative Remote Sensing of Land Surfaces*; John Wiley & Sons: Hoboken, NJ, USA, 2005; pp. 1–528.

85. Brutsaert, W. Aspect of bulk atmospheric boundary layer similarity under free-convective conditions. *Rev Geophys.* **1999**, *37*, 439–451. [CrossRef]

86. Monin, A.S.; Obukhov, A.M.F. Basic laws of turbulent mixing in the surface layer of the atmosphere. *Contrib. Geophys. Inst. Acad. Sci. USSR* **1954**, *24*, 163–187.

87. Beljaars, A.C.M.; Holtslag, A.A.M. Flux parameterization over land surfaces for atmospheric models. *J. Appl. Meteorol.* **1991**, *30*, 327–341. [CrossRef]

88. Gupta, R.K.; Prasad, T.S.; Vijayan, D. Estimation of roughness length and sensible heat flux from WiFS and NOAA AVHRR data. *Adv. Space Res.* **2002**, *29*, 33–38. [CrossRef]

89. Su, Z.; Schmugge, T.; Kustas, W.P.; Massman, W.J. An evaluation of two models for estimation of the roughness height for heat transfer between the land surface and the atmosphere. *J. Appl. Meteorol.* **2001**, *40*, 1933–1951. [CrossRef]

90. Choudhury, B.J.; Monteith, J.L. A four-layer model for the heat budget of homogeneous land surfaces. *Q. J. R. Meteorol. Soc.* **1988**, *114*, 373–398. [CrossRef]

91. Gowda, P.H.; Chavez, J.L.; Colaizzi, P.D.; Howell, T.A.; Schwartz, R.C.; Marek, T.H. Relationship between LAI and Landsat TM spectral vegetation indices in the Texas Panhandle. In Proceedings of the American Society of Agricultural and Biological Engineers Annual Meeting, Minneapolis, MI, USA, 17–20 June 2007.

92. Monteith, J.L. *Evaporation and the Environment: The State and Movement of Water in Living Organism*; XIXth Symposium; Cambridge University Press: Swansea, UK, 1965.

93. Monteith, J.L. Evaporation and surface temperature. *Q. J. R. Meteorol. Soc.* **1981**, *107*, 1–27. [CrossRef]

94. Fischer, M.L.; Torn, M.S.; Billesbach, D.P.; Doyle, G.; Northup, B.; Biraud, S.C. Carbon, water, and heat flux responses to experimental burning and drought in a tallgrass prairie. *Agric. For. Meteorol.* **2012**, *166*, 169–174. [CrossRef]

95. Billesbach, D.; Bradford, J.; Margaret, T. *AmeriFlux US-AR2 ARM USDA UNL OSU Woodward Switchgrass 2*; U.S. Department of Agriculture: Washington, DC, USA; University of Nebraska: Lincoln, NE, USA, 2015.

96. Twine, T.E.; Kustas, W.P.; Norman, J.M.; Cook, D.R.; Houser, P.R.; Meyers, T.P.; Prueger, J.H.; Starks, P.J.; Wesely, M.L. Correcting Eddy-Covariance Flux Underestimates over a Grassland. *Agric. For. Meteorol.* **2000**, *103*, 279–300. [CrossRef]

97. Mu, Q.; Heinsch, F.A.; Zhao, M.; Running, S.W. Development of a global evapotranspiration algorithm based on MODIS and global meteorology data. *Remote Sens. Environ.* **2007**, *111*, 519–536. [CrossRef]

98. Mu, Q.; Zhao, M.; Running, S.W. Improvements to a MODIS global terrestrial evapotranspiration algorithm. *Remote Sens. Environ.* **2011**, *115*, 1781–1800. [CrossRef]

99. Gokmen, M.; Vekerdy, Z.; Verhoef, A.; Verhoef, W.; Batelaan, O.; Van der Tol, C. Integration of soil moisture in SEBS for improving evapotranspiration estimation under water stress conditions. *Remote Sens. Environ.* **2012**, *121*, 261–274. [CrossRef]

100. Gibson, L.A. The Application of the Surface Energy Balance System Model to Estimate Evapotranspiration in South Africa. Ph.D. Thesis, University of Cape Town, Cape Town, South Africa, 2013.

101. Paul, G.; Gowda, P.H.; Prasad, P.V.; Howell, T.A.; Aiken, R.M.; Neale, C.M. Investigating the influence of roughness length for heat transport (zoh) on the performance of SEBAL in semi-arid irrigated and dryland agricultural systems. *J. Hydrol.* **2014**, *509*, 231–244. [CrossRef]

102. Bhattarai, N.; Mallick, K.; Brunsell, N.A.; Sun, G.; Jain, M. Regional evapotranspiration from image-based implementation of the Surface Temperature Initiated Closure (STIC1. 2) model and its validation across an aridity gradient in the conterminous United States. *Hydrol. Earth Syst. Sci.* **2018**, *22*, 2311–2341. [CrossRef]

103. Khan, I.S.; Hong, Y.; Vieux, B.; Liu, W. Development and evaluation of an actual evapotranspiration estimation algorithm using satellite remote sensing and meteorological observational network in Oklahoma. *Int. J. Remote Sens.* **2010**, *31*, 3799–3819. [CrossRef]

104. Liaqat, U.W.; Choi, M. Accuracy comparison of remotely sensed evapotranspiration products and their associated water stress footprints under different land cover types in Korean peninsula. *J. Clean. Prod.* **2017**, *155*, 93–104. [CrossRef]

105. Yang, Z.; Zhang, Q.; Yang, Y.; Hao, X.; Zhang, H. Evaluation of evapotranspiration models over semi-arid and semi-humid areas of China. *Hydrol. Process.* **2016**, *30*, 4292–4313. [CrossRef]

106. Li, Y.; Huang, C.; Hou, J.; Gu, J.; Zhu, G.; Li, X. Mapping daily evapotranspiration based on spatiotemporal fusion of ASTER and MODIS images over irrigated agricultural areas in the Heihe River Basin, Northwest China. *Agric. For. Meteorol.* **2017**, *244*, 82–97. [CrossRef]

107. Huang, C.; Li, Y.; Gu, J.; Lu, L.; Li, X. Improving estimation of evapotranspiration under water-limited conditions based on SEBS and MODIS data in arid regions. *Remote Sens.* **2015**, *7*, 16795–16814. [CrossRef]

108. Kustas, W.P.; Norman, J.M. Evaluating the effects of subpixel heterogeneity on pixel average fluxes. *Remote Sens. Environ.* **2000**, *74*, 327–342. [CrossRef]

109. Van der Kwast, J.; Timmermans, W.; Gieske, A.; Su, Z.; Olioso, A.; Jia, L.; Elbers, J.; Karssenberg, D.; de Jong, S. Evaluation of the Surface Energy Balance System (SEBS) applied to ASTER imagery with flux-measurements at the SPARC 2004 site (Barrax, Spain). *Hydrol. Earth Syst. Sci. Discuss.* **2009**, *6*, 1165–1196. [CrossRef]

110. Liaqat, U.W.; Choi, M.; Awan, U.K. Spatio-temporal distribution of actual evapotranspiration in the Indus Basin Irrigation System. *Hydrol. Process.* **2015**, *29*, 2613–2627. [CrossRef]

111. Wang, Y.; Li, X.; Tang, S. Validation of the SEBS-derived sensible heat for FY3A/VIRR and TERRA/MODIS over an alpine grass region using LAS measurements. *Int. J. Appl. Earth Obs. Geoinf.* **2013**, *23*, 226–233. [CrossRef]

112. Gibson, L.A.; Münch, Z.; Engelbrecht, J. Particular uncertainties encountered in using a pre-packaged SEBS model to derive evapotranspiration in a heterogeneous study area in South Africa. *Hydrol. Earth Syst. Sci.* **2011**, *15*, 295–310. [CrossRef]
113. Su, H.; Wood, E.F.; McCabe, M.F.; Su, Z. Evaluation of remotely sensed evapotranspiration over the CEOP EOP-1 reference sites. *J. Meteorol. Soc. Jpn.* **2007**, *85*, 439–459. [CrossRef]
114. Hu, G.; Jia, L.; Menenti, M. Comparison of MOD16 and LSA-SAF MSG evapotranspiration products over Europe for 2011. *Remote Sens. Environ.* **2015**, *156*, 510–526. [CrossRef]
115. Feng, X.M.; Sun, G.; Fu, B.J.; Su, C.H.; Liu, Y.; Lamparski, H. Regional effects of vegetation restoration on water yield across the Loess Plateau, China. *Hydrol. Earth Syst. Sci.* **2012**, *16*, 2617–2628. [CrossRef]
116. Svoboda, M.; LeComte, D.; Hayes, M.; Heim, R.; Gleason, K.; Angel, J.; Rippey, B.; Tinker, R.; Palecki, M.; Stooksbury, D.; et al. The drought monitor. *Bull. Am. Meteorol. Soc.* **2002**, *83*, 1181–1190. [CrossRef]

Article

An Improved Spatio-Temporal Adaptive Data Fusion Algorithm for Evapotranspiration Mapping

Tong Wang [1,2], Ronglin Tang [1,2,*], Zhao-Liang Li [3,4], Yazhen Jiang [2,3], Meng Liu [1,2] and Lu Niu [1,2]

[1]   State Key Laboratory of Resources and Environment Information System, Institute of Geographic Sciences and Natural Resources Research, Chinese Academy of Sciences, Beijing 100101, China; tongwang_whu@163.com (T.W.); xinyin_liumeng@163.com (M.L.); niul.17s@igsnrr.an.cn (L.N.)

[2]   College of Resources and Environment, University of Chinese Academy of Sciences, Beijing 100049, China; jiangyazhen15@mails.ucas.ac.cn

[3]   ICube, UdS, CNRS; 300 Boulevard Sebastien Brant, CS10413, 67412 Illkirch, France; lizl@igsnrr.ac.cn

[4]   Key Laboratory of Agricultural Remote Sensing, Ministry of Agriculture/Institute of Agricultural Resources and Regional Planning, Chinese Academy of Agricultural Sciences, Beijing 100081, China

*   Correspondence: tangrl@lreis.ac.cn; Tel.: +86-106-488-8172

Received: 25 February 2019; Accepted: 25 March 2019; Published: 29 March 2019

Abstract: Continuous high spatio-temporal resolution monitoring of evapotranspiration (ET) is critical for water resource management and the quantification of irrigation water efficiency at both global and local scales. However, available remote sensing satellites cannot generally provide ET data at both high spatial and temporal resolutions. Data fusion methods have been widely applied to estimate ET at a high spatio-temporal resolution. Nevertheless, most fusion methods applied to ET are initially used to integrate land surface reflectance, the spectral index and land surface temperature, and few studies completely consider the influencing factor of ET. To overcome this limitation, this paper presents an improved ET fusion method, namely, the spatio-temporal adaptive data fusion algorithm for evapotranspiration mapping (SADFAET), by introducing critical surface temperature (the corresponding temperature to decide soil moisture), importing the weights of surface ET-indicative similarity (the influencing factor of ET, which is estimated from remote sensing data) and modifying the spectral similarity (the differences in spectral characteristics of different spatial resolution images) for the enhanced spatial and temporal adaptive reflectance fusion model (ESTARFM). We fused daily Moderate Resolution Imaging Spectroradiometer (MODIS) and periodic Landsat 8 ET data in the SADFAET for the experimental area downstream of the Heihe River basin from April to October 2015. The validation results, based on ground-based ET measurements, indicated that the SADFAET could successfully fuse MODIS and Landsat 8 ET data (mean percent error: -5%), with a root mean square error of 45.7 W/m^2, whereas the ESTARFM performed slightly worse, with a root mean square error of 50.6 W/m^2. The more physically explainable SADFAET could be a better alternative to the ESTARFM for producing ET at a high spatio-temporal resolution.

Keywords: evapotranspiration; fusion; multi-source satellite data; Landsat 8; MODIS; SADFAET

---

1. Introduction

Evapotranspiration (ET), including soil evaporation and vegetation transpiration, is defined as the movement of water from the land surface into air and continuously acquiring ET at a high spatio-temporal resolution at field or sub-field scales is of critical significance for agricultural and hydrological cycle modelling, irrigation water efficiency quantification and agricultural water resource management [1]. Satellite remote sensing has already been considered a reliable and efficient tool for monitoring spatially distributed ET over large zones. However, a single satellite cannot provide ET at a high spatio-temporal resolution due to the trade-off between the spatial and temporal resolutions of thermal bands in current satellite sensors. For instance, Landsat series imagery can be used to map ET at a relatively high spatial resolution, while ET dynamics cannot be observed, due to the 16-day revisit interval of a single Landsat platform. In contrast, moderate-resolution sensors, such as the Moderate Resolution Imaging Spectroradiometer (MODIS), provide imagery for ET estimation on a daily basis but the relatively coarse spatial resolution hinders the application of ET for the quantification of irrigation water efficiency and water resource management at field, local or basin scales.

To overcome this limitation, previous studies have proposed several methods, mainly including traditional downscaling methods and data fusion methods [2]. Downscaling methods comprise a scaling process of converting coarse spatial resolution images into finer spatial resolution images; however, these methods cannot simultaneously enhance the temporal resolution of the sensor [3] and have rarely been used in recent years [4–6]. Data fusion methods use two or more images to obtain fine spatial resolution images; thus, they can simultaneously improve the spatial resolution and temporal coverage. To date, several data fusion methodologies, which were originally developed to fuse land surface reflectance and spectral index data, have been utilized to attempt the estimation of ET at a fine spatio-temporal resolution from ET at a high spatial resolution and high temporal resolution [7–24]. These data fusion methods can be divided into two categories: (1) fuse intermediate variables to estimate ET and (2) fuse ET data. For the first category, ET at a high spatio-temporal resolution is estimated by different ET models using intermediate variables that are closely related to ET, such as the reference ET fraction (ET_rF), normalized differential vegetation index (NDVI) and land surface temperature (LST). These intermediate variables can be fused using a regression model [12,25] or different spatio-temporal data fusion models [23,26,27]. For the second category, different spatio-temporal data fusion models are directly applied to fuse ET. For example, Ke et al. [26] and Ma et al. [28] applied the spatial and temporal adaptive reflectance fusion model (STARFM) and the enhanced spatial and temporal adaptive reflectance fusion model (ESTARFM), respectively, to fuse multi-source ET data, while a research group at the U.S. Department of Agriculture (USDA) used the STARFM to fuse ET at the GEOS Imager-derived 3 km–10 km, MODIS 1 km and Landsat 30 m scales from the Atmosphere-Land Exchange Inverse (ALEXI) model and the associated flux disaggregation model (DisALEXI) [7–11,13,18,29,30]. For either the first or second category, the most widely and successfully used spatio-temporal data fusion models are the STARFM and ESTARFM, respectively. However, it should be noted that existing methods also have the following limitations: most fusion methods applied to ET are initially used to integrate the land surface reflectance, spectral index and LST; thus, these methods cannot completely consider the influencing factor of ET including remote sensing and atmospheric characteristics [31] (especially some critical issues, such as soil moisture [32] and vegetation distribution) [33,34].

The objectives of this paper are twofold: (1) to develop a spatio-temporal adaptive data fusion algorithm for evapotranspiration mapping (SADFAET) for producing ET at a fine resolution and (2) to fuse ET using Landsat 8 and MODIS image data and validate the fused ET data with ground-based measured data collected downstream of the Heihe River basin. The SADFAET improves on the original ESTARFM algorithm by introducing the critical surface temperature (the corresponding temperature to decide soil moisture) to select similar pixels, importing the weight for surface ET-indicative similarity (the influencing factor of ET which is estimated from remote sensing data) and modifying the spectral similarity (the differences in spectral characteristics between different spatial resolution images) by incorporating the shortwave infrared bands. Section 2 presents the background and the methodology on how the SADFAET was derived. Section 3 describes the study site, the ground-based meteorological and energy flux measurements, satellite data and data preprocessing. Section 4 provides the results and discussion. The conclusions are finally made in Section 5.

2. Methods

2.1. A Brief Overview of the ESTARFM

The ESTARFM was developed to fuse the land surface reflectance by Zhu et al. [35], which was based on the STARFM [36] and the two methods are the most common remote sensing data fusion methods that obtain fine spatial resolution images on a specified date from coarse spatial resolution images on the same date and one or more additional fine-coarse resolution pair. These methods can accurately predict fine-resolution images for heterogeneous landscapes and are the most common remote sensing data fusion methods. In ESTARFM, unknown fine-resolution reflectance on the predicted date t_p can be fused using the known fine-resolution reflectance acquired on the reference date t_m (right before t_p) and t_n (right after t_p) together with the corresponding coarse-resolution reflectance on t_m, t_n and t_p by introducing similar pixels (i.e., spectrally similar and homogeneous neighboring pixels within the moving window), a conversion coefficient and a weighting coefficient. Although the ESTARFM and the STARFM were originally designed to produce reflectance data, they were also used to produce high spatial resolution NDVI, LST and ET data [7–9,13,18,26,28,37,38]. Similar to the ESTARFM, Weng et al. [3] proposed a spatio-temporal adaptive data fusion algorithm for temperature mapping (SADFAT), corresponding to deriving the ESTARFM by nonlinear methods, which was proven to accurately predict fine-resolution radiance products. Since both reflectance and radiance contribute to the variation in ET and the ESTARFM was proven to accurately fuse reflectance and radiance data, this study developed a spatio-temporal adaptive data fusion algorithm for evapotranspiration mapping (SADFAET) by introducing ET, which has a greater influence, into the ESTARFM.

2.2. Theoretical Basis of the SADFAET

The SADFAET, which improves on the original ESTARFM algorithm by introducing the critical surface temperature (T^*) the corresponding temperature when surface soil moisture availability in the upper soil layer for each pixel decreases to 0 and vegetation becomes soil water-stressed, which is defined in the work of Tang & Li [39] to decide soil moisture, importing the weight for surface ET-indicative similarity and modifying the spectral similarity by incorporating the shortwave infrared bands (representing soil moisture), is developed to fuse different spatial-temporal resolution ET products to estimate fine-resolution ET data. In the SADFAET, the calculation of fine-resolution ET is modified from that of the ESTARFM algorithm and the ET over a fine-resolution pixel at the prediction time can be calculated by the sum of the fine-resolution ET data at the reference time and coarse-resolution ET data at the reference and prediction times. Similar to the ESTARFM, the moving window is used to search similar pixels within the window and information of similar pixels is then integrated into fine-resolution ET calculation. Unknown fine-resolution ET data of the central pixel $(x_{w/2}, y_{w/2})$ on the predicted date t_p in the SADFAET can be computed using the known fine-resolution ET data acquired on the reference date t_m and t_n together with the corresponding coarse-resolution ET data on t_m, t_n and t_p:

$$ET_F{}^{w/2,w/2,t_p} = T_m \times ET_{F,m}{}^{w/2,w/2,t_p} + T_n \times ET_{F,n}{}^{w/2,w/2,t_p} \tag{1}$$

$$ET_{F,k}{}^{w/2,w/2,t_p} = ET_F{}^{w/2,w/2,t_k} + \Gamma(W_{i,j,t_k} \times V_{i,j,t_k} \times (ET_C{}^{i,j,t_p} - ET_C{}^{i,j,t_k})), (k = m, n) \tag{2}$$

$$T_k = \frac{1/\left|\Gamma(ET_C{}^{i,j,t_k} - ET_C{}^{i,j,t_p})\right|}{\Sigma_{k=m,n}\left(1/\left|\Gamma(ET_C{}^{i,j,t_k} ET_C{}^{i,j,t_p})\right|\right)} \tag{3}$$

where the subscript C represents the coarse-resolution pixel and the subscript F represents the fine-resolution pixel. w represents the side length of the moving window, which is used to search similar pixels and is determined by the spatial resolution of the input image. The co-ordinate location of the similar pixel is (i, j) and $(w/2, w/2)$ is the coordinate location of the central pixel. $W_{i,j,tk}$ represents the weight of the similar pixel computed from the surface ET-indicative similarity, the improved spectral similarity and the distance weight between the similar pixel and the central pixel. $V_{i,j,tk}$ represents the conversion coefficient of the similar pixel, which can be obtained by linear regression analysis for each similar pixel on the reference date t_m and t_n [35]. T_k represents the temporal weight. The operator $\Gamma(X)$ represents $\sum_{i=1}^{w} \sum_{j=1}^{w} (X)$.

There are 4 major steps, including the selection of similar neighboring pixels, the calculation of the weights of similar pixels, the calculation of the conversion coefficient and the calculation of the temporal weight for ET fusion. The SADFAET improves the selection of similar neighboring pixels and the calculation of the weight of similar pixels but follows the work of Zhu et al. [35] to determine the window size, temporal weight and conversion coefficient.

2.2.1. Selection of Similar Neighboring Pixels

According to the original ESTARFM models, two methods were used to obtain similar pixels, including the setting threshold and unsupervised classification. However, these two methods mainly consider spectral similarity rather than the ET similarity between the central pixel and other pixels within the search window, while the number of classes limits the automated fusion processing and reduces the accuracy of the similar pixel selection. Note that soil moisture is an important environmental factor that can significantly influence ET. In the SADFAET, the new method is therefore to take into account soil moisture in order to select similar neighboring pixels more reasonably by first introducing a critical surface temperature (T^*) and then judging how T^* and the remotely sensed surface temperature vary. The T^* corresponds to the surface temperature when the surface soil moisture availability in the upper soil layer decreases to 0 and the vegetation becomes soil water-stressed [39] and it can be estimated using the theoretical surface temperature at 2 hypothesized end-members, T_{sd} with no soil water availability at the upper layer and with zero evaporation and T_{vw} with well-watered vegetation with potential transpiration, from the following equations.

$$T_{sd} = \frac{r_{as}(R_{n,s} - G_s)}{\rho C_p} + T_a \tag{4}$$

$$T_{vw} = \frac{r_{av}R_{n,v}}{\rho C_p} \frac{\gamma(1 + r_{vw}/r_{av})}{\Delta + \gamma(1 + r_{vw}/r_{av})} - \frac{VPD}{\Delta + \gamma(1 + r_{vw}/r_{av})} + T_a \tag{5}$$

$$F_v = \left(\frac{NDVI - NDVI_{MIN}}{NDVI_{MAX} - NDVI_{MIN}}\right)^2 \tag{6}$$

$$T^* = [T_{sd}^4(1 - F_v) + T_{vw}^4 F_v]^{1/4} \tag{7}$$

where T_{sd} and the T_{vw} represent the theoretical surface temperature of the dry soil and well-watered vegetation, respectively. C_p represents the specific heat capacity at constant pressure (J/(m·K)). ρ represents the density of air (kg/m^3). Δ represents the slope of saturated vapor pressure versus air temperature (kPa/°C). VPD represents the vapor pressure deficit of the air (kPa). T_a represents the near-surface air temperature (K). γ represents the psychrometric (kPa/°C). r_{vw} represents the canopy resistances at the well-watered vegetation (s/m). r_{as} represents the aerodynamic resistance at dry soil (s/m).

When the actual LST (retrieved by remote sensing) for the pixel is lower than or equal to T^*, the vegetation component is considered to be potentially transpiring and the soil component is considered to be evaporating at a rate between zero and its maximum value (i.e., the potential rate); otherwise, no water is available to be evaporated for the soil component and the vegetation component transpires with a certain degree of soil water stress. Below are the procedures of the proposed method for selecting similar pixels.

(1). For each fine-resolution pixel at t_m and t_n, record T^* and the LST retrieved from remote sensing data (T_R).

(2). For a given pixel, if the remotely sensed T_R at a fine resolution at t_m is equal to or greater than T^* and T_R at t_n is equal to or greater than T^* as well, this pixel is considered to fall into CLASS 1, where no water is available to be evaporated for the soil component and the vegetation component transpires with a certain degree of soil water stress between t_m and t_n.

(3). If T_R at t_m is greater than T^* but T_R at t_n is less than T^* at the same time, this pixel is considered to fall into CLASS 2, where the surface soil moisture is increasing, and the vegetation transpiration is increasing to the potential transpiration amount between t_m and t_n.

(4). If T_R at t_m is less than T^* but T_R at t_n is greater than T^* at the same time, the pixel is considered to fall into CLASS 3, where the surface soil moisture decreases to zero while the vegetation component transpires from a maximum value (i.e., the potential transpiration) to a certain degree of soil water stress between t_m and t_n.

(5). If T_R at t_m is less than T^* and T_R at t_n is less than T^* as well, this pixel is considered to fall into CLASS 4, where the vegetation component transpires potentially, and the surface soil moisture is between zero and a maximum value between t_m and t_n.

(6). Finally, compare the class (CLASS 1 through CLASS 4) of the central pixel with the given neighboring pixel. If the two pixels fall into the same class, the given neighboring pixel is considered to be a similar neighboring pixel.

2.2.2. Calculation of the Weight of the Similar Pixel

After finishing the selection of a similar pixel, its weight (W) is calculated. However, the original W cannot completely represent the ET; thus, W_i is modified in this paper and the improved weight is computed from the spectral similarity, the surface ET-indicative similarity and the distance between the similar pixel and the central pixel in the SADFAET.

Calculations of the spectral similarity need to consider the effects of surface soil moisture and several indices derived from optical remote sensing observations have been proposed for measuring soil moisture [40], such as the Shortwave Infrared Water Stress Index ($SIWSI$) [41], the Shortwave-infrared Perpendicular Drought Index ($SPDI$) [42] and the Visible and Shortwave-infrared Drought Index ($VSDI$) [43]. We can see that all of these indices use shortwave infrared bands. However, in the ESTARFM and SADFAT, the spectral similarity is calculated using only shortwave and thermal infrared bands; thus, we make an improvement to introduce the shortwave infrared bands into the calculation of spectral similarity to indicate the changes in soil moisture and ET. Here, the spectral similarity ($S_{i,j,tk}$) is determined by the correlation coefficient of the spectral vector, including shortwave bands (representing vegetation cover), shortwave infrared bands (representing soil moisture) and thermal infrared bands (representing LST), between each similar pixel and its corresponding coarse-resolution pixel with the following equations:

$$S_{i,j,t_k} = \frac{E\left[F_{ijk} - E(F_{ijk}))(C_{ijk} - E(C_{ijk}))\right]}{\sqrt{D(F_{ijk})} \cdot \sqrt{D(C_{ijk})}} \tag{8}$$

$$F_{ijk} = \left\{ B_{S,F}{}^{i,j,t_m}, B_{S,F}{}^{i,j,t_m}, B_{TIR,F}{}^{i,j,t_m}, B_{S,F}{}^{i,j,t_n}, B_{SWIR,F}{}^{i,j,t_n}, B_{TIR,F}{}^{i,j,t_n} \right\} \tag{9}$$

$$C_{ijk} = \left\{ B_{S,C}{}^{i,j,t_m}, B_{S,C}{}^{i,j,t_m}, B_{TIR,C}{}^{i,j,t_m}, B_{S,C}{}^{i,j,t_n}, B_{SWIR,C}{}^{i,j,t_n}, B_{TIR,C}{}^{i,j,t_n} \right\} \tag{10}$$

where F_{ijk} and C_{ijk} represent the coarse and fine spatial resolution image spectral vectors respectively at t_k, including the reflectance of shortwave infrared bands (B_{SWIR}), shortwave bands (B_S) and the radiance of thermal infrared bands (B_{NIR}). For example, with the Landsat 8 data in this study, the B_S covers bands 2–5 (blue, green, red and NIR, respectively), B_{SWIR} covers bands 6 and 7 (SWIR 1 and 2, respectively) and the B_{NIR} covers bands 10 and 11 (TIRS 1 and 2, respectively). $E()$ represents the expected value and $D()$ represents the variance value. Combining SWIR information can introduce soil moisture into the fusion model and improve the accuracy of fused ET data.

The surface ET indicators include NDVI, LST, soil moisture and ET [44,45]. We can calculate the changes in surface ET indicators for both the central pixel and the selected similar pixel between t_m and t_n and determine the weights of different similar pixels to avoid the influence of the selected useless pixel. The weight for the surface ET-indicative similarity of the *ith* similar pixel ($WSC_{i,j,tk}$) can be calculated from the correlation coefficient of the surface ET indicator vector as follows:

$$WSC_{i,j,t_k} = \frac{E[SC_{ijk} - E(SC_{ijk}))(SC_{central} - E(SC_{central}))]}{\sqrt{D(SC_{ijk})} \cdot \sqrt{D(SC_{central})}} \tag{11}$$

$$SC_{ijk} = (NDVI_F^{i,j,t_k}, LST_F^{i,j,t_k}, SM_F^{i,j,t_k}, ET_F^{i,j,t_k}) \tag{12}$$

$$SC_{center} = (NDVI_F^{w/2,w/2,t_k}, LST_F^{w/2,w/2,t_k}, SM_F^{w/2,w/2,t_k}, ET_F^{w/2,w/2,t_k}) \tag{13}$$

where the subscripts "*central*" represent central pixels, respectively. *SM* represents the soil moisture. The weight of a similar pixel ($W_{i,j,tk}$) is calculated can be given as follows:

$$W_{i,j,t_k} = (1/D_{i,j,t_k})/\Gamma(1/D_{i,j,t_k}) \tag{14}$$

$$D_{i,j,t_k} = (1 - S_{i,j,t_k})(1 - WSC_{i,j,t_k}) \times d_{i,j,t_k} \tag{15}$$

$$d_{i,j,t_k} = 1 + \sqrt{(w/2 - i)^2 + (w/2 - j)^2}/(w/2) \tag{16}$$

where $S_{i,j,tk}$ represents the spectral similarity between fine- and coarse-resolution pixels for the similar pixels (i, j). $WSC_{i,j,tk}$ represents the weight of the surface ET-indicative similarity between the similar pixels (i, j) and central pixel $(w/2, w/2)$. $d_{i,j,tk}$ represents the geographic distance between the *ith* similar pixel and central pixel.

Figure 1 presents a schematic diagram of the SADFAET. This algorithm requires two pairs of fine- and coarse-resolution images on the same date and a set of coarse-resolution images for the prediction dates. Before implementing the SADFAET, the fine-resolution LST, NDVI, soil moisture and ET data at the reference time must be retrieved. There are 6 major steps in the SADFAET implementation. First, fine-resolution LST and T^* data are used to select similar neighboring pixels. Second, the fine- and coarse-resolution shortwave infrared bands, the shortwave bands and the radiance in the thermal infrared bands are used to calculate the spectral similarity. Third, surface ET-indicative similarity is estimated by fine-resolution LST, NDVI, soil moisture and ET data. Fourth, the geographic distance and temporal weight are calculated. Fifth, the conversion coefficient is determined by linear regression. Finally, the above weight, similarity and coefficient are used to calculate the ET at a fine resolution at the predicted time.

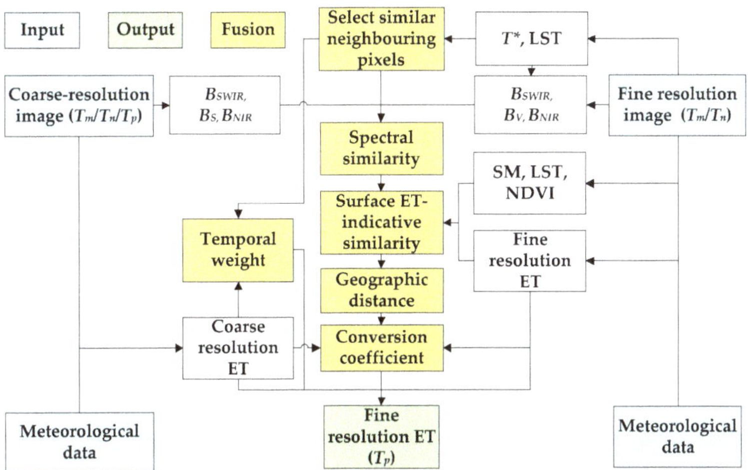

Figure 1. Schematic diagram of the spatio-temporal adaptive data fusion algorithm for evapotranspiration mapping (SADFAET).

2.3. End-Member-Based Soil and Vegetation Energy Partitioning Model

The end-member-based soil and vegetation energy partitioning (ESVEP) model [39] was developed for estimating soil and vegetation ET from remote sensing data by considering the differing responses of the soil water content in the upper surface layer to soil evaporation and those in the deeper root zone layer to vegetation transpiration. This model defines four hypothesized end-members, including dry soil (i.e., no soil water availability in the upper layer and zero evaporation), dry vegetation (i.e., no soil water availability and zero transpiration), wet soil (i.e., potential evaporation) and well-watered vegetation (i.e., potential transpiration). Based on the difference in vegetation height in different partially vegetated pixels, four end-member temperatures can be estimated pixel-by-pixel. The actual surface temperature for the partially vegetated pixel consequentially falls between the maximum and minimum of its hypothesized end-member temperature. When the vegetation temperature and energy are separated from their soil components, the ESVEP abides by two principles: (1) when soil evaporation is greater than zero, the vegetation transpiration equals the maximum (potential transpiration) and (2) when vegetation is soil-water stressed, the soil evaporation equals zero.

There are three major steps in the ESVEP model implementation. First, a physical algorithm [46] is used to estimate the divergence in the surface net radiation and soil heat flux. Second, the theoretical surface temperature at each of the four hypothesized end-members is calculated by the surface energy balance equation and the Penman–Monteith equation. Finally, when the actual LST (retrieved by remote sensing) for the pixel is lower than or equal to the critical surface temperature, as presented in Equation (7), the vegetation component is considered to be potentially transpiring and the soil component is considered to be evaporating at a rate between zero and the maximum value (i.e., the potential rate). The vegetation and soil component latent heat flux (LE_v and LE_s, respectively) under this condition are estimated as:

$$LE_v = LE_{vw} \tag{17}$$

$$LE_s = \frac{T_{sd} - T_s}{T_{sd} - T_{sw}} LE_{sw} \tag{18}$$

where T_s and T_{sw} represent the soil component temperature and theoretical surface temperature of the saturated soil, respectively. LE_{vw} and LE_{sw} represent the theoretical latent heat flux in the well-watered vegetation and saturated soil, respectively.

Otherwise, when no water is available to be evaporated in the soil component and the vegetation component transpires with a certain degree of soil water stress, the LE_v and LE_s are estimated as:

$$LE_v = \frac{T_{vd} - T_v}{T_{vd} - T_{vw}} LE_{vw} \tag{19}$$

$$LE_s = LE_{sd} = 0 \tag{20}$$

where T_v and T_{vd} represent the vegetation component temperature and the theoretical surface temperature of dry vegetation, respectively. LE_{sd} represents the theoretical LE of the dry soil.

The ESVEP model is demonstrated to be no more sensitive to meteorological, vegetation and remote sensing inputs than other ET models and has great potential for producing reasonably good surface energy fluxes. In our study, the fine- and coarse-resolution ET data at the reference time and the fine-resolution ET data at the prediction time were estimated by the ESVEP model [39].

2.4. Validation of the SADFAET

The SADFAET algorithm was applied to the actual Landsat 8 and MODIS images, which aided in understanding its accuracy and reliability. To avoid the impact of ET retrieval, the accuracy of the ET estimates from the ESVEP was first evaluated. The performance of the SADFAET was then assessed through retrieved ET from Landsat 8 data and ground-based measurements. The mean, maximum and minimum ET, standard deviation and histogram of differences of SADFAET and Landsat 8 ET were also used to validate the spatial patterns of fused ET. The fused ET was further compared with the ground-based eddy covariance (EC) measurement to validate the accuracy of the SADFAET based on the mean bias (MB), mean percent error (MPE) and root mean square error (RMSE). The MPE can be calculated as:

$$MPE = \frac{1}{N} \times \sum_{i=1}^{N} B_i / O_i \times 100\% \tag{21}$$

where B_i represents the bias of the *ith* datum. O_i represents the observation of the *ith* data. N represents the number of data.

3. Materials

3.1. Test Sites and Ground-Based Data

The study area is downstream of the Heihe River basin (LU: 42.0333°N, 101.1°E, RL: 41.9667°N, 101.1667°E), which is an endorheic basin located in the arid and semiarid regions of Northwest China. The annual mean temperature, relative humidity and wind speed are 9.4 °C, 33.7% and 3.2 m/s, respectively and the annual precipitation is approximately 50 mm. There are five ground sites, namely, Sidaoqiao, Populus euphratica, Mixed Forest, Barren land and Cropland, in the study area and the landscapes are comprised of Tamarix, populus euphratica, Populus euphratica and Tamarix, bare land and melon, respectively (Table 1 & Figure 2). Ground-based half-hourly average atmospheric variables (air temperature, wind speed, atmospheric pressure and relative humidity) and downward solar radiation, which were collected from April 2015 to October 2015 at the five ground sites, are used for this study and can be obtained from the Heihe Integrated Observatory Network (http://www.heihedata.org/) [47,48]. The EC system and a large aperture scintillometer (LAS) provided the measurements of turbulent fluxes downstream of the Heihe River basin, while the sample frequencies are 10 Hz (for EC) and 1 min (for LAS) and the heights are 3.5 m (for EC at the Cropland and Barren Land sites), 22 m (for EC at the Mixed Forest and Populus euphratica sites), 8 m (for EC at the Sidaoqiao site) and 22.5 m (for LAS). The LAS and EC flux data were processed and screened according to the criteria suggested in Liu et al. [47]. The instantaneous ET was validated by using 10-min flux measurements as ground-truth data.

Table 1. Attributes of the test sites in the study area (Sidaoqiao).

Ground Sites	Landscape	Longitude	Latitude	Elevation (m)	Observation Instrument
Sidaoqiao	tamarix	101.1374 E	42.0012 N	873	LAS/EC
Populus euphratica	populus euphratica	101.1239 E	41.9932 N	876	EC
Mixed Forest	populus euphratica and tamarix	101.1335 E	41.9903 N	874	EC
Barren Land	bare land	101.1326 E	41.9993 N	878	EC
Cropland	melon	101.1338 E	42.0048 N	875	EC

3.2. Satellite Data

Multi-source remote sensing data at different spatial and temporal resolutions were obtained from Landsat 8 and Terra MODIS (Table 2). The spatial resolutions of Landsat 8 and MODIS are 30 m and 500–1000 m, respectively, while the temporal resolutions are 8–16 days and 1 day, respectively. Fifteen predominantly clear scenes (day of year [doy]: 96, 103, 135, 144, 160, 176, 199, 215, 231, 240, 247, 256, 288, 295 and 304) of Landsat 8 data from Path 133, Row 31 and Path 134, Row 31 acquired for the period of April to October 2015 were collected from the United States Geological Survey (USGS) (https://landsat.usgs.gov/). The shortwave reflectance band and the thermal band of the 15 scenes were atmospherically corrected using the ENVI. The corresponding MODIS (Terra) datasets, including the 1 km calibrated radiance product (MOD021KM), geolocation product (MOD03), precipitable water product (MOD05_L2), surface reflectance product (MOD09GA), land surface temperature/emissivity product (MOD11A1) and vegetation index product (MOD13A2), were obtained from the Land Processes Distributed Active Archive Center (LP DAAC) (http://glovis.usgs.gov/).

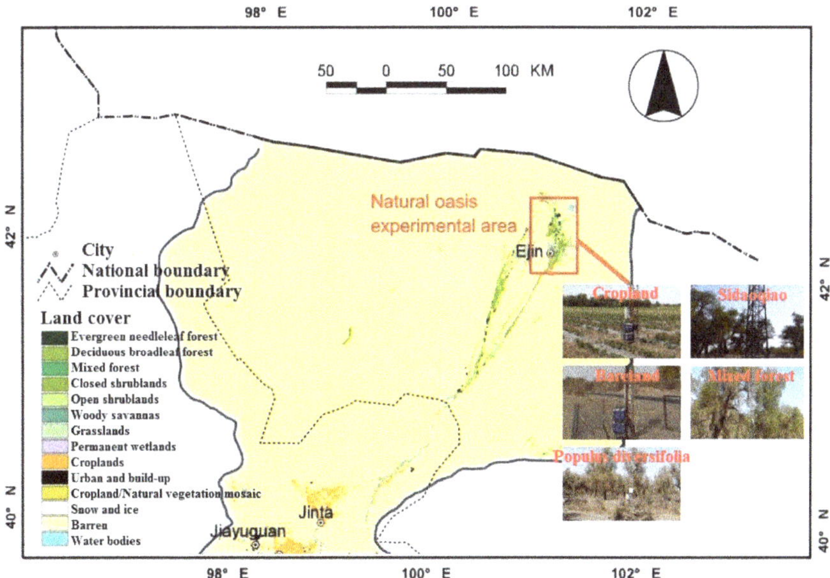

Figure 2. The location of the test area over the Heihe Watershed Allied Telemetry Experimental Research (HiWATER) experimental areas [49].

Table 2. Terra, Moderate Resolution Imaging Spectroradiometer (MODIS) and Landsat 8 data (April to October 2015) used in this study.

Data Type	MODIS (Horizontal 25, Vertical 04)		Landsat 8 (Path 133/134, Row 031)	
	MOD09GA	MOD021KM etc.	OLI	TIRS
Resolution	500 m	1000 m	30 m	30 m (resample)
Day of year (DOY)	096–304		096, 103, 135, 144, 160, 176, 199, 215, 231, 240, 247, 256, 288, 295, 304,	

Before the implementation of the SADFAET, both Landsat 8 (OLI/TIRS) and MODIS data need to be calibrated over the same coordinate system (Universal Transverse Mercator projection, UTM) and resampled at the same spatial resolution (30 m). MODIS Reprojection Tools (MRTs) were used to resample and re-project the MODIS data to the Landsat resolution and extent. Landsat 8 data were calibrated and atmospherically corrected for the shortwave reflectance band using the ENVI. Considering the landscape and land cover types of this area, the size of the search window was set as 30 Landsat 8 pixels.

Based on the 15 original Landsat-MODIS images, the dates with missing remote sensing or meteorological data were excluded and the fusion strategy is shown in Table 3.

Table 3. Fusion strategy: day of year (DOY) of the input data and results.

DOY of MODIS ET (1 km)	DOY of Landsat 8 ET (30 m)	DOY of Fusion Results and Validation (30 m)
96/103/135	96/135	103
103/135/144	103/144	135
135/144/176	135/176	144
144/176/199	144/199	176
176/199/231	176/231	199
199/231/240	199/240	231
231/240/247	231/247	240
240/247/256	240/256	247
247/256/288	247/288	256
256/288/295	256/295	288

The input parameters at a spatial resolution of 30 m for the ESVEP model and the SADFAET were derived from the Landsat 8 shortwave reflectance and thermal bands, while the input parameters at a spatial resolution of 1 km were derived from MODIS products. Specifically, for the 1 km spatial resolution data, MOD021KM was used to calculate the spectral similarity, MOD03, MOD09GA, MOD11A1 and MOD13A2 were used to estimate the coarse-resolution ET data and MOD05_L2 provided the atmospheric water vapor content. For 30 m spatial resolution data, the 30 m resolution NDVI was calculated using corrected OLI Band 4 and Band 5 reflectance data of Landsat 8. Soil moisture at a 30 m resolution was retrieved from OLI Band 7 of Landsat 8 and the NDVI using the OPtical TRApezoid Model (OPTRAM) [50]:

$$W = \frac{i_d + s_d NDVI - STR}{i_d - i_w + (s_d - s_w)NDVI} \tag{22}$$

where STR can be estimated by SWIR reflectance; i_d and s_d represent the intercept and slope of the dry edge in the $STR-NDVI$ space, respectively; and i_w and s_w represent the intercept and slope of the wet edge in the $STR-NDVI$ space, respectively. In this research, the dry edge and wet edge were determined by the method in Sadeghi et al. [50].

The LST at a 30 m resolution was retrieved from TIRS Band 10 of Landsat 8 using the widely applied mono-window algorithm [51]:

$$LST = \left[K_2(\varphi_1 + \varphi_2)T_{10} + (1 - \varphi_1 - \varphi_2)T_{10}^2 - K_2\varphi_2 T_a \right] / K_2\varphi_1 \tag{23}$$

$$\varphi_1 = \varepsilon_{10}\tau_{10}, \varphi_2 = (1 - \tau_{10})[1 + (1 - \varepsilon_{10})\tau_{10}] \tag{24}$$

where T_{10} represents the brightness temperature at Landsat 8 band 10. The surface emissivity (ε) is estimated following an NDVI threshold method [52]. τ represents the atmospheric transmittance. T_a represents the average atmospheric operating temperature, K. K_2 is a constant (1321.08).

The main input data sources used in the SADFAET are outlined in Table 4.

Table 4. Summary of the primary inputs into the SADFAET (Landsat and MODIS).

Variable	MODIS	Landsat 8
NDVI	MOD13A2	OLI Band 4, Band 5
LST	MOD11A1	TIRS Band 10, mono-window algorithm [51]
SM	-	OLI Band 7, OPTRAM [50]

Landsat 8 OLI bands 4 and 5 were used to estimate the NDVI, band 7 was used to estimate the soil moisture and band 10 was used to estimate the LST, while ET at a fine resolution, surface ET-indicative similarity and spectral similarity can be estimated by these parameters.

4. Results and Discussion

4.1. Validation of ET Estimated by the ESVEP Model

Accurate estimates of ET from fine- and coarse-resolution satellite data are a prerequisite for producing reliable fusion results. The accuracy of the ET estimates from the ESVEP was evaluated by validating the LE (latent heat flux) retrievals against the EC/LAS observations at the five test sites. At the Sidaoqiao site, the LAS measurement was used to validate the MODIS-based retrievals, while at the other sites, the EC measurement was used to validate the Landsat-based retrievals. Figure 3 and Table 5 show that the retrieved LEs agreed well with local measurements, with an overall low root mean square error (RMSE) of 40.9 W/m² at all sites and the average retrieved LEs were slightly lower than the average observations; thus, the retrieved instantaneous ET is reliable as the input parameter for data fusion.

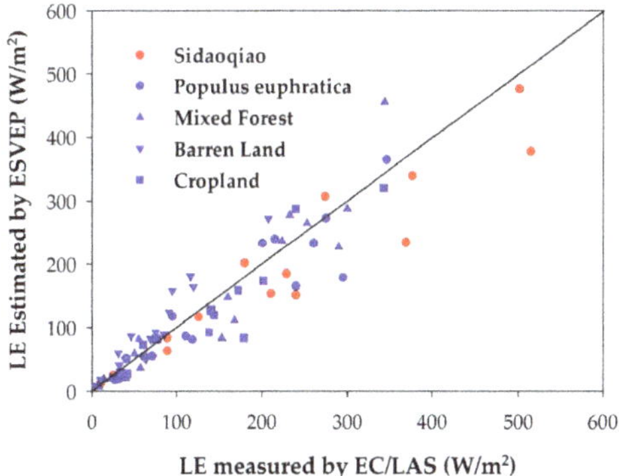

Figure 3. Validation of the instantaneous LE estimations from the end-member-based soil and vegetation energy partitioning (ESVEP) model for MODIS- and Landsat-based retrievals against ground-based LAS (red symbols) and EC (blue symbols) measurements.

Table 5. Error metrics of instantaneous LE estimates from MODIS and Landsat 8 data. Mean O: mean observation; Mean P: mean prediction; MB: mean bias; MPE: mean percent error; RMSE: root mean square error.

Ground Sites	Mean O (W/m^2)	Mean P (W/m^2)	MB (W/m^2)	MPE (%)	RMSE (W/m^2)
Sidaoqiao	217.5	189.7	−27.8	−16	42.7
Populus euphratica	160.7	144.7	−16.0	−12	40.4
Mixed Forest	155.4	154.2	−1.2	−3	44.8
Barren Land	70.5	86.7	16.2	22	24.6
Cropland	135.1	120.3	−14.8	−14	31.8
Average	147.8	139.1	−8.7	−5	40.9

Statistics are also provided in Table 5 for different land cover types. The MPEs of the Sidaoqiao, Populus euphratica, Mixed Forest, Barren Land, Cropland sites are −16%, −12%, −3%, 22% and −14%, respectively. Only the LE at the Bareland site was overestimated, while the LEs at the other sites were underestimated. The largest MPE (22%) appeared for the estimations of LE at the Barren Land site, which was likely because the fractional vegetation cover of the bare land was too low, resulting in large errors for the four end-members in the ESVEP.

4.2. Evaluation of the Spatial Pattern of ET Fused with the SADFAET Model

The performance of the SADFAET was first evaluated through the comparison of fused ET data with the retrieved ET data by the ESVEP model using the Landsat 8 data on the 10 selected fusion dates. Three typical pairs of the spatial pattern of fused ET and retrieved ET data over the study area on May 15, July 18 and August 28, together with a histogram of the ET difference, are illustrated in Figure 4. Table 6 illustrates the statistical measures of ET fused by the SADFAET and that retrieved using the Landsat 8 data over the typical days. It could be found that over the study area, the fused ET data were close to the retrieved ET data, with a smaller absolute value of the percent difference varying between 1.0% and 5.6%. The fused ET data also had a standard deviation (SD) magnitude similar to that of compared to the retrieved ET on each of the three typical days. The SD value over the study area varied between 16.2 W/m^2 and 48.2 W/m^2 for the fused ET and between 17.6 W/m^2 and 48.8 W/m^2 for the retrieved ET. The mean values over the study area varied between 30.1 W/m^2 and 74.6 W/m^2 for the fused ET and between 28.5 W/m^2 and 76.7 W/m^2 for the retrieved ET.

It is clear that the image fused by the SADFAET was similar to that predicted directly using the Landsat 8 image in terms of the overall spatial patterns of ET. The difference between SADFAET ET and Landsat 8 ET is mostly ±10 W/m^2. The fused ET images contained most of the spatial details found in the Landsat 8 ET images, including surface features, such as buildings, barren lands (with low ET values) and croplands (with high values).

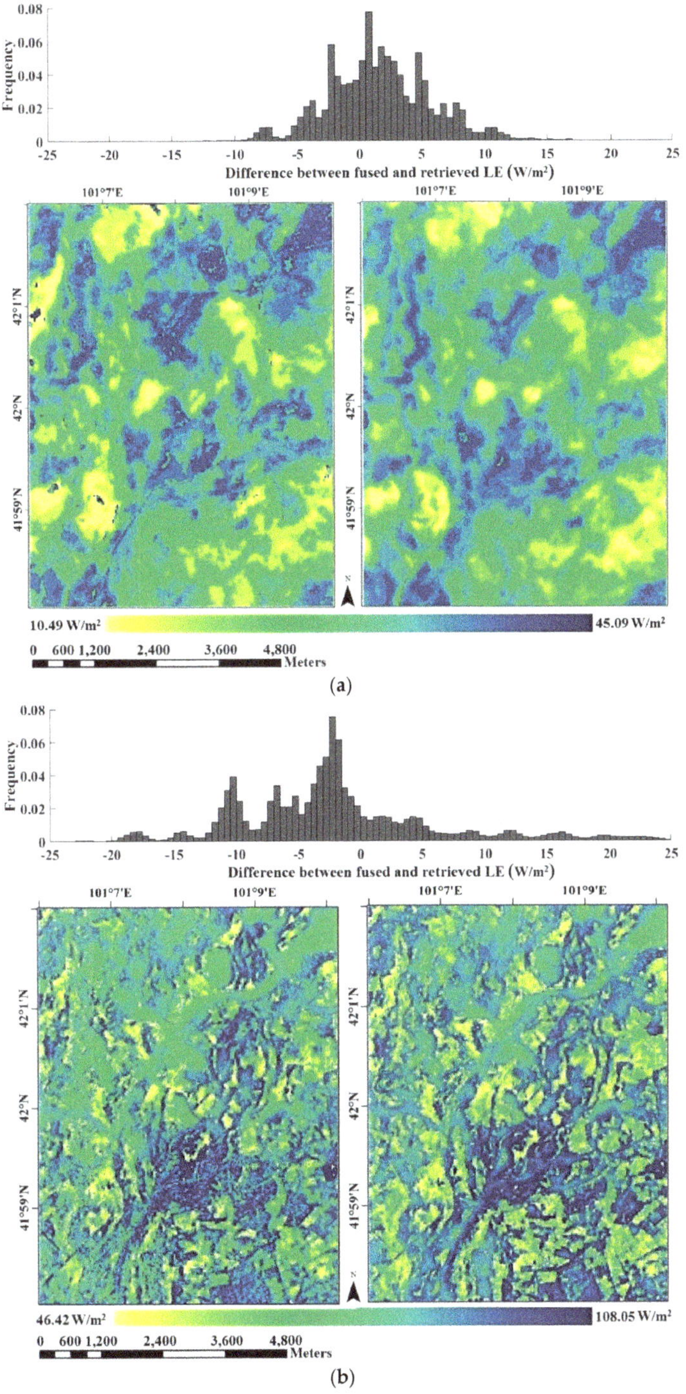

Figure 4. *Cont.*

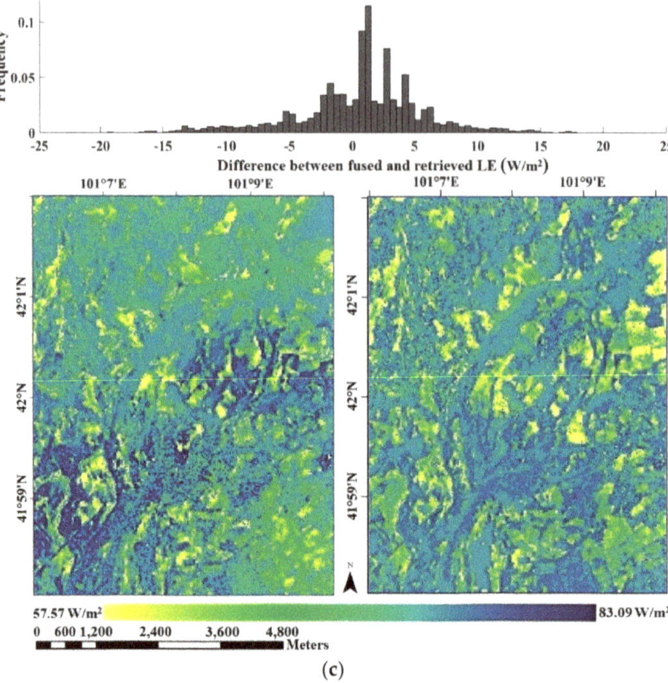

Figure 4. Histogram (top panel) of the difference between fine-resolution LE data fused by the SADFAET (right panel) and the estimates directly retrieved by the ESVEP model using Landsat 8 data (left panel) on (**a**) May 15, (**b**) July 18 and (**c**) August 28.

Table 6. Comparison between instantaneous ET fused by the SADFAET and that directly retrieved by the ESVEP model from Landsat 8 data on typical days (in spring, summer and autumn) over the study area. Max: maximum ET; Min: minimum ET; SD: standard deviation; R: correlation coefficient.

DOY	Item	Mean (W/m^2)	Max (W/m^2)	Min (W/m^2)	SD (W/m^2)	R
May 15	Fused ET	30.06	45.09	10.49	16.17	0.47
	Retrieved ET	28.48	40.84	12.76	17.58	
July 18	Fused ET	74.61	99.26	38.29	38.85	0.49
	Retrieved ET	76.69	108.05	34.32	40.55	
August 28	Fused ET	69.58	67.21	32.61	48.21	0.46
	Retrieved ET	68.90	83.09	29.21	48.78	

4.3. Validation of ET Data Fused by the SADFAET Model Using Ground-Based Measurements

The instantaneous LE data fused by the SADFAET were further compared with the ground-based EC measurement collected at the 5 test sites over the 10 selected days (see Figure 5). Table 7 illustrates the statistical parameters between the fused LEs and measured LEs on the selected days for different land cover types. The results show an overall reasonably good agreement between the instantaneously fused LEs and measured LEs, with a slight underestimation of 13.1 W/m^2 and a RMSE of 45.7 W/m^2. An overestimation of 18.0 W/m^2 was found at the Barren Land site, with a RMSE of 25.1 W/m^2, while underestimations were shown at the other four sites, with mean errors varying from -27.5 W/m^2 to -18.3 W/m^2 and RMSE varying from 41.4 W/m^2 to 60.6 W/m^2. The verification results at the site scale show that the SADFAET can accurately fuse ET data over most surfaces, while the relatively larger error at the Barren Land site was possibly due to the poor accuracy of the instantaneously retrieved LE.

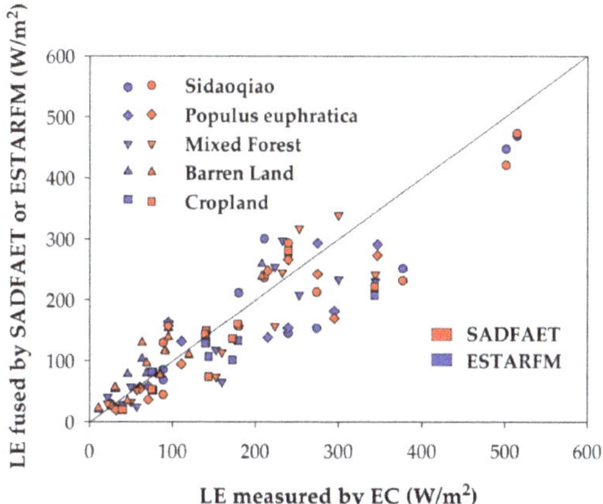

Figure 5. Comparison of instantaneous LE predictions using the SADFAET and ESTARFM with LE measured by EC.

Table 7. Error metrics of the LE fused by the SADFAET with the ground-based EC measurement.

Ground Sites	Mean O (W/m^2)	MB (W/m^2)	MPE (%)	RMSE (W/m^2)
Sidaoqiao	250.0	−27.5	−11	60.6
Populus euphratica	174.2	−18.3	−11	50.7
Mixed Forest	179.6	−19.5	−11	50.5
Barren Land	81.9	18.0	22	25.1
Cropland	155.1	−18.6	−12	41.4
Average	168.2	−13.1	−5	45.7

4.4. Comparison of ET Data Fused by the SADFAET and ESTARFM

We compared the SADFAET with the original ESTARFM to verify the improvement in the SADFAET. The scatter plots in Figure 5 also show the measured LE and ESTARFM LE at the 5 sites. According to Figure 5, we can see that both the SADFAET and ESTARFM can accurately predict ET at a high resolution, while the SADFAET LE data in the scatter plots fall closer to the 1:1 line. For all the sites, the prediction precision of the SADFAET is higher than that of the original ESTARFM (MPE: −5% vs. −8%; MB: −13.1 W/m^2 vs. −18.6 W/m^2; RMSE: 45.7 W/m^2 vs. 50.6 W/m^2). Statistics are provided in Figure 6 for different sites for the ESTARFM and SADFAET. Similar to the SADFAET, only fused ET data by the ESTARFM at the Barren Land site were overestimated, while fused ET data at the other sites were underestimated. The greatest improvement in the SADFAET, in comparison to the ESTARFM, appeared at the Cropland site (MPE: −12% vs. −17%), while at the Barren Land site, there was almost no improvement (MPE: 22% vs. 22%). This may be because the SADFAET considers the variation in soil moisture, while the soil moisture of the Bareland site has a low value and the soil moisture of the Cropland site has significant variation during the crop growth season.

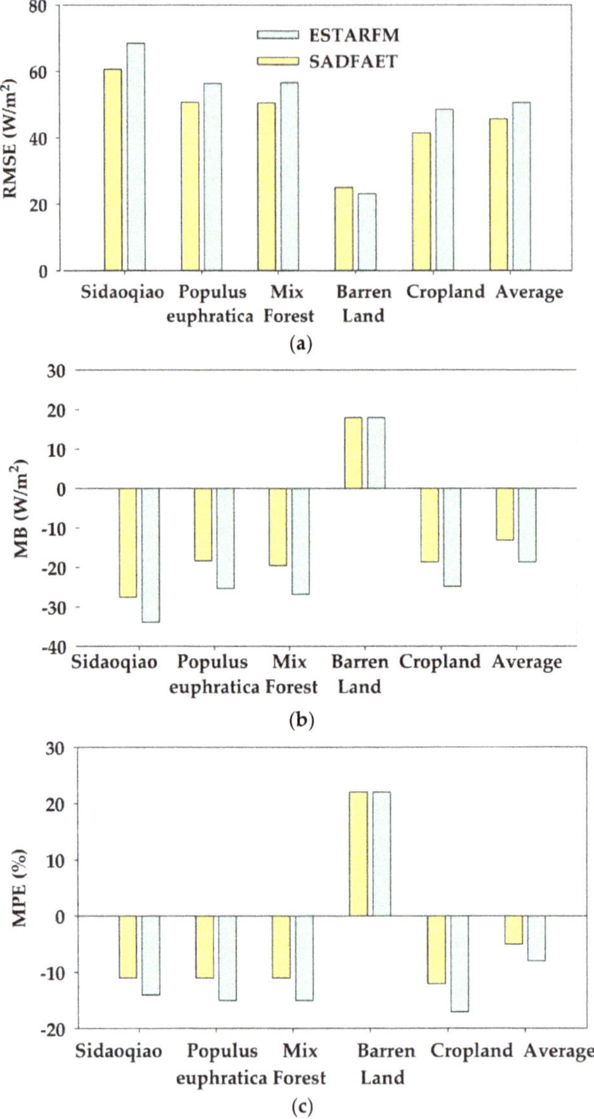

Figure 6. Statistical parameters between the instantaneous LE predictions using ESTARFM/ SADFAET and the EC values: (**a**) RMSE; (**b**) MB; (**c**) MPE.

4.5. Discussion

4.5.1. Uncertainties in the SADFAET

Our results indicate that the SADFAET can accurately estimate ET at a high spatio-temporal resolution by fusing ET at a high spatial resolution with that at a high temporal resolution. However, there are some uncertainties are associated with the use of the algorithm and the uncertainty in the fused ET method primarily results from (1) the retrieval of surface ET-indicative variables, (2) the fine- and coarse-resolution ET data at the reference and prediction times and (3) the fusion algorithm itself, including the selection of similar neighbouring pixels, the calculation of the weights of these pixels and the determination of the conversion coefficient and the temporal weight. For example, the uncertainties in the remotely sensed and meteorological data could affect the calculation of instantaneous ET from the ESVEP model. Validation results indicated that LE was underestimated at the Sidaoqiao site (16%), Mixed Forest site (3%), Populus euphratica site (12%) and Cropland site (14%), while LE was overestimated at the Barren Land site (22%). The uncertainty in the calculation of the critical surface temperature could cause the inaccurate classification of similar pixels and the computation of the weight of the similar pixel. In addition, the inherent uncertainties in the EC and LAS validation data and the mismatch in the spatial scales between the predicted and fused ET and EC measurements could also impact the fusion accuracy.

4.5.2. Improvements and Limitations of the SADFAET

The SADFAET has made several improvements to the original ESTARFM when it is used for ET fusion. The most significant improvement is the introduction of the critical surface temperature (T^*) to consider the impact of soil moisture for the selection of similar pixels, which can solve the problem that determining the number of classes limits automated processing and reduces the accuracy of similar pixel selection in the ESTARFM. In the original ESTARFM, shortwave bands are used to select similar pixels by setting a threshold and unsupervised classification without considering the soil moisture. However, our SADFAET introduced T^* to calculate the soil moisture variation between the predicted time (t_p) and reference times $(t_m$ and $t_n)$ of different pixels, which can ensure that the correct similar pixels with similar ET variations are selected. Second, for the weight calculation of each similar pixel, the SADFAET introduces surface ET-indicative similarity to consider the influences of NDVI, LST, soil moisture and ET at t_m and t_n. Finally, ET may change rapidly with time and using more bands will contribute to accurately computing the weight of the similar pixel; thus, we introduced the shortwave bands (representing vegetation cover), shortwave infrared bands (representing soil moisture) and thermal infrared bands (representing LST) between each similar pixel and its corresponding coarse-resolution pixel for the calculation of spectral similarity. The ET fusion estimates in the present study yielded an MPE of -11% to 22%, which was slightly better than that from previous studies. For example, Ma et al. [28] reported MPE values of -14% to 29% in the midstream region of the Heihe River basin when the ESTARFM was used to fuse ET retrieved from Landsat 7 ETM+ and MODIS data. Semmens et al. [13] and Yang et al. [16] employed the original STARFM combined with a multi-scale ET retrieval algorithm to compute 30 m resolution ET over a mixed forested/agricultural landscape in North Carolina, USA and the mean absolute percent errors of the fused ET were 20% to 23% and 19% to 30% respectively. For the same study area, the prediction precision of the SADFAET is higher than that of the original ESTARFM (-14% to 22%) and the fused ET images contained most of the spatial details found in the Landsat 8 ET images.

There are several limitations and constraints when using the SADFAET. First, similar to the ESTARFM and STARFM [35,36], the SADFAET cannot accurately predict shape changes and may cause boundary blurring. Second, when the soil moisture maintains a low value or exhibits slight variation, the accuracy of the SADFAET may decrease. Finally, although the SADFAET solves the problem whereby determining the number of classes limits automated processing in the ESTARFM, the size of the moving window remains a restriction.

5. Conclusions

This research proposed a spatio-temporal adaptive data fusion algorithm for evapotranspiration mapping (SADFAET) to estimate evapotranspiration (ET) at high spatial resolutions (30 m) and at a high temporal frequency (daily). An experiment that predicted ET at a 30 m spatial resolution on 10 dates in 2015 from April to October downstream of the Heihe River basin was performed using Landsat 8 and MODIS datasets. The main advantage of SADFAET is the consideration of soil moisture by introducing the critical surface temperature (T^*) during the selection of similar pixels.

The other improvement is the use of multiple spectral bands, including shortwave bands (representing vegetation cover), shortwave infrared bands (representing soil moisture) and thermal infrared bands (representing LST) and the introduction of the surface ET-indicative similarity to calculate the weights of similar pixels.

The results showed an overall reasonably good agreement between the fused ET and measured ET with a slight underestimation of 13.1 W/m^2 and a RMSE of 45.7 W/ m^2. The difference between SADFAET ET and Landsat 8 ET was mostly $\pm$10 W/m^2 and the SADFAET ET images preserved most of the original spatial details. The prediction precision of the SADFAET was higher than that of the original ESTARFM (MPE: -5% vs. -8%; MB: -13.1 W/m^2 vs. -18.6 W/m^2; RMSE: 45.7 W/m^2 vs. 50.6 W/m^2). Therefore, our proposed SADFAET could effectively fuse ET at high and low spatial resolutions and the method was better than that from previous studies and other common fusion algorithms. Future work will be performed on the evaluation of the SADFAET on other regions that are covered by different vegetation types and characterized by different climates.

Author Contributions: R.T. and T.W. designed the experiments; T.W. and R.T. analyzed the data and wrote the paper; Z.L, Y.J., M.L. and L.N. revised the paper.

Funding: This research was funded by the National Key R&D Program of China under Grant 2018YFA0605401, the Beijing Municipal Science and Technology Project under Grant Z181100005318003, the Youth Innovation Promotion Association CAS under 2015039, the National Natural Science Foundation of China under Grant 41571351 and 41571367.

Acknowledgments: The staff members in the Heihe Watershed Allied Telemetry Experimental Research (HiWATER) experimental areas are acknowledged for their hard work on the setup and maintenance of the ground-based instruments and data collection.

Conflicts of Interest: The authors declare no conflict of interest.

References

1. Li, A.; Zhao, W.; Deng, W. A Quantitative Inspection on Spatio-Temporal Variation of Remote Sensing-Based Estimates of Land Surface Evapotranspiration in South Asia. *Remote Sens.* **2015**, *7*, 4726–4752. [CrossRef]
2. Mahour, M.; Tolpekin, V.; Stein, A.; Sharifi, A. A comparison of two downscaling procedures to increase the spatial resolution of mapping actual evapotranspiration. *ISPRS J. Photogramm. Remote Sens.* **2017**, *126*, 56–67. [CrossRef]
3. Weng, Q.; Fu, P.; Gao, F. Generating daily land surface temperature at Landsat resolution by fusing Landsat and MODIS data. *Remote Sens. Environ.* **2014**, *145*, 55–67. [CrossRef]
4. Guo, L.J.; Moore, J.M. Pixel block intensity modulation: Adding spatial detail to TM band 6 thermal imagery. *Int. J. Remote Sens.* **1998**, *19*, 2477–2491. [CrossRef]
5. Nichol, J. An emissivity modulation method for spatial enhancement of thermal satellite images in urban heat island analysis. *Photogramm. Eng. Remote Sens.* **2009**, *75*, 547–556. [CrossRef]
6. Hong, S.-H.; Hendrickx, J.M.H.; Borchers, B. Down-scaling of SEBAL derived evapotranspiration maps from MODIS (250 m) to Landsat (30 m) scales. *Int. J. Remote Sens.* **2011**, *32*, 6457–6477. [CrossRef]
7. Anderson, M.C.; Kustas, W.P.; Norman, J.M.; Hain, C.R. Mapping daily evapotranspiration at field to global scales using geostationary and polar orbiting satellite imagery. *Hydrol. Earth Syst. Sci.* **2011**, *15*, 223–239. [CrossRef]
8. Anderson, M.C.; Kustas, W.P.; Alfieri, J.G.; Gao, F.; Hain, C.; Prueger, J.H.; Evett, S.; Colaizzi, P.; Howell, T.; Chávez, J.L. Mapping daily evapotranspiration at Landsat spatial scales during the BEAREX'08 field campaign. *Adv. Water Resour.* **2012**, *50*, 162–177. [CrossRef]
9. Anderson, M.C.; Allen, R.G.; Morse, A.; Kustas, W.P. Use of Landsat thermal imagery in monitoring evapotranspiration and managing water resources. *Remote Sens. Environ.* **2012**, *122*, 50–65. [CrossRef]
10. Cammalleri, C.; Anderson, M.C.; Gao, F.; Hain, C.R.; Kustas, W.P. A data fusion approach for mapping daily evapotranspiration at field scale. *Water Resour. Res.* **2013**, *49*, 4672–4686. [CrossRef]
11. Cammalleri, C.; Anderson, M.C.; Gao, F.; Hain, C.R.; Kustas, W.P. Mapping daily evapotranspiration at field scales over rainfed and irrigated agricultural areas using remote sensing data fusion. *Agric. For. Meteorol.* **2014**, *186*, 1–11. [CrossRef]

12. Bhattarai, N.; Quackenbush, L.J.; Dougherty, M.; Marzen, L.J. A simple Landsat–MODIS fusion approach for monitoring seasonal evapotranspiration at 30 m spatial resolution. *Int. J. Remote Sens.* **2015**, *36*, 115–143.

13. Semmens, K.A.; Anderson, M.C.; Kustas, W.P.; Gao, F.; Alfieri, J.G.; Mckee, L.; Prueger, J.H.; Hain, C.R.; Cammalleri, C.; Yang, Y.; et al. Monitoring daily evapotranspiration over two California vineyards using Landsat 8 in a multi-sensor data fusion approach. *Remote Sens. Environ.* **2015**, *185*, 155–170. [CrossRef]

14. Ke, Y.; Im, J.; Park, S.; Gong, H. Downscaling of MODIS One Kilometer Evapotranspiration Using Landsat-8 Data and Machine Learning Approaches. *Remote Sens.* **2016**, *8*, 215. [CrossRef]

15. Bastiaanssen, W.; Noordman, E.; Pelgrum, H.; Davids, G.; Thoreson, B.; Allen, R. SEBAL model with remotely sensed data to improve water-resources management under actual field conditions. *J. Irrig. Drain. Eng.* **2005**, *131*, 85–93. [CrossRef]

16. Allen, R.G.; Robison, C.W.; Trezza, R.; Garcia, M.; Kjaersgaard, J. Comparison of evapotranspiration images from MODIS and Landsat along the Middle Rio Grande of New Mexico. In Proceedings of the 17th William T. Pecora Memorial Remote Sensing Symposium Proceedings, Denver, CO, USA, 18–20 November 2008.

17. Yang, Y.; Shang, S.; Jiang, L. Remote sensing temporal and spatial patterns of evapotranspiration and the responses to water management in a large irrigation district of North China. *Agric. For. Meteorol.* **2012**, *164*, 112–122. [CrossRef]

18. Yang, Y.; Anderson, M.C.; Gao, F.; Hain, C.R.; Semmens, K.A.; Kustas, W.P.; Noormets, A.; Wynne, R.H.; Thomas, V.A.; Sun, G. Daily Landsat-scale evapotranspiration estimation over a forested landscape in North Carolina, USA, using multi-satellite data fusion. *Hydrol. Earth Syst. Sci.* **2017**, *21*, 1017–1037. [CrossRef]

19. Wu, P.; Shen, H.; Zhang, L.; Göttsche, F.M. Integrated fusion of multi-scale polar-orbiting and geostationary satellite observations for the mapping of high spatial and temporal resolution land surface temperature. *Remote Sens. Environ.* **2015**, *156*, 169–181. [CrossRef]

20. Zayed, I.S.A.; Elagib, N.A.; Ribbe, L.; Heinrich, J. Satellite-based evapotranspiration over Gezira Irrigation Scheme, Sudan: A comparative study. *Agric. Water Manag.* **2016**, *177*, 66–76. [CrossRef]

21. Allen, R.; Tasumi, M.; Trezza, R.; Kjaersgaard, J. *Mapping Evapotranspiration at High Resolution, Applications Manual for Landsat Satellite Imagery*; University of Idaho: Kimberly, ID, USA, 2010; Volume 2, p. 248.

22. Yi, Z.; Zhao, H.; Jiang, Y.; Yan, H.; Yin, C.; Huang, Y.; Zhen, H. Daily Evapotranspiration Estimation at the Field Scale: Using the Modified SEBS Model and HJ-1 Data in a Desert-Oasis Area, Northwestern China. *Water* **2018**, *10*, 640. [CrossRef]

23. Yang, G.; Weng, Q.; Pu, R.; Gao, F.; Sun, C.; Li, H.; Zhao, C. Evaluation of ASTER-Like Daily Land Surface Temperature by Fusing ASTER and MODIS Data during the HiWATER-MUSOEXE. *Remote Sens.* **2016**, *8*, 75. [CrossRef]

24. Bai, Y.; Wong, M.S.; Shi, W.-Z.; Wu, L.-X.; Qin, K. Advancing of Land Surface Temperature Retrieval Using Extreme Learning Machine and Spatio-Temporal Adaptive Data Fusion Algorithm. *Remote Sens.* **2015**, *7*, 4424–4441. [CrossRef]

25. Oliveraguerra, L.; Mattar, C.; Merlin, O.; DuránAlarcón, C.; SantamaríaArtigas, A.; Fuster, R. An operational method for the disaggregation of land surface temperature to estimate actual evapotranspiration in the arid region of Chile. *ISPRS J. Photogramm. Remote Sens.* **2017**, *128*, 170–181. [CrossRef]

26. Ke, Y.; Im, J.; Park, S.; Gong, H. Spatiotemporal downscaling approaches for monitoring 8-day 30 m actual evapotranspiration. *ISPRS J. Photogramm. Remote Sens.* **2017**, *126*, 79–93. [CrossRef]

27. Yan, L.; Huang, C.; Hou, J.; Gu, J.; Zhu, G.; Xin, L. Mapping daily evapotranspiration based on spatiotemporal fusion of ASTER and MODIS images over irrigated agricultural areas in the Heihe River Basin, Northwest China. *Agric. For. Meteorol.* **2017**, *244*, 82–97.

28. Ma, Y.; Liu, S.; Song, L.; Xu, Z.; Liu, Y.; Xu, T.; Zhu, Z. Estimation of daily evapotranspiration and irrigation water efficiency at a Landsat-like scale for an arid irrigation area using multi-source remote sensing data. *Remote Sens. Environ.* **2018**, *216*, 715–734. [CrossRef]

29. Najmaddin, P.M.; Whelan, M.J.; Balzter, H. Estimating Daily Reference Evapotranspiration in a Semi-Arid Region Using Remote Sensing Data. *Remote Sens.* **2017**, *9*, 779. [CrossRef]

30. Stefan, V.; Merlin, O.; Er-Raki, S.; Escorihuela, M.-J.; Khabba, S. Consistency between in situ, model-derived and high-resolution-image-based soil temperature endmembers: Towards a robust data-based model for multi-resolution monitoring of crop evapotranspiration. *Remote Sens.* **2015**, *7*, 10444–10479. [CrossRef]

31. Kuriqi, A. Assessment and quantification of meteorological data for implementation of weather radar in mountainous regions. *MAUSAM* **2016**, *67*, 789–802.

32. Martínez-de la Torre, A.; Blyth, E.M.; Robinson, E.L. Evaluation of drydown processes in global land surface and hydrological models using flux tower evapotranspiration. *Water* **2019**, *11*, 356. [CrossRef]

33. Small, E.; Badger, A.; Abolafia-Rosenzweig, R.; Livneh, B. Estimating Soil Evaporation Using Drying Rates Determined from Satellite-Based Soil Moisture Records. *Remote Sens.* **2018**, *10*, 1945. [CrossRef]

34. Bruin, H.A.; Trigo, I.F. A New Method to Estimate Reference Crop Evapotranspiration from Geostationary Satellite Imagery: Practical Considerations. *Water* **2019**, *11*, 382. [CrossRef]

35. Zhu, X.L.; Jin, C.; Feng, G.; Chen, X.H.; Masek, J.G. An enhanced spatial and temporal adaptive reflectance fusion model for complex heterogeneous regions. *Remote Sens. Environ.* **2010**, *114*, 2610–2623. [CrossRef]

36. Gao, F.; Masek, J.; Schwaller, M.; Hall, F. On the blending of the Landsat and MODIS surface reflectance: Predicting daily Landsat surface reflectance. *IEEE Trans. Geosci. Remote Sens.* **2006**, *44*, 2207–2218.

37. Emelyanova, I.V.; Mcvicar, T.R.; Niel, T.G.V.; Li, L.T.; van Dijk, A.I. Assessing the accuracy of blending Landsat–MODIS surface reflectances in two landscapes with contrasting spatial and temporal dynamics: A framework for algorithm selection. *Remote Sens. Environ.* **2013**, *133*, 193–209. [CrossRef]

38. Bai, L.; Cai, J.; Liu, Y.; Chen, H.; Zhang, B.; Huang, L. Responses of field evapotranspiration to the changes of cropping pattern and groundwater depth in large irrigation district of Yellow River basin. *Agric. Water Manag.* **2017**, *188*, 1–11. [CrossRef]

39. Tang, R.; Li, Z.L. An End-Member-Based Two-Source Approach for Estimating Land Surface Evapotranspiration From Remote Sensing Data. *IEEE Trans. Geosci. Remote Sens.* **2017**, *55*, 5818–5832. [CrossRef]

40. Vicente-Serrano, S.M.; Cabello, D.; Tomás-Burguera, M.; Martín-Hernández, N.; Beguería, S.; Azorin-Molina, C.; Kenawy, A.E. Drought Variability and Land Degradation in Semiarid Regions: Assessment Using Remote Sensing Data and Drought Indices (1982–2011). *Remote Sens.* **2015**, *7*, 4391–4423. [CrossRef]

41. Fensholt, R.; Sandholt, I. Derivation of a shortwave infrared water stress index from MODIS near-and shortwave infrared data in a semiarid environment. *Remote Sens. Environ.* **2003**, *87*, 111–121. [CrossRef]

42. Ghulam, A.; Li, Z.-L.; Qin, Q.; Tong, Q.; Wang, J.; Kasimu, A.; Zhu, L. A method for canopy water content estimation for highly vegetated surfaces-shortwave infrared perpendicular water stress index. *Sci. China Ser. D Earth Sci.* **2007**, *50*, 1359–1368. [CrossRef]

43. Zhang, N.; Hong, Y.; Qin, Q.; Liu, L. VSDI: A visible and shortwave infrared drought index for monitoring soil and vegetation moisture based on optical remote sensing. *Int. J. Remote Sens.* **2013**, *34*, 4585–4609. [CrossRef]

44. Talsma, C.; Good, S.; Miralles, D.; Fisher, J.; Martens, B.; Jimenez, C.; Purdy, A. Sensitivity of Evapotranspiration Components in Remote Sensing-Based Models. *Remote Sens.* **2018**, *10*, 1601. [CrossRef]

45. Campo-Bescós, M.A.; Muñoz-Carpena, R.; Southworth, J.; Zhu, L.; Waylen, P.R.; Bunting, E. Combined Spatial and Temporal Effects of Environmental Controls on Long-Term Monthly NDVI in the Southern Africa Savanna. *Remote Sens.* **2013**, *5*, 6513–6538. [CrossRef]

46. Kustas, W.P.; Norman, J.M. A two-source energy balance approach using directional radiometric temperature observations for sparse canopy covered surfaces. *Agron. J.* **2000**, *92*, 847–854. [CrossRef]

47. Liu, S.M.; Xu, Z.W.; Wang, W.; Jia, Z.; Zhu, M.; Bai, J.; Wang, J. A comparison of eddy-covariance and large aperture scintillometer measurements with respect to the energy balance closure problem. *Hydrol. Earth Syst. Sci.* **2011**, *15*, 1291–1306. [CrossRef]

48. Liu, S.; Li, X.; Xu, Z.; Che, T.; Xiao, Q.; Ma, M.; Liu, Q.; Jin, R.; Guo, J.; Wang, L.; et al. The Heihe Integrated Observatory Network: A basin-scale land surface processes observatory in China. *Vadose Zone J.* **2018**, *17*. [CrossRef]

49. Li, X.; Cheng, G.; Liu, S.; Xiao, Q.; Ma, M.; Jin, R.; Che, T.; Liu, Q.; Wang, W.; Qi, Y. Heihe Watershed Allied Telemetry Experimental Research (HiWATER): Scientific Objectives and Experimental Design. *Bull. Am. Meteorol. Soc.* **2013**, *94*, 1145–1160. [CrossRef]

50. Sadeghi, M.; Babaeian, E.; Tuller, M.; Jones, S.B. The optical trapezoid model: A novel approach to remote sensing of soil moisture applied to Sentinel-2 and Landsat-8 observations. *Remote Sens. Environ.* **2017**, *198*, 52–68. [CrossRef]

51. Qin, Z.; Karnieli, A.; Berliner, P. A mono-window algorithm for retrieving land surface temperature from Landsat TM data and its application to the Israel-Egypt border region. *Int. J. Remote Sens.* **2001**, *22*, 3719–3746. [CrossRef]
52. Sobrino, J.A.; Jimenez-Muoz, J.C.; Sòria, G.; Romaguera, M.; Guanter, L.; Moreno, J.; Plaza, A.; Martínez, P. Land surface emissivity retrieval from different VNIR and TIR sensors. *IEEE Trans. Geosci. Remote Sens.* **2008**, *46*, 316–327. [CrossRef]

Article

Uncertainties in Evapotranspiration Estimates over West Africa

Hahn Chul Jung [1,2,*], Augusto Getirana [1,3], Kristi R. Arsenault [1,4], Thomas R.H. Holmes [1] and Amy McNally [1,3]

[1]   Hydrological Sciences Laboratory, NASA Goddard Space Flight Center, Greenbelt, MD 20771, USA; augusto.getirana@nasa.gov (A.G.); kristi.r.arsenault@nasa.gov (K.R.A.); thomas.r.holmes@nasa.gov (T.R.H.H.); amy.l.mcnally@nasa.gov (A.M.)
[2]   Science Systems and Applications, Inc., Lanham, MD 20706, USA
[3]   Earth System Science Interdisciplinary Center, University of Maryland, College Park, MD 20740, USA
[4]   Science Applications International Corporation, McLean, VA 22102, USA
*   Correspondence: hahnchul.jung@nasa.gov; Tel.: +1-301-614-5563

Received: 18 March 2019; Accepted: 10 April 2019; Published: 12 April 2019

Abstract: An evapotranspiration (ET) ensemble composed of 36 land surface model (LSM) experiments and four diagnostic datasets (GLEAM, ALEXI, MOD16, and FLUXNET) is used to investigate uncertainties in ET estimate over five climate regions in West Africa. Diagnostic ET datasets show lower uncertainty estimates and smaller seasonal variations than the LSM-based ET values, particularly in the humid climate regions. Overall, the impact of the choice of LSMs and meteorological forcing datasets on the modeled ET rates increases from north to south. The LSM formulations and parameters have the largest impact on ET in humid regions, contributing to 90% of the ET uncertainty estimates. Precipitation contributes to the ET uncertainty primarily in arid regions. The LSM-based ET estimates are sensitive to the uncertainty of net radiation in arid region and precipitation in humid region. This study serves as support for better determining water availability for agriculture and livelihoods in Africa with earth observations and land surface models.

Keywords: evapotranspiration; uncertainty; land surface model; West Africa

---

1. Introduction

Accurately estimating evapotranspiration (ET) over West Africa is particularly important for water resources management, weather monitoring and climate change impact assessment on agriculture and food security due to a strong land-atmosphere coupling [1,2]. Also, ET, governed by the surface water and energy budgets, plays an important role in West African monsoon development [3,4]. High uncertainty in ET estimates over West Africa from global models is an obstacle to investigate temporal and spatial variability in the regional hydrology, especially in the context of climate change [5,6]. Understanding ET uncertainties can improve estimates of water availability for agriculture and livelihoods in Africa. However, such a task is still a challenge in data-sparse regions [7–9]. Previously, Kato et al. [10] performed a sensitivity study of land surface model (LSM) simulations, including ET, under the Coordinated Enhanced Observing Period (CEOP) initiative, but none of the four reference sites were located in West Africa.

Several intercomparisons of different model- and/or satellite-based ET estimates have been carried out at different scales. At the global scale, the LandFlux-EVAL initiative [11,12] presented the benchmark synthesis products based on the analyses of multiple global scale ET estimates from LSMs, diagnostic (i.e., observation-based) datasets, and reanalyses. Other studies focused on Asia [13], South America [14], Africa [8], West Africa [9,15,16], and smaller regions, such as the Volta basin [7,17] and the Ouémé River basin [18]. In particular, the African Monsoon Multidisciplinary Analysis

Land Surface Intercomparison Project (ALMIP) computed water budget components over the whole West Africa in its first phase (ALMIP-1) [4] and smaller areas within the region in its second phase (ALMIP-2) [19,20]. However, little is known about seasonal and regional ET uncertainty variations and the impacts of the choice of LSM, radiation forcing datasets, and precipitation on the ET uncertainty over West Africa. Few studies have only attempted to estimate annual ET uncertainties of regional watershed basins (e.g., the Lake Chad basin and the Niger River basin) [5] and investigate uncertainties and trends of global ET estimates using different combinations of ET models and meteorological forcing datasets [6].

The ensemble of ALMIP-1 models revealed that total annual ET corresponds to 77% of the total annual precipitation in West Africa and 85% in the Sahel [21]. The annual precipitation cycle is highly subject to the West African monsoon [22]. Overall, the average precipitation rates increase southwards. In the north, the hyper-arid, arid, and semi-arid regions are located within the Sahel region with a single peak rainy season from July to September [8,23]. In the south, due to a north-south migration of the Inter Tropical Convergence Zone (ITCZ), the sub-humid and humid regions are characterized by two rainy seasons: the first rainy season from May to July and the second rainy season from August to October. Although advances in climate modeling indicate oceans as the main contributor to the recent drought persistence in the Sahel [24], Tian and Peters-Lidard [25] found that West African land shows higher precipitation uncertainties associated with the ITCZ migration than its oceanic counterpart. Also, Vinukollu et al. [6] revealed high ET uncertainties in the Sahel due to the high variability of precipitation, radiation and other meteorological variables, leading to large differences among the models.

Satellite-based ET datasets share some features with physical energy and water balance descriptions of the LSM's but are explicitly constrained by 'diagnostic' observations of surface states (e.g., surface temperature, soil moisture, vegetation water contents or relative humidity) [26]. For better understanding of the characteristics of the ET uncertainties, we investigate 36 LSM-based ET estimates and four diagnostic ET datasets from 2007 to 2011 during which all datasets overlap. Using the diagnostic ET uncertainties to shed some additional light on the model ensemble uncertainty can help give more insight to the variances and spread found among the ensemble of LSM-based ET results, which can be useful in a multi-model ensemble approach. Also, in terms of individual model spread across the different forcing data types, this study can offer some insights for assimilating ET or soil moisture fields. In this study, evaluation of model performance and identification of the primary sources of error in the ET estimates related to inaccurate forcing datasets and limited model parameterizations are beyond the scope of this work. Ground-based ET observations are not representative of LSM-based estimates due to spatial scale differences. The use of the satellite-based reference datasets are more appropriate scale for the LSM ensemble and uncertainty evaluation

The objectives of this study are to: (1) quantify the uncertainty range of net radiation and precipitation datasets as input parameters to estimate ET over West Africa; (2) compare spatial and temporal characteristics between LSM-based and diagnostic ET estimates; (3) analyze the sensitivity of LSMs in simulating ET to the uncertainty of net radiation and precipitation; and (4) investigate the impact of the choice of the model parameterization, meteorological forcing dataset, and precipitation on the LSM-based ET uncertainty. Because the annual precipitation has a latitudinal gradient in West Africa, we divided the domain between 18°W–25°E and 5°N–27°N into five climate regions (highlighted in Figure 1a). The classification of climate regions was based on similar aridity conditions for the 1950–2000 period, as suggested in [27,28]. This study serves to offer further insight on improved land surface modeling designs and better monitoring of water and energy budgets in West Africa.

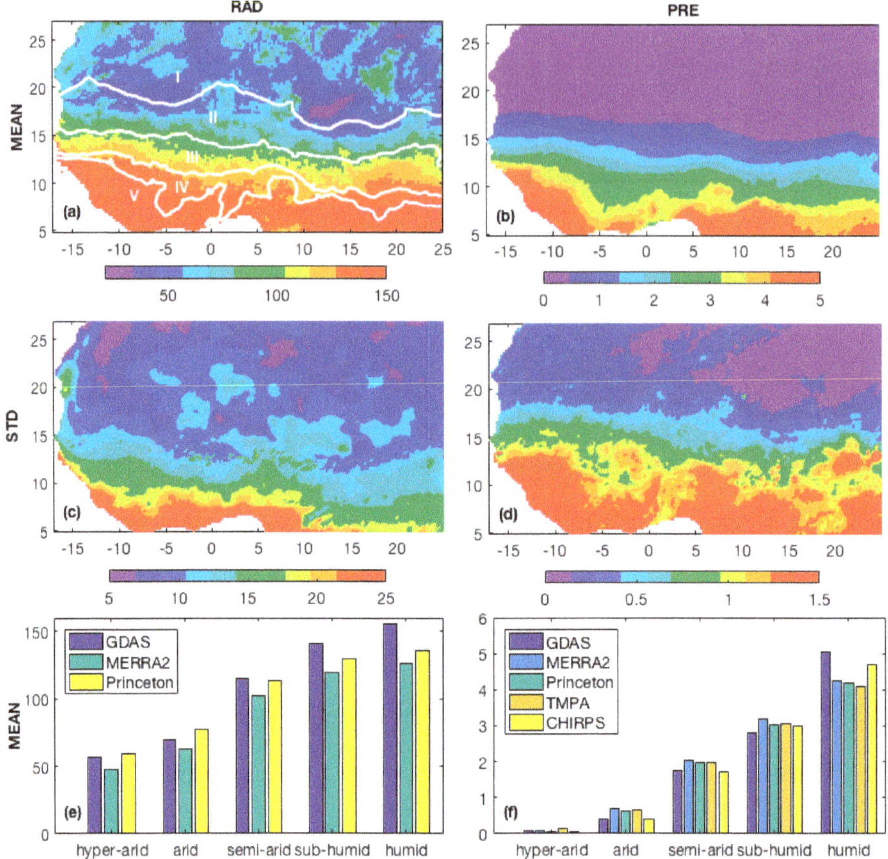

Figure 1. (**a**,**b**) Mean, (**c**,**d**) standard deviation (STD) maps, and (**e**,**f**) regional averages of the mean values from three net radiation (RAD, unit: W·m^{-2}) and five precipitation datasets (PRE, unit: mm· day^{-1}) for years, 2007–2011. The white lines in panel (**a**) delineate the five climate regions (i.e., I-hyper-arid, II-arid, III-semi-arid, IV-sub-humid, V-humid, going from north to south).

2. Data and Methods

2.1. LSM-based ET Datasets

All LSM-based ET outputs were produced daily on a 0.25° spatial resolution domain. The NASA Land Information System (LIS) [29] was used as the modeling platform. Table 1 lists the details of the 36 model experiments.

Table 1. Overview of ET datasets used in this study. An "x" indicates the input not used by the model.

Category	Name (Reference)	PET Scheme	Met. Forcing	Precip.	Spatial Resolution	Temporal Resolution	Time Period
Land Surface Models	Noah [30,31]	Penman	GDAS	GDAS	0.25 deg	Daily	01/2007–12/2011
			GDAS	TMPA			
			GDAS	CHIRPS			
			MERRA2	MERRA2			
			MERRA2	TMPA			
			MERRA2	CHIRPS			
			Princeton	Princeton			
			Princeton	TMPA			
			Princeton	CHIRPS			
	NoahMP [32,33]	Penman–Monteith	*	**			
	VIC [34]	Penman–Monteith	*	**			
	CLSMF [35,36]	Penman–Monteith	*	**			
Diagnostics	GLEAM [37]	Priestley-Taylor	CFS-R	MSWEP	0.25 deg	Daily	01/1980–12/2015
	ALEXI [38]	Priestley-Taylor	ERA-Interim	X	0.05 deg	Weekly	01/2007–12/2015
	MOD16 [39]	Penman–Monteith	GMAO	X	0.01 deg	Monthly	01/2000–12/2012
	FLUXNET [40]	X	X	X	0.50 deg	Monthly	01/1982–12/2011

*, ** Each of the four LSMs was forced with three meteorological forcing datasets and two additional observation-based precipitation datasets, generating a total of 9 experiments.

2.1.1. Land Surface Models

Four LSMs are used in this study to simulate the water and energy fluxes in West Africa, including the Noah land surface model (Noah), version 3.3 [30,31], Noah land surface model with Multi-Parameterization (NoahMP), version 3.6 [32,33], the Variable Infiltration Capacity (VIC), version 4.1.2 [34], and Catchment Land Surface Model (CLSM), version Fortuna 2.5 [35,36]. CLSM and NoahMP were selected in this study because of their ability to explicitly represent groundwater and high performance of data assimilation framework within the NASA LIS system [41,42]. Noah and VIC were included to investigate the impact of different LSM parameterizations on modeling ET estimates and to increase the number of ensemble members included in the uncertainty analysis

All four LSMs calculate ET as the sum of water loss from bare soil and canopy intercepted water (i.e., evaporation) and transpiration via the canopy leaves. In each model, the actual ET is computed as a modification of a potential value, using the Penman equation [43] for potential ET in Noah or the Penman–Monteith [44] in the other three LSMs. The Noah model applies additional resistance factors as a variation on Penman–Monteith [30]. The Penman–Monteith equation calculates actual ET typically through scaling coefficients related to different vegetation or crop types. The Penman–Monteith is more sensitive to vegetation specific parameters and allows for a composited plant stomatal resistance to vapor transport, whereas the Penman equation assumes a continuously available water source and no canopy resistance.

In addition to the ET formulation, model inputs and physics directly impacting ET estimates include meteorological forcing data, precipitation, land cover, soil type, infiltration rates, and drainage in the soil column. In this study, ET uncertainties attributable to these LSM formulations and parameters in the modeled ET rates, except for meteorological forcing and precipitation, are considered as ET uncertainty due to LSMs. For land cover classes, Noah and NoahMP use the International Geosphere-Biosphere Programme (IGBP) classification from the NASA's Terra Moderate Resolution Imaging Spectroradiometer (MODIS) observations [45]. VIC and CLSM used the University of Maryland classification from the Advanced Very High-Resolution Radiometer (AVHRR) observations [46].

2.1.2. Meteorological Forcings

LSMs were driven with three meteorological forcing datasets: NCEP's Global Data Assimilation System (GDAS) [47], NASA's Modern-Era Retrospective analysis for Research and Applications, version 2 (MERRA2) [48], and the Princeton global meteorological forcing datasets (Princeton) [49]. As the sum of net short wave and long wave radiation, net radiation has the largest effect on LSM-based ET estimates in relation to energy balance terms. In Figure 1, the mean of net radiation from the three meteorological forcing datasets ranges from 50 W/m^2 for the hyper-arid to 150 W/m^2 for the humid region. Standard deviations in net radiation are generally 10–20 W/m^2, but coastal areas at longitudes between $-13°$ and $10°$ show higher variations. When considering the three meteorological datasets, MERRA2 has the lowest net radiation of these datasets, whereas Princeton has the highest in both hyper-arid and arid regions. GDAS has the highest values in semi-arid, sub-humid, and humid regions.

2.1.3. Precipitation

Additional experiments were conducted replacing the precipitation fields from the aforementioned meteorological reanalysis forcings with two other precipitation datasets: the Tropical Rainfall Measuring Mission Multi-Satellite Precipitation Analysis product 3B42, version 7 (TMPA) [50] and the Climate Hazards group InfraRed Precipitation with Stations (CHIRPS) [51]. In Figure 1, the mean and standard deviation of the five precipitation datasets increase linearly from north to south, associated with the ITCZ precipitation band. The mean and standard deviation maps show a local effect at longitude 10° and latitude 5°, showing higher values (than 10 mm/day in mean and 5 mm/day in standard deviation; not shown in Figure 1). A comparison of the five precipitation datasets illustrates that GDAS and CHIRPS provides lower precipitation rates in both arid and semi-arid regions, but higher rates in the humid region.

2.2. Diagnostic ET Datasets

The four diagnostic ET datasets are used in this study, including the Global Land Evaporation Amsterdam Model, v3.0a (GLEAM) [37], the Atmosphere-Land Exchange Inverse (ALEXI) [38], the Moderate Resolution Imaging Spectroradiometer (MODIS) land ET product 16 (MOD16) [39], and FLUXNET [40]. Our study time period is limited to the 5-year overlap between these datasets from January 2007 to December 2011. Most of arid and hyper-arid regions in the ALEXI and MOD16 datasets correspond to missing data as they are deemed unreliable. Diagnostic ET datasets were averaged to monthly values and re-scaled to a 0.25-degree spatial resolution, whenever necessary.

For PET estimation, ALEXI and GLEAM use the Priestley-Taylor equation, whereas MOD16 uses the Penman–Monteith equation. The Priestley-Taylor equation estimates evaporation from an extensive wet surface under conditions of minimum advection by removing the aerodynamic terms from the Penman–Monteith equation and adding an empirically derived constant factor [52]. Forcing inputs for the Penman–Monteith equation (used in MOD16) include vapor pressure deficit, air temperature, net solar radiation, wind speed and air pressure whereas the Priestly-Taylor equation (used in ALEXI and GLEAM) uses only net radiation or solar irradiance. GLEAM includes a running water balance, using precipitation input (i.e., the Multi-Source Weighted-Ensemble Precipitation) [53], assimilating soil moisture observations, and combining them with microwave vegetation optical depth to parameterize evaporative stress. ALEXI uses observed morning changes in temperature to diagnose the partitioning of the surface energy balance between sensible and latent heats. MOD16 estimates the surface resistance based on Leaf Area Index (LAI). FLUXNET is based on upscaling eddy-covariance flux measurements of ET with satellite-based vegetation indices. The FLUXNET database is composed of regional and global analysis of observations from over 750 micrometeorological tower sites, including two flux towers over West Africa, starting in 2010, with a limited time period as provided with the latest version of the FLUXNET 2015 Tier 1 dataset.

2.3. Uncertainty Analysis

The evapotranspiration uncertainty analysis is performed separately on LSM-based and diagnostic datasets to compare their spatial and temporal characteristics and quantify their uncertainty ranges. The analysis is also performed separately for the five West African climate regions.

To compare the mean and standard deviation between LSM-based and diagnostic ET datasets, the relative difference (RD) of the two ET products are computed as:

$$RD = \frac{E_1 - E_2}{\overline{\overline{E}}} \times 100 \ (\%) \tag{1}$$

where E_1 and E_2 represent ET^{LSM} and ET^{DIAG}, respectively, and $\overline{E}$ is the mean of the products E_1 and E_2. Also, to incorporate the bias-insensitive and multiple component nature of evaluation metrics for a comprehensive comparison, the spatial efficiency metrics (SPAEF) [54,55] are computed as:

$$SPAEF = 1 - \sqrt{(\alpha - 1)^2 + (\beta - 1)^2 + (\gamma - 1)^2}$$
$$\alpha = corr(E1, E2), \ \ \beta = \left(\frac{\sigma_{E1}}{\mu_{E1}}\right) / \left(\frac{\sigma_{E2}}{\mu_{E2}}\right) \ and \ \ \gamma = \frac{\sum_{j=1}^{n} \min(K_j, L_j)}{\sum_{j=1}^{n} K_j} \tag{2}$$

where α is the Pearson correlation coefficient, β is the fraction of the coefficient of variation and γ is the histogram intersection for the given histograms K and L of the E1 and E2 patterns, respectively.

The impact of the choice of LSM, meteorological forcing datasets (MET), and precipitation-only (PRE) on the uncertainty in the LSM-based ET estimates (ET^{LSM}) is estimated through the standard deviation attributable to each of the three components at each grid cell location as follows:

$$\sigma^x_{ET(LSM)} = sqrt\left(\frac{\sum_{t=1}^{T} \sum_{l=1}^{L} \sigma^2\left[ET^{x,t,l}_{LSM1}, \ldots, ET^{x,t,l}_{LSM4}\right]}{T \times L}\right) \tag{3}$$

$$\sigma^x_{ET(MET)} = sqrt\left(\frac{\sum_{t=1}^{T} \sum_{m=1}^{M} \sigma^2\left[ET^{x,t,m}_{MET1}, \ldots, ET^{x,t,m}_{MET3}\right]}{T \times M}\right) \tag{4}$$

$$\sigma^x_{ET(PRE)} = sqrt\left(\frac{\sum_{t=1}^{T} \sum_{n=1}^{N} \sigma^2\left[ET^{x,t,n}_{PRE1}, \ldots, ET^{x,t,n}_{PRE5}\right]}{T \times N}\right) \tag{5}$$

where σ^2 is variance, x is the grid cell, T (=60) is the total number of monthly time step t, for years 2007–2011. A number of the combined experiments ($L = 9$ from 3 METs times 3 PREs, $M = 8$ from 4 LSMs times 2 additional PREs, $N = 12$ from 4 LSMs times 3 METs) with the four LSMs, the three METs, and the five PREs are used for each of the ET uncertainties $\sigma^x_{ET(LSM)}$, $\sigma^x_{ET(MET)}$, and $\sigma^x_{ET(PRE)}$, respectively. To separate the temporal ET variability from the uncertainty analysis, uncertainties are estimated by averaging standard deviations over the ET datasets at the same time, t, and then calculating the temporal average of the standard deviations.

3. Results and Discussion

3.1. Comparison of LSM-Based and Diagnostic ET Estimates

Figure 2 shows the spatial distribution maps of the mean, standard deviation, and relative difference from the LSM-based and diagnostic ET datasets over West Africa for years, 2007–2011. Overall, arid regions at the higher latitudes show lower values in mean ET rates and standard deviations than for the humid regions at lower latitudes. Specifically, lower standard deviations are found in the hyper-arid and arid regions when mean ET values are less than 1 mm/day. This threshold is reached in the arid region, resulting in standard deviations of 0.2 mm/day and greater below 15°N. Compared to diagnostic ET datasets, LSM-based ET datasets show the similar mean ET rates in West Africa except at longitudes between −5° and 25° in the hyper arid region, ranging from −30 to −30 % in relative difference. Overall, the LSM-based ET dataset has higher standard deviation values with positive relative differences over West Africa, except for a part of the hyper-arid region. Comparing the four LSMs demonstrates that Noah and NoahMP provides lower mean ET rates than CLSM and VIC for all climate regions (Figure 2g). The order of the spatially averaged mean ET rates, from the highest to the lowest, is CLSM, VIC, Noah, NoahMP. These results are likely due to the differences of the model physics and parameters, including land cover datasets of IGBP classification (used in CLSM and VIC) and UMD classification (used in Noah and NoahMP). For the diagnostic model datasets (Figure 2h), ALEXI has the highest ET rate (except in the hyper-arid region), whereas MOD16 has the lowest ET rates (except in humid region) for all climate regions. Also, it is noteworthy that GLEAM, in model structure such as a running water balance driven with precipitation dataset, is more consistent with FLUXNET-based observations. On the other hand, ALEXI and MOD16 are more pure diagnostic estimates that do not use antecedent information (e.g., water balance estimates).

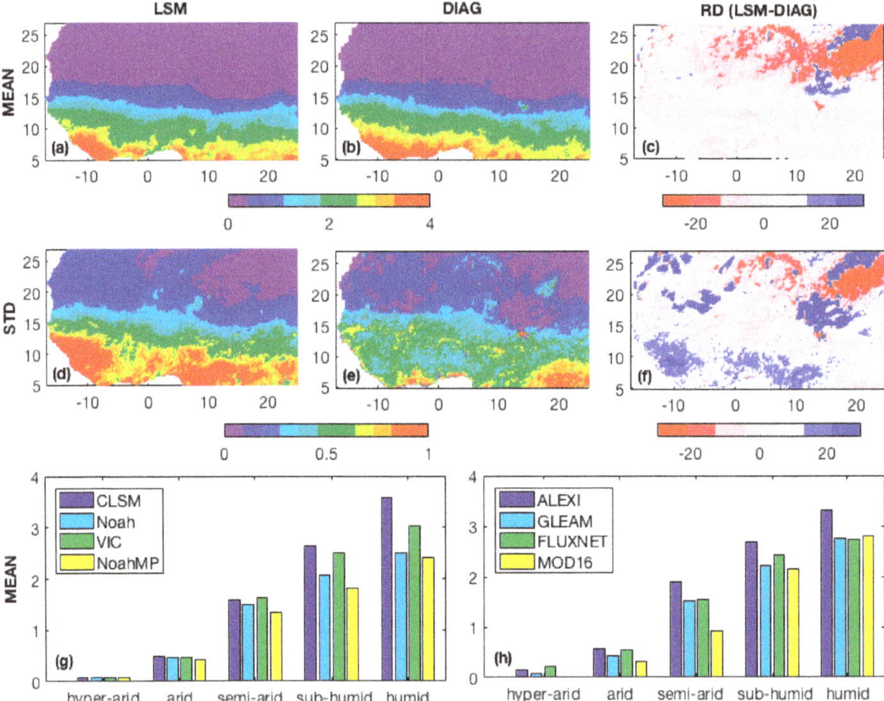

Figure 2. (**a,b**) Mean (mm·day^{-1}), (**d,e**) standard deviation (mm·day^{-1}), (**c,f**) relative difference (%) maps, and (**g,h**) regional averages of the mean values of LSM-based and diagnostic (i.e., DIAG) ET datasets combined.

Figure 3 shows monthly climatologies of mean, standard deviation and relative difference values for the five West African climate regions. The LSM-based and diagnostic ET datasets show the similar temporal variations of their mean ET rates. Compared to diagnostic datasets, LSM-based mean ET rates show lower values during winter and spring, slightly higher during summer, and similar in fall. This finding is consistent with that the LSMs tend to underestimate ET during winter and spring seasons when local rainfall is not the primary source of water available for ET [26]. The hyper-arid region shows low mean ET rates over the year, whereas the arid region presents a range from 0.3 to 1.5 mm/day, mainly from June to October. The humid region shows a bimodal ET cycle with two peaks with high ET rates (>3 mm/day) from April to November, whereas semi-arid and sub-humid regions show one peak in September.

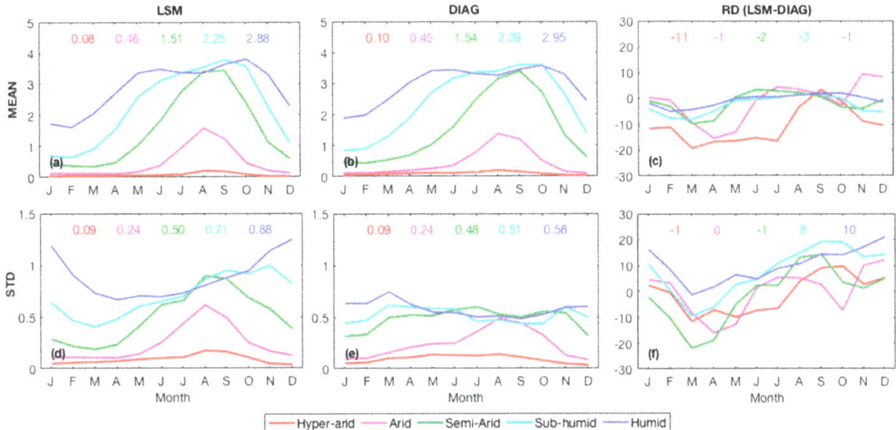

Figure 3. Monthly climatologies of the (**a,b**) mean (mm·day^{-1}), (**d,e**) standard deviation (mm·day^{-1}), and (**c,f**) relative difference (%) from the LSM-based and diagnostic ET datasets for the five West African climate regions. Values represent the average for each climate region.

For the standard deviation of monthly climatologies, the diagnostic ET datasets (Figure 3e) show values lower than LSM-based ET estimates (Figure 3d), particularly in the semi-arid, sub-humid and humid regions. The hyper-arid and arid regions show the standard deviation increasing with mean ET and peaking in late summer or early fall. The semi-arid region shows more similar patterns of ET uncertainty in the monthly climatology as humid regions than the other arid regions. This suggests that the behavior of ET is dependent on water and energy budget variables, which can lead to different characteristics than climate regions determined by similar annual precipitation cycle or aridity condition. Compared to the LSM-based ET datasets, the diagnostic ET datasets show lower seasonal variation in their standard deviations for the semi-arid, sub-humid, and humid regions. LSM-based ET datasets show higher standard deviations for all four seasons than the diagnostic ET datasets in humid region. Also, it is noteworthy that uncertainties of the LSM-based ET estimates increase during winter for the humid region. This can be explained by the fact that CLSM generates the low peaks in ET rates for April, lagging by two months behind the low peaks of the other LSMs, which occur in February for the humid region.

Figure 4 shows SPAEF-based comparison of mean and standard deviation of the LSM-based and diagnostic ET datasets combined. Compared to metrics of standard deviation (Figure 4b,d), all three metrics (i.e., correlation coefficient, coefficient of variation, histogram match) and SPAFE values of the spatially averaged mean ET rates (Figure 4a,c) are closer to the optimal condition, one. This supports that larger difference between LSM-based and diagnostic ET datasets exists in their standard deviation than the mean ET rates. Monthly climatologies of the mean ET rates (Figure 4e) show lower histogram match between LSM-based and diagnostic ET datasets than those of the standard deviation (Figure 4f), which leads to reduced SPAEF values despite high correlation coefficient and coefficient of variation for all four seasons. This implies that seasonal variation of the mean ET rates between two different ET datasets are not consistent as much as one spatially averaged for the whole time period 2007–2011.

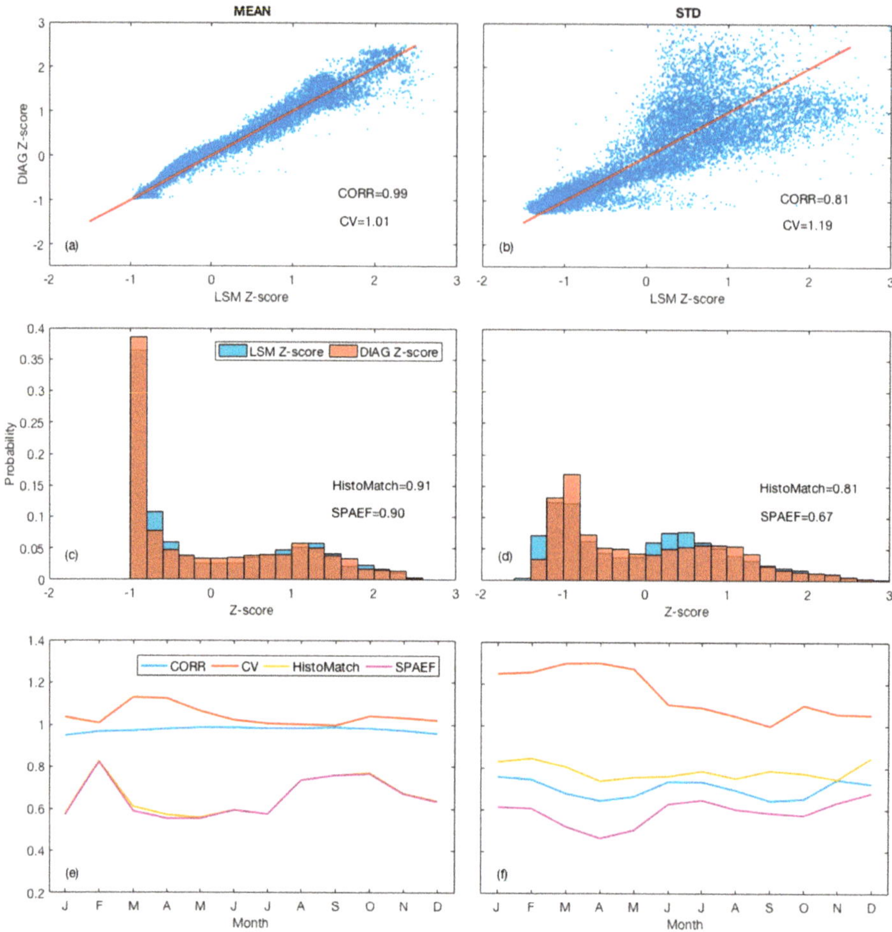

Figure 4. Comprehensive comparison of (**a,c,e**) mean and (**b,d,f**) standard deviation of the LSM-based and diagnostic (i.e., DIAG) ET datasets. Correlation coefficient (CORR), coefficient of variation (CV), histogram match (HistoMatch), SPAEF values are spatially averaged for (**a–d**) the 2007–2011 period and (**e–f**) monthly climatologies.

3.2. Uncertainty Analysis of LSM-Based ET Estimates

Figure 5 shows the spatially distributed impacts of model parameterization, meteorological forcing datasets, and precipitation uncertainties on the LSM-based ET uncertainty, as defined in Equations (3)–(5). Overall, LSM-based ET uncertainties attributable to LSM, MET, and PRE increase from north to south in West Africa. The ET uncertainties are mostly attributed to LSM, or the differences in model physics and parameterizations, particularly in the two humid climate regions plus the semi-arid region. This is consistent with a previous sensitivity study of land surface simulations against 1-year field measurements [10]. Arid regions with little precipitation have ET values much lower than PET, whereas humid regions have ET values closer to PET values. This implies that the choice of PET equations in LSMs provide more diversity in our ET estimations over West Africa as compared to the forcing data. This is also supported by the evidence that the ET uncertainties between the potential evaporation products are higher than ones between the different actual evaporation products over the African continent [8].

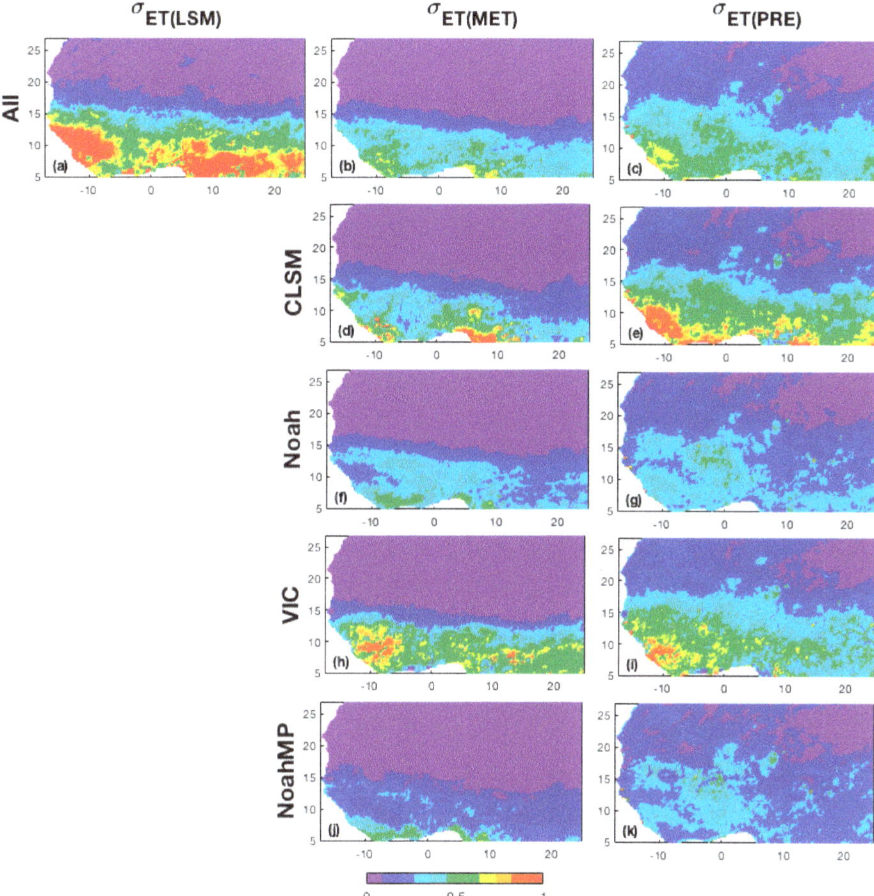

Figure 5. LSM-based ET uncertainties are calculated from standard deviations (mm·day^{-1}) attributable to (**a**) LSM, (**b**) MET, and (**c**) PRE for each grid cell location. The impacts of MET and PRE uncertainties on LSM-specific ET estimates are shown for (**d,e**) CLSM, (**f,g**) Noah, (**h,i**) VIC, and (**j,k**) NoahMP, respectively.

The impacts of MET and PRE uncertainties on LSM-specific ET estimates are also investigated. In arid regions, the four LSMs show the similar impact of MET and PRE uncertainties on the modeled ET estimates. However, CLSM and VIC in humid regions show higher effects from MET and PRE than Noah and NoahMP with higher mean ET rates (see Figure 2g). MET and PRE uncertainties generate a local effect in the modeled ET uncertainties, having higher values along the coast for CLSM and at longitudes between −12° and −8° in humid region for VIC. This is related to the fact that CLSM and VIC are more sensitive to high uncertainties of radiation and precipitation in these areas (shown in Figure 1c,d) than Noah and NoahMP.

Figure 6 shows the sensitivity of ET uncertainties to the uncertainties of both radiation and precipitation. The sensitivity is calculated by dividing the ET uncertainties (Figure 5b,c) by the uncertainty of the radiation (Figure 1c) and precipitation (Figure 1d). From north to south, the sensitivity of ET uncertainties to the uncertainties of radiation decreases, whereas the sensitivity of ET uncertainties to precipitation increases. This indicates that ET uncertainties in arid regions with little water are more sensitive to the uncertainty of net radiation, whereas humid regions have a higher sensitivity of ET uncertainties to the uncertainty of precipitation.

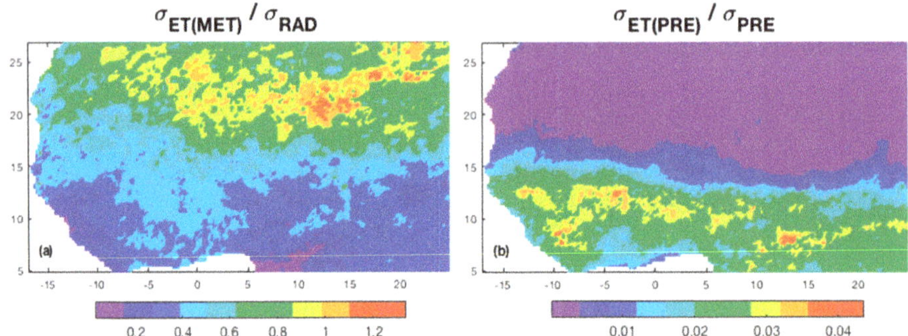

Figure 6. Sensitivity of modeled ET uncertainties to the uncertainties of the meteorological based (**a**) net radiation (mm·day^{-1}/W·m^{-2}) and (**b**) precipitation (mm·day^{-1}/mm·day^{-1}).

When examining the monthly climatology of the impacts of the LSM, MET, and PRE uncertainties on the LSM-based ET uncertainty, LSM shows the largest effect in Figure 7. This leads to the fact that seasonal variation of standard deviations due to LSM shows the most similar seasonal variation of the modeled ET uncertainty (shown in Figure 3d) with lower uncertainty values for all climate regions. The standard deviations attributable to meteorological forcing datasets show clear seasonal variations except in hyper-arid region. Interestingly, the standard deviations attributable to precipitation show lesser seasonal variation in humid region with different peak seasons in June for sub-humid region and in August for the other three arid regions.

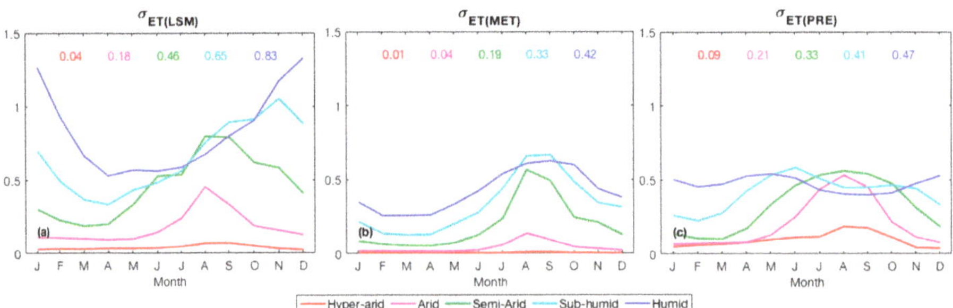

Figure 7. Monthly climatology (mm·day^{-1}) of the impacts of the LSM, MET, and PRE uncertainties on the LSM-based ET uncertainty for the five West African climate regions. Values represent the average for each climate region.

Figure 8 shows that LSM physics and parameters contribute to more than 90% of the uncertainty of the LSM-based ET estimates for semi-arid (91%), sub-humid, (95%) and humid (97%) regions. The impact of the choice of LSM includes the calibration process on inaccurate input data including meteorological forcing datasets and precipitation. Thus, it implies that standard deviations attributable to meteorological forcing dataset and precipitation can be reduced by the LMS physics. In arid regions, precipitation forcing contributes to the ET uncertainties (>90%) primarily, and the meteorological forcing datasets have the smallest impact on the ET uncertainty. This can be explained by the fact that

ET estimates in arid regions are more governed by water availability (e.g., precipitation) than energy availability (e.g., net radiation).

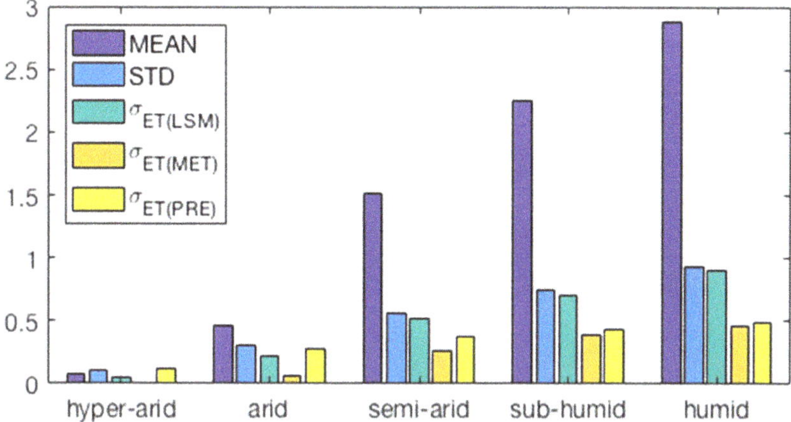

Figure 8. Regional average (unit: mm·day^{-1}) of mean ET rates, standard deviations, and impacts of LSM uncertainties ($\delta_{ET(LSM)}$), MET ($\delta_{ET(MET)}$), and PRE ($\delta_{ET(PRE)}$) on the LSM-based ET uncertainty.

4. Summary and Conclusions

We quantified evapotranspiration uncertainties for five West African climate regions for the 2007–2011 period. Uncertainty is defined as the standard deviations, and the analysis is performed using four LSM-based and four diagnostic ET estimates. Results show clear regional ET uncertainty variations, increasing southward from the hyper-arid to humid regions. Diagnostic ET datasets show lower uncertainties and smaller seasonal variations than the LSM-based ET datasets, particularly in the humid climate regions and semi-arid region. This suggests that assimilating diagnostic ET datasets into LSMs or hydrological models could improve ET simulations. This finding is supported by a recent study that assimilated MODIS-based actual ET data and showed improved simulated discharge [56]. The NASA LIS system incorporates a data assimilation (DA) framework and supports a variety of model and DA type approaches, which could support future ET assimilation studies, accounting for uncertainty in the model and observational ET reference-type datasets (e.g., MODIS products).

In addition to the ET uncertainty quantification, we demonstrated that LSMs have the biggest impact in humid regions, contributing to more than 90% of simulated ET uncertainties. Also, the seasonal variation of the ET uncertainties is mostly affected by the uncertainties attributable to LSMs. Humid regions have ET values close to PET. This can be explained by the fact that the model physics differentiate most in their parametrization of evaporative stress when conditions are closer to potential ET. The precipitation uncertainty has higher influence on ET uncertainty in all West African climate regions than the net radiation uncertainty. Specifically, in both hyper-arid and arid regions, precipitation has the biggest effect on the modeled ET estimates. The sensitivity analysis reveals that the modeled ET estimates are more sensitive to the uncertainty of net radiation in arid regions and precipitation in humid regions, respectively.

Major rivers, wetlands, lakes, and floodplains flowing through arid regions, such as the Niger River and, in particular, its inner delta, located in Mali, are main sources of evaporation. However, this process is neglected in all LSM-based ET estimates used in this study. As a result, the actual ET rates over West Africa could be higher than what has been reported in the literature and also in this study. These limit our approach to quantify ET uncertainty and analyze error sources of our ET datasets. Also, further studies based on more diversified LSM attributed from a choice of different land cover datasets, vegetation parameters, and soil physics are recommended to better quantify the uncertainties of the model ET estimates.

Multiple land surface models are used in Land Data Assimilation Systems (LDAS) such as Global LDAS (GLDAS), North American LDAS (NLDAS), and FEWS NET LDAS (FLDAS) to increase simulation skill along with the use of hybrid forcing ensembles [57,58]. Also, a multi-model ensemble framework is used to develop estimates of model uncertainty and understand the level of similarity and dissimilarity between the constituent models. The utility of the multi-model ensemble can be increased when sufficient dissimilarity among the constituent models is guaranteed. The results from this study could suggest that by including certain LSMs to these LDAS, additional information would be provided for enhancing the water availability monitoring for agriculture and livelihoods in Africa.

Author Contributions: H.C.J.: Conceptualization; Investigation; Methodology; Visualization; Writing-original draft. A.G.: Project Administration; Conceptualization; Methodology; Writing-review and editing. K.R.A.: Investigation; Writing-review and editing. T.R.H.H.: Methodology; Writing-review and editing. A.M.: Investigation; Writing-review and editing.

Funding: This study was funded by NASA SERVIR Applied Sciences Program in Earth Science Division (NNH15ZDA001N-SERVIR).

Acknowledgments: Computing was supported by the resources at the NASA Center for Climate Simulation (NCCS). We acknowledge C. Hain and M. Anderson for use of the ALEXIS ET data. The MOD16 data are publicly available from http://files.ntsg.umt.edu/data/NTSG_Products/MOD16/, which was supported by the NASA Earth Observing System MODIS project. The GLEAM data are freely accessed from http://www.GLEAM.eu. The FLUXNET data are available from https://www.bgc-jena.mpg.de/geodb/.

Conflicts of Interest: The authors declare no conflict of interest.

References

1. Koster, R.D.; Suarez, M.J.; Ducharne, A.; Stieglitz, M.; Kumar, P. A catchment based approach to modeling land surface processes in a general circulation model: 1. Model structure. *J. Geophys. Res.-Atmos.* **2000**, *105*, 24809–24822. [CrossRef]

2. Lei, F.; Crow, W.T.; Holmes, T.R.H.; Hain, C.; Anderson, M.C. Global investigation of soil moisture and latent heat flux coupling strength. *Water Resour. Res.* **2018**, *54*, 8196–8215. [CrossRef]

3. Xue, Y.; Juang, H.-M.H.; Li, W.-P.; Prince, S.; DeFries, R.; Jiao, Y.; Vasic, R. Role of land surface processes in monsoon development: East Asia and West Africa. *J. Geophys. Res.* **2004**, *109*, D03105. [CrossRef]

4. Boone, A.; De Rosnay, P.; Balsamo, G.; Beljaars, A.; Chopin, F.; Decharme, B.; Delire, C.; Ducharne, A.; Gascoin, S.; Grippa, M.; et al. The AMMA Land Surface Model Intercomparison Project. *Bull. Am. Meteorol. Soc.* **2009**, *90*, 1865–1880. [CrossRef]

5. Li, K.Y.; Coe, M.T.; Ramankutty, N. Investigation of hydrological variability in West Africa using land surface models. *J. Clim.* **2005**, *18*, 3173–3188. [CrossRef]

6. Vinukollu, R.K.; Meynadier, R.; Sheffield, J.; Wood, E.F. Multi-model, multi-sensor estimates of global evapotranspiration: Climatology, uncertainties and trends. *Hydrol. Process.* **2011**, *25*, 3993–4010. [CrossRef]

7. Andam-Akorful, S.A.; Ferreira, V.G.; Awange, J.L.; Forootan, E.; He, X.F. Multi-model and multi-sesnor estimates of evapotranspiration over the Volta Basin, West Africa. *Int. J. Climatol.* **2015**, *35*, 3132–3145. [CrossRef]

8. Trambauer, P.; Dutra, E.; Maskey, S.; Werner, M.; Pappenberger, F.; van Beek, L.P.H.; Uhlenbrook, S. Comparison of different evaporation estimates over the African continent. *Hydrol. Earth Syst. Sci.* **2014**, *18*, 193–212. [CrossRef]

9. McNally, A.; Husak, G.J.; Brown, M.; Carroll, M.; Funk, C.; Yatheendradas, S.; Arsenault, K.; Peters-Lidard, C.; Verdin, J.P. Calculating Crop Water Requirement Satisfaction in the West Africa Sahel with Remotely Sensed Soil Moisture. *J. Hydrometeorol.* **2015**. [CrossRef]

10. Kato, H.; Rodell, M.; Beyrich, F.; Cleugh, H.; van Gorsel, E.; Liu, H.; Meyers, T.P. Sensitivity of land surface simulations to model physics, land characteristics, and forcings at four CEOP sites. *J. Meteorol. Soc. Jpn.* **2007**, *85*, 187–204. [CrossRef]

11. Mueller, B.; Seneviratne, S.I.; Jimenez, C.; Corti, T.; Hirschi, M.; Balsamo, G.; Ciais, P.; Dirmeyer, P.; Fisher, J.B.; Guo, Z.; et al. Evaluation of global observations-based evapotranspiration datasets and IPCC AR4 simulations. *Geophys. Res. Lett.* **2011**, *38*, L06402. [CrossRef]

12. Mueller, B.; Hirschi, M.; Jimenez, C.; Ciais, P.; Dirmeyer, P.A.; Dolman, A.J.; Fisher, J.B.; Jung, M.; Ludwig, F.; Maignan, F.; et al. Benchmark products for land evapotranspiration: LandFlux-EVAL multi-data set synthesis. *Hydrol. Earth Syst. Sci.* **2013**, *17*, 3707–3720. [CrossRef]

13. Khan, M.S.; Liaqat, U.W.; Baik, J.; Choi, M. Stand-alone uncertainty characterization of GLEAM, GLDAS and MOD16 evapotranspiration products using an extended triple collocation approach. *Agric. For. Meteorol.* **2018**, *252*, 256–268. [CrossRef]

14. Sorensson, A.A.; Ruscica, R.C. Intercomparison and uncertainty assessment of nine evapotranspiration estimates over South America. *Water Resour. Res.* **2018**, *54*, 2891–2908. [CrossRef]

15. Guichard, F.; Asencio, N.; Peugeot, C.; Bock, O.; Redelsperger, J.L.; Cui, X.; Garvert, M.; Lamptey, B.; Orlandi, E.; Sander, J.; et al. An Intercomparison of Simulated Rainfall and Evapotranspiration Associated with a Mesoscale Convective System over West Africa. *Weather Forecast.* **2010**, *25*, 37–60. [CrossRef]

16. Peugeot, C.; Guichard, F.; Bock, O.; Bouniol, D.; Chong, M.; Boone, A.; Cappelaere, B.; Gosset, M.; Besson, L.; Lemaitre, Y.; et al. Mesoscale water cycle within the West African Monsoon. *Atmos. Sci. Lett.* **2011**, *12*, 45–50. [CrossRef]

17. Opoku-Duah, S.; Donoghue, D.N.M.; Burt, T.P. Intercomparison of Evapotranspiration Over the Savannah Volta Basin in West Africa Using Remote Sensing Data. *Sensors* **2008**, *8*, 2736–2761. [CrossRef]

18. Getirana, A.; Boone, A.; Peugeot, C. Evaluating LSM-based water budgets over a West African basin assisted with a river routing scheme. *J. Hydrometeorol.* **2014**. [CrossRef]

19. Boone, A.; Getirana, A.C.V.; Demarty, J.; Cappelaere, B.; Galle, S.; Grippa, M.; Lebel, T.; Mougin, E.; Peugeot, C.; Vischel, T. AMMA Land Surface Model Intercomparison Project Phase 2, (ALMIP-2). *Gewex News* **2009**, *9*, 9–10.

20. Getirana, A.; Boone, A.; Peugeot, C. Streamflows over a West African basin from the ALMIP-2 model ensemble. *J. Hydrometeorol.* **2017**. [CrossRef]

21. Grippa, M.; Kergoat, L.; Frappart, F.; Araud, Q.; Boone, A.; de Rosnay, P.; Lemoine, J.M.; Gascoin, S.; Balsamo, G.; Ottle, C.; et al. Land water storage variability over West Africa estimated by Gravity Recovery and Climate Experiment (GRACE) and land surface models. *Water Resour. Res.* **2011**, *47*, W05549. [CrossRef]

22. Wolters, E.L.L.A.; Van den Hurk, B.J.J.M.; Roebeling, R.A. Evaluation of rainfall retrievals from SEVIRI reflectances over West Africa using TRMM-PR and CMORPH. *Hydrol. Earth Syst. Sci.* **2011**, *15*, 437–451. [CrossRef]

23. Sylla, M.B.; Coppola, E.; Mariotti, L.; Giorgi, F.; Ruti, P.; Dell'Aquila, A.; Bi, X. Multiyear simulation of the African climate using a regional climate model (RegCM3) with the high resolution ERA-interim reanalysis. *Clim. Dynam.* **2010**, *35*, 231–247. [CrossRef]

24. Giannini, A.; Biasutti, M.; Verstraete, M.M. A climate model-based review of drought in the Sahel: Desertification, the re-greening and climate change. *Glob. Planet. Chang.* **2008**, *64*, 119–128. [CrossRef]

25. Tian, Y.; Peters-Lidard, C.D. A global map of uncertainties in satellite-based precipitation measurements. *Geophys. Res. Lett.* **2010**, *37*, L24407. [CrossRef]

26. Yilmaz, M.T.; Anderson, M.C.; Zaitchik, B.; Hain, C.R.; Crow, W.T.; Ozdogan, M.; Chun, J.A.; Evans, J. Comparison of prognostic and diagnostic surface flux modeling approaches over the Nile River basin. *Water Resour. Res.* **2014**, *50*, 386–408. [CrossRef]

27. UNEP (United Nations Environment Programme). *World Atlas of Desertification*; UNEP: Nairobi, Kenya, 1997.

28. Zomer, R.J.; Trabucco, A.; Bossio, D.A.; van Straaten, O.; Verchot, L.V. Climate Change Mitigation: A Spatial Analysis of Global Land Suitability for Clean Development Mechanism Afforestation and Reforestation. *Agric. Ecosyst. Environ.* **2008**, *126*, 67–80. [CrossRef]

29. Kumar, S.V.; Peters-Lidard, C.D.; Tian, Y.; Geiger, J.; Houser, P.R.; Olden, S.; Lighty, L.; Eastman, J.L.; Dirmeyer, P.; Doty, B.; et al. LIS—An interoperable framework for high resolution land surface modeling. *Environ. Model. Softw.* **2006**, *21*, 1402–1415. [CrossRef]

30. Chen, F.; Mitchell, K.; Schaake, J.; Xue, Y.; Pan, H.; Koren, V.; Duan, Y.; Ek, M.; Betts, A. Modeling of land-surface evaporation by four schemes and comparison with FIFE observations. *J. Geophys. Res.* **1996**, *101*, 7251–7268. [CrossRef]

31. Ek, M.B.; Mitchell, K.E.; Lin, Y.; Rogers, E.; Grummann, P.; Koren, V.; Gayno, G.; Tarpley, J.D. Implementation of Noah land surface model advances in the National Centers for Environmental Prediction operational Mesoscale Eta Model. *J. Geophys. Res.* **2003**, *108*, 8851. [CrossRef]

32. Niu, G.-Y.; Yang, Z.-L.; Mitchell, K.E.; Chen, F.; Ek, M.B.; Barlage, M.; Longuevergne, L.; Kumar, A.; Manning, K.; Niyogi, D.; et al. The community Noah land surface model with multiparam-eterization options (Noah-MP): 1. Model description and evaluation with local-scale measurements. *J. Geophys. Res.* **2011**. [CrossRef]

33. Cai, X.; Yang, Z.-L.; David, C.H.; Niu, G.-Y.; Rodell, M. Hydrological evaluation of the Noah-MP land surface model for the Mississippi River Basin. *J. Geophys. Res. Atmos.* **2014**, *119*, 23–38. [CrossRef]

34. Liang, X.; Lettenmaier, D.P.; Wood, E.F.; Burges, S.J. A simple hydrologically based model of land surface water and energy fluxes for general circulation models. *J. Geophys. Res.* **1994**, *99*, 14415–14428. [CrossRef]

35. Koster, R.D.; Dirmeyer, P.A.; Guo, Z.; Bonan, G.; Chan, E.; Cox, P.; Gordon, C.T.; Kanae, S.; Kowalczyk, E.; Lawrence, D.; et al. Regions of strong coupling between soil moisture and precipitation. *Science* **2004**, *305*, 1138–1140. [CrossRef]

36. Reichle, R.H.; Koster, R.D.; De Lannoy, G.J.M.; Forman, B.A.; Liu, Q.; Mahanama, S.P.P.; Toure, A. Assessment and enhancement of MERRA land surface hydrology estimates. *J. Clim.* **2011**, *24*, 6322–6338. [CrossRef]

37. Martens, B.; Miralles, D.G.; Lievens, H.; van der Schalie, R.; de Jeu, R.A.M.; Fernández-Prieto, D.; Beck, H.E.; Dorigo, W.A.; Verhoest, N.E.C. GLEAM v3.0: Satellite-based land evaporation and root-zone soil moisture. *Geosci. Model Dev. Discuss.* **2016**. [CrossRef]

38. Hain, C.R.; Anderson, M.C. Estimating Morning Change in Land Surface Temperature from MODIS Day/Night Land Surface Temperature: Applications for Surface Energy Balance Modeling. *Geophys. Res. Lett.* **2017**. [CrossRef]

39. Mu, Q.; Zhao, M.; Running, S.W. Improvements to a MODIS global terrestrial evapotranspiration algorithm. *Remote Sens. Environ.* **2011**, *115*, 1781–1800. [CrossRef]

40. Jung, M.; Reichstein, M.; Margolis, H.A.; Cescatti, A.; Richardson, A.D.; Arain, M.A.; Arneth, A.; Bernhofer, C.; Bonal, D.; Chen, J.; et al. Global patterns of land-atmosphere fluxes of carbon dioxide, latent heat, and sensible heat derived from eddy covariance, satellite, and meteorological observations. *J. Geophys. Res.* **2011**, *116*, G00J07. [CrossRef]

41. Houborg, R.; Rodell, M.; Li, B.; Reichle, R.; Zaitchik, B.F. Drought indicators based on model-assimilated gravity recovery and climate experiment (GRACE) terrestrial water storage observations. *Water Resour. Res.* **2012**, *48*, W07525. [CrossRef]

42. Getirana, A.; Kumar, S.; Girotto, M.; Rodell, M. Rivers and floodplains as key components of global terrestrial water storage variability. *Geophys. Res. Lett.* **2017**, *44*. [CrossRef]

43. Penman, H.L. Natural evaporation from open water, bare soil, and grass. *Proc. R. Soc. Lond.* **1948**, *193*, 120–146.

44. Monteith, J.L. Evaporation and environment. In *Proceedings of the 19th Symposium of the Society for Experimental Biology*; Cambridge University Press: New York, NY, USA, 1965; pp. 205–233.

45. Friedl, M.A.; Sulla-Menashe, D.; Tan, B.; Schneider, A.; Ramankutty, N.; Sibley, A.; Huang, X. MODIS Collection 5 global land cover: Algorithm refinements and characterization of new datasets. *Remote Sens. Environ.* **2010**, *114*, 168–182. [CrossRef]

46. Hansen, M.; DeFries, R.; Townshend, J.R.; Sohlberg, R. Global land cover classification at 1 km spatial resolution using a classification tree approach. *Int. J. Remote Sens.* **2000**, *21*, 1331–1364. [CrossRef]

47. Rodell, M.; Houser, P.R.; Jambor, U.; Gottschalck, J.; Mitchell, K.; Meng, C.-J.; Arsenault, K.; Cosgrove, B.; Radakovich, J.; Bosilovich, M.; et al. The Global Land Data Assimilation System. *Bull. Am. Meteorol. Soc.* **2004**, *85*, 381–394. [CrossRef]

48. Reichle, R.H.; Liu, Q.; Koster, R.D.; Draper, C.S.; Mahanama, S.P.P.; Partyka, G.S. Land Surface Precipitation in MERRA2. *J. Clim.* **2017**, *30*, 1643–1664. [CrossRef]

49. Sheffield, J.; Goteti, G.; Wood, E.F. Development of a 50-year high-resolution global dataset of meteorological forcings for land surface modeling. *J. Clim.* **2006**, *19*, 3088–3111. [CrossRef]

50. Huffman, G.; Adler, R.; Bolvin, D.; Gu, G.; Nelkin, E.; Bowman, K.; Hong, Y.; Stocker, E.; Wolff, D. The TRMM multisatellite precipitation analysis (TCMA): Quasi-global, multiyear, combined-sensor precipitation estimates at fine scales. *J. Hydrometeorol.* **2007**, *8*, 38–55. [CrossRef]

51. Funk, C.; Peterson, P.; Landsfeld, M.; Pedreros, D.; Verdin, J.; Shukla, S.; Husak, G.; Rowland, J.; Harrison, L.; Hoell, A.; et al. The climate hazards infrared precipitation with stations—A new environmental record for monitoring extremes. *Sci. Data* **2015**, *2*, 150066. [CrossRef]
52. Priestly, C.H.B.; Taylor, R.J. On the assessment of surface heat flux and evaporation using large-scale parameters. *Mon. Weather Rev.* **1972**, *100*, 81–91. [CrossRef]
53. Beck, H.E.; van Dijk, A.I.J.M.; Levizzani, V.; Schellekens, J.; Miralles, D.G.; Martens, B.; de Roo, A. MSWEP: 3-hourly 0.25° global gridded precipitation (1979–2015) by merging gauge, satellite, and reanalysis data. *Hydrol. Earth Syst. Sci.* **2017**, *21*, 589–615. [CrossRef]
54. Koch, J.; Demirel, M.C.; Stisen, S. On the importance of multiple-component evaluation of spatial patterns for optimization of earth system models. *Geosci. Model Dev.* **2018**, *11*, 1873–1886. [CrossRef]
55. Demirel, M.C.; Mai, J.; Mendiguren, G.; Koch, J.; Samaniego, L.; Stisen, S. Combining satellite data and appropriate objective functions for improved spatial pattern performance of a distributed hydrologic model. *Hydrol. Earth Syst. Sci.* **2018**, *22*, 1299–1315. [CrossRef]
56. Hartato, I.M.; van der Kwast, J.; Alexandridis, T.K.; Almeida, W.; Song, Y.; van Andel, S.J.; Solomatine, D.P. Data assimilation of satellite-based actual evapotranspiration in a distributed hydrological model of a controlled water system. *Int. J. Appl. Earth Obs. Geoinf.* **2017**, *57*, 123–135. [CrossRef]
57. Kumar, S.V.; Wang, S.; Mocko, D.M.; Peters-Lidard, C.D.; Xia, Y. Similarity assessment of land surface model outputs in the North American Land Data Assimilation System. *Water Resour. Res.* **2017**, *53*. [CrossRef]
58. Kumar, S.V.; Dong, J.; Peters-Lidard, C.D.; Mocko, D.; Gómez, B. Role of forcing uncertainty and background model error characterization in snow data assimilation. *Hydrol. Earth Syst. Sci.* **2017**, *21*, 2637–2647. [CrossRef]

Article

Differences among Evapotranspiration Products Affect Water Resources and Ecosystem Management in an Australian Catchment

Zhixiang Lu [1,2,†], Yan Zhao [2,3,†], Yongping Wei [2], Qi Feng [1,*] and Jiali Xie [4,5]

[1] Key Laboratory of Ecohydrology of Inland River Basin, Northwest Institute of Eco-Environment and Resources, Chinese Academy of Sciences, Lanzhou 730000, China; lzhxiang@lzb.ac.cn

[2] School of Earth and Environmental Sciences, The University of Queensland, Brisbane 4067, Australia; yan.zhao@uq.edu.au (Y.Z.); yongping.wei@uq.edu.au (Y.W.)

[3] Institute of Geographic Sciences and Natural Resources Research, Chinese Academy of Sciences, Beijing 100101, China

[4] Key Laboratory of Desert and Desertification, Northwest Institute of Eco-Environment and Resources, Chinese Academy of Sciences, Lanzhou 730000, China; xiejl@lzb.ac.cn

[5] University of Chinese Academy of Sciences, Beijing 100049, China

* Correspondence: qifeng@lzb.ac.cn

† The authors contributed equally to this work.

Received: 18 March 2019; Accepted: 17 April 2019; Published: 22 April 2019

Abstract: Evapotranspiration (ET) is a critical component of the water and energy balance of climate–soil–vegetation interactions and can account for a water loss of about 90% in arid regions. It is recognized that there are differences among different ET products, but it is not known what the range of this difference is and to what extent it impacts on water resources and ecosystem management. In this study, we assess the effects of value differences of five representative ET products on water resources and ecosystem management in the Murrumbidgee River catchment in Australia. The results show there are obvious differences in the annual and monthly ET values among these five ET products, which lead to huge differences on the estimations of mean annual runoff, soil water storage changes, and yearly irrigation water per area. Meanwhile, they result in different relationships between the annual gross primary productivity and ET and different water-use efficiency values for both forest and grassland, but the influence of ET variations on forest is less obvious than on grassland. The effects of the variations among the ET products on water resources and ecosystem management are remarkable and need to be the subject of more attention.

Keywords: evapotranspiration; remote sensing; Murrumbidgee River catchment; water resources management; ecosystem management

---

1. Introduction

Water scarcity is one of the most serious global challenges [1,2]. Evidence shows that the changing and uncertain climate in the future could lead to more uncertainty in water availability [3,4]. This creates greater pressure on water resources and ecosystem management in arid and semi-arid regions where there is often poor infrastructure and a fragile ecological environment [2]. Thus, to obtain accurate information of water use is important for water resources and ecosystem management in these regions [2,5].

Evapotranspiration (ET) is an essential component in catchment water balance since it is almost the only water-extracting constituent in arid and semi-arid regions [6,7]. ET is the loss of water when mitigating vegetation stress responses and the key variable for linking ecosystem functioning, carbon and climate feedbacks [5]. Therefore, ET is commonly considered as a reference and indicator for water resources and ecosystem management. It is used for: the determination of irrigation requirements of crops, evaluation of the hydrological effects of afforestation and deforestation, and the estimation of the productivity and water-use efficiencies of ecosystems [5,8–10].

A wide range of global and regional ET products have been developed in recent decades for monitoring water use and assisting in planning management. They can be broadly classified into three categories: remote sensing (RS)-based ET products using vegetation index-based data or land surface temperature (LST) (e.g., the Moderate Resolution Imaging Spectroradiometer (MODIS) ET, Advanced Very High Resolution Radiometer (AVHRR) ET, Priestly-Taylor Jet Propulsion Laboratory (PT-JPL) ET and Global Land Evaporation Amsterdam Model (GLEAM) ET), ET output from land surface models (LSMs) (e.g., Global Land Data Assimilation System (GLDAS) ET, Modern-Era Retrospective analysis for Research and Applications (MERRA) ET and European Centre for Medium-Range Weather Forecasts Re-analysis (ERA)-interim ET), and ET from a water budget with measured precipitation, runoff and total water storage change (e.g., TerraClimate ET and Gravity Recovery and Climate Experiment (GRACE)-inferred ET) [5,11,12]. These ET products have been used to a varying degree for irrigation water management [8], vegetation productivity estimation [9,10], calibration and optimization of hydrological models and prediction of runoff in the ungauged basins [13,14], and water management modes and water rights regimes [2,15,16]. However, these different ET products are of different spatial-temporal scales, and differ in accuracy or uncertainty [11,17,18].

As ET is the largest water-use component in the catchment water balance in arid and semi-arid regions, any variation of ET will cause corresponding errors in the estimation of runoff and soil water storage, thus also large errors in the determination of water demand of irrigated agriculture and in the estimation of productivity of water use in an ecosystem [19,20]. To date, most hydrologic studies, as well as river basin management proposals, have tended to focus on supply elements e.g., precipitation, snow, runoff, water storage soil moisture and groundwater, but have largely ignored the demand side (e.g., ET) (Figure 1) [5]. This is mainly because of insufficient knowledge of ET at local to global scales. The ground observations for ET (eddy covariance towers, Bowen ratio systems, and weighing lysimeters) are poorly covered in many regions, furthermore there are uncertainties about the ET from a water budget perspective which accumulate from errors in other hydrological components, and uncertainties of the ET from land surface models and RS-based approaches due to differences in the structures, resolutions, and inputs of the models and approaches. However, the range of these errors caused by the variations of ET products is not yet clear. Without such knowledge, our capacity for accurate water resource management, economic development and sustainable ecosystems would be seriously compromised.

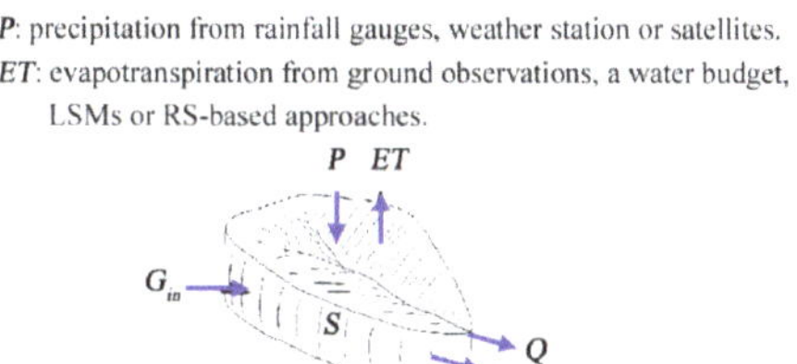

Figure 1. The access methods of the water balance components within a drainage basin.

This paper aims to analyse the chain effects of variations of several widely used ET products on water resource and ecosystem management in which the Murrumbidgee River catchment (hereafter

MRC) in Australia, a semi-arid and arid catchment with the intensified agriculture and declining ecosystems, is taken as a case study area. The water balance in the MRC obtained from the Australian Water Availability Project (AWAP) datasets from 2001–2016 will be used as the "reference" and to compare the performance differences of four global representative ET products selected from dozens of products: a RS-based ET product from MODIS, a LSMs-based ET product from GLDAS, a water budget-based ET product from TerraClimate and the ET product developed by the Australian indigenous research institute, the Commonwealth Scientific and Industrial Research Organisation (CSIRO), and their effects on the other water balance components, irrigation, and gross primary productivity (GPP) and water-use efficiency of forest and grassland are analysed.

2. Methods and Data

2.1. Study Area

The Murray–Darling Basin (hereafter MDB) is the most productive agricultural area in Australia and a globally recognized epitome of successful water management. It covers around 14% of Australia's land, but produces about 40% of Australia's gross value of agricultural production. As one of the most important catchments in the MDB, the MRC has an area of 84,000 square kilometers and accounts for about 7.9% of the MDB (Figure 2). It encompasses all of the Australian Capital Territory. The Murrumbidgee River rises in the Snowy Mountains and flows 1690 km from east to west. The average annual runoff is about 37.84×10^8 m^3. The highest precipitation occurs in the eastern mountains of the catchment and averages 1500 mm yr^{-1}, while the average annual rainfall in the lower catchment is about a fifth of this figure. The annual potential ET ranges from 1250 mm to 1500 mm. Most of the precipitation losses are in the form of ET and account for up to 90%.

The MRC was opened to settlement in the 1830s and soon became an important farming area, and now its irrigated area accounts for about 15% of the total in the MDB, especially the irrigated rice area which is up to about two third of the total rice area in the MDB according to the statistical data from Australian Bureau of Agricultural and Resource Economics and Sciences (ABARE). In the MRC, agricultural land, grassland, and forest are the main land cover types, which accounts for about 60%, 15%, and 20%, respectively [21]. The Murrumbidgee River is one of the most regulated rivers in Australia on which there are 26 water storage structures. Irrigation for agriculture is the greatest user of water in the MRC, and the total volume of irrigation entitlements is approximately 28.0×10^8 m^3, which accounts for 74.0% of the average annual runoff [22]. However, with the development of the farming practices, there is increasing deterioration of river health, thus, protection of the environment is on the political agenda, along with a commitment not only to return water to rivers to nurse them back to health, but also to help agricultural industries rise up to the challenge of a drier climate [23]. In such a catchment, precise and reliable water resources management is extremely important for agricultural development and ecosystem management.

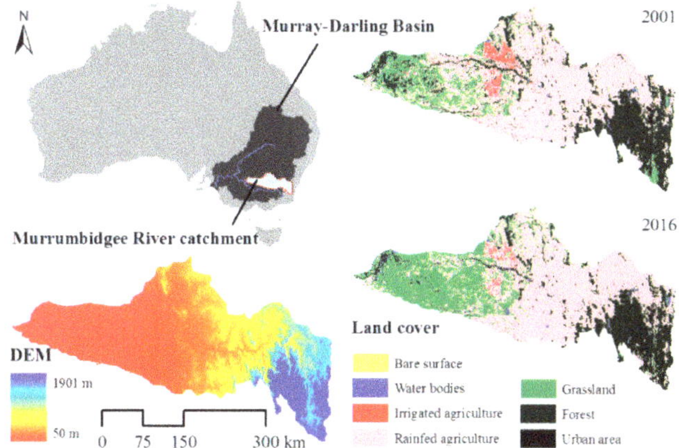

Figure 2. The location of the Murrumbidgee River catchment (MRC) and its topography and land cover.

2.2. Data Sources and Processing

The AWAP datasets concern the terrestrial water balance across continental Australia and are developed by the Australian Water Availability Project (AWAP, http://www.csiro.au/awap) Team and the CSIRO Marine and Atmospheric Research using the WaterDyn model. They include a long-term historical monthly time series (data set "Run 26m", 1900 to 2017) of the conventional water balance components at a spatial resolution of 0.05° [24]. These datasets have been widely used in research on water resource management in Australia and can be viewed as referenced or observational data to evaluate the other types of water balance components [25]. The water balance components (in mm mon^{-1}) of the MRC from 2001 to 2016, including precipitation, ET, runoff, and soil water storage changes were obtained from these datasets.

The four other ET products selected for comparative analysis include ET from CSIRO, GLDAS, MODIS, and TerraClimate. The CSIRO ET is a monthly global ET product developed by the CSIRO [26]. The estimates are computed through the observation-driven Penman–Monteith–Leuning (PML) model. The GLDAS ET is the GLDAS-2.1 ET product, which is one of two components of the GLDAS Version 2 (GLDAS-2) dataset and is developed with the advanced land surface modeling and data assimilation techniques utilising satellite and ground-based observational data products [27]. The MODIS ET is the MOD16A2 Version 6 Evapotranspiration/Latent Heat Flux product and is an 8-day composite product produced at 500-metrepixel resolution [28]. The algorithm used for the MOD16 data product collection is based on the logic of the Penman–Monteith equation, which includes inputs of daily meteorological reanalysis data along with MODIS remotely sensed data products such as vegetation property dynamics, albedo, and land cover. The TerraClimate ET is the actual evapotranspiration of TerraClimate dataset and is derived using a one-dimensional soil water balance model [29]. TerraClimate is a dataset of monthly climate and climatic water balance for global terrestrial surfaces. It uses climatically aided interpolation, combining high-spatial resolution climatological normals from the WorldClim dataset, with coarser spatial resolution of 2.5°.

Other data used in this study include irrigation information, ecosystem data, and land cover data. The irrigation information was obtained from farm survey reports from ABARE, which is the only national farm survey information source in Australia. As one of the most important indicators of ecosystem function and closely related to water use, the GPP was selected for ecosystem assessment. The GPP data is the MOD17A3 V055 product, which provides information about annual GPP at 1 km pixel resolution [30]. In addition, we used the current release of the second version of the Dynamic Land Cover Dataset (DLCDv2.1) as a basis for water balance analyses [21]. The land cover classification scheme conforms to the 2007 International Standards Organisation (ISO) land cover standard (19144-2).

In this dataset Australian land covers are clustered into 22 classes, ranging from cultivated and managed land cover to natural land cover. The data presents land cover information for every 250 m by 250 m area of the country at a two-year interval. In accord with our research aim in this study, we combined these 22 land cover types into 7 types: bare surface, water bodies/wet land, irrigated agriculture, rain-fed agriculture, grassland, forest, and urban area. Then, we uniformly resampled and adjusted resolution of the spatial data (except AWAP data set and CSIRO ET) to 500 m and projection and temporal resolutions (annual and monthly) to make the data products comparable using the Google Earth Engine (GEE). Manipulation of the AWAP data set and CSIRO ET, and reclassification and combination of land cover data were conducted in the ArcGIS platform. We made the analysis based on the land cover types in monthly and annual scales respectively using SPSS and Microsoft Excel software.

All data used in this study including the water balances components from AWAP, the ET products from CSIRO, GLDAS, MODIS, and TerraClimate, irrigation information, land cover, and GPP are summarized in Table 1.

Table 1. The information of data used in this study.

Data Name	Sources	Spatial Resolution	Spatial Extent	Temporal Resolution	Time Span	Provider
Precipitation, runoff, and water storage changes	AWAP [a]	0.05°	Australian	Monthly	1900–2017	CSIRO
Evapotranspiration (ET)	1. AWAP	0.05°	Australian	Monthly	1900–2017	CSIRO
	2. CSIRO [b]	0.5°	Global	Monthly	1981–2012	CSIRO
	3. GLDAS [c]	0.25°	Global	3 h	2000–August 2018	NASA
	4. MODIS [d]	500 m	Global (−60° to 80° Latitude; −180° to 180° Longitude)	8 days	2001–September 2018	NASA LP DAAC at the USGS EROS Center
	5. TerraClimate	2.5°	Global	Monthly	1958–2017	University of Idaho
Irrigation	ABARE [e]	Catchment and administrative area	Australian	Yearly	2005–2017	ABARE
GPP [f]	MOD17A3.055	1000 m	Global	Yearly	2000–2014	NASA LP DAAC at the USGS EROS Center
Land cover	DLCDv2.1 [g]	250 m	Australian	Yearly	2001–2015	www.ga.gov.au

Note: [a] Australian Water Availability Project (AWAP), [b] Commonwealth Scientific and Industrial Research Organisation (CSIRO), [c] Global Land Data Assimilation System (GLDAS), [d] Moderate Resolution Imaging Spectroradiometer (MODIS), [e] Australian Bureau of Agricultural and Resource Economics and Sciences (ABARE), [f] gross primary productivity (GPP), and [g] Dynamic Land Cover Dataset v2.1 (DLCDv2.1).

2.3. Comparison among These Five Evapotranspiration (ET) Products

The variations of ET among the products were quantified by comparing the four ET products (comparison value) with the AWAP ET component (reference value). The percent bias (PBIAS) and mean absolute error (MAE) were used [31] to assess the difference between the reference and comparison ET values. PBIAS measures the average tendency of comparison values. Positive values show overestimation and negative values indicate underestimation; low-magnitude values of PBIAS are preferred and the optimal value is 0. MAE means the average value of the absolute errors between the reference and comparison ET values and can better reflect the actual level of the comparison value error. The smaller the MAE, the better the comparison value. They are calculated as follows:

$$\text{PBIAS} = \frac{\sum_{i=1}^{n}\left(Q_i^{sim} - Q_i^{obs}\right)}{\sum_{i=1}^{n} Q_{i=1}^{obs}} \times 100\% \tag{1}$$

$$\text{MAE} = \frac{1}{n}\sum_{i=1}^{n}\left|Q_i^{sim} - Q_i^{obs}\right| \tag{2}$$

The root mean square error (RMSE) and root mean square error observation standard deviation ratio (RSR) were then used to derive statistical goodness of fit of the ET comparison values and to evaluate the performance of the data product against reference values. RMSE is very sensitive to the large errors in a group of data. RSR is used to standardise RMSE and integrate the advantages of error statistics, and RSR ranges from 0 to 1. They are computed using the following equations:

$$\text{RMSE} = \sqrt{\frac{\sum_{i=1}^{n}\left(Q_i^{sim} - Q_i^{obs}\right)^2}{n}} \tag{3}$$

$$\text{RSR} = \frac{\sqrt{\sum_{i=1}^{n}\left(Q_i^{sim} - Q_i^{obs}\right)^2}}{\sqrt{\sum_{i=1}^{n}\left(Q_i^{obs} - Q_{avg}^{obs}\right)^2}} \tag{4}$$

Following on, the coefficient of determination (R^2) value was computed to evaluate the linear relationship between reference and comparison ET data. Along with R^2, we also reported slope and intercept values to provide useful information of the degree of bias at higher ET and intercept for the magnitude of bias affecting lower comparison ET. R^2 ranges from 0 to 1, and this represents the proportion of the total variance in the reference data that can be explained, with higher R^2 values indicating better performance.

$$R^2 = \frac{\left(\sum_{i=1}^{n}\left(Q_i^{obs} - Q_{avg}^{obs}\right)\left(Q_i^{sim} - Q_{avg}^{sim}\right)\right)^2}{\sum_{i=1}^{n}\left(Q_i^{obs} - Q_{avg}^{obs}\right)^2 \sum_{i=1}^{n}\left(Q_i^{sim} - Q_{avg}^{sim}\right)^2} \tag{5}$$

Finally, the Nash–Sutcliffe efficiency (NSE) [32] was used to measure how well comparison values represent the reference data, relative to a prediction made by using the average reference value. It is a more stringent measure than R^2. NSE ranges from $-\infty$ to 1, with NSE = 1 being the optimal value. NSE is calculated as:

$$\text{NSE} = 1 - \frac{\sum_{i=1}^{n}\left(Q_i^{sim} - Q_i^{obs}\right)^2}{\sum_{i=1}^{n}\left(Q_i^{obs} - Q_{avg}^{obs}\right)^2} \tag{6}$$

where Q_i^{sim} is the comparison ET (other four ET products), and Q_i^{obs} is the reference ET (AWAP ET) at time step i, respectively, whereas Q_{avg}^{obs} and Q_{avg}^{sim} are the average reference and comparison ET values, and n is the total number of data, and subscript i represents the time step (month or year).

2.4. The Effects of ET Variations on the Runoff and Water Storage Changes

The catchment water balance approach was used to analyse the effects of value differences in these ET products on the runoff and water storage changes. The water balance approach, routinely used for estimating mean annual ET, is based on the catchment water budget [33]. For a given watershed, the instantaneous equation of the water mass balance is:

$$P = \Delta W + ET + R \tag{7}$$

where P, ΔW, ET, and R are precipitation, water mass storage changes, evapotranspiration, and runoff respectively. These terms are generally expressed in terms of water mass (mm of equivalent water height) per day. There are many transformations for the Equation (7) for different purposes and based on different available data.

To assess the effects of ET variations on runoff, other terms of the equations were set at AWAP levels with ET values from the other four products were used respectively, and the equation can be transferred into the following:

$$R = P - ET - \Delta W \tag{8}$$

To assess the impacts of ET variations on water storage changes, other terms of the equations were set at AWAP levels with ET values from the four products were used respectively, and the equation can be transferred into the following:

$$\Delta W = P - ET - R \tag{9}$$

2.5. The Chain Effects of ET Variations on Irrigation

For water catchments with intensive irrigation, irrigation water (I) as an external source should be added to Equation (7). Then the equation of water mass balance is given as:

$$P + I = \Delta W + ET + R \tag{10}$$

$$I = \Delta W + ET + R - P \tag{11}$$

Because the detailed groundwater observational data is lacking and the main irrigation water resources in this catchment is surface water, and the recharge to the groundwater through infiltration of irrigation water is very small according to data drawn from ABARE's farm survey reports, two assumptions were made: only when the sum of ET and R was more than P, I was calculated by dividing the P from the sum of the ET and R, and ΔW was viewed as 0 in Equation (12); and when the sum of the ET and R was less than or equal to P, there was no irrigation, and ΔW was the difference between P and the sum of the ET and R in Equation (13).

$$I = ET + R - P, \quad \text{when } ET + R > P \text{ and } \Delta W = 0 \tag{12}$$

$$I = 0, \quad \text{when } ET + R < P \text{ and } \Delta W = P - ET - R \tag{13}$$

2.6. The Chain Effects of ET Variations on Productivity and Water-Use Efficiency of Ecosystems

The large water consumption sectors in the ecological systems in the MRC include the grasslands and forest. *GPP*, the most important indicator related to both the water cycle and carbon cycle, is used to measure the effects of *ET* variations on terrestrial ecosystems because of the linear relationship between GPP and ET at a regional scale [34]. In the MDB, it was modified from a linear relationship to a quadratic relationship [10]. The function between annual *GPP* and *ET* is given in Equation (14).

$$GPP = a * ET^2 + b * ET + c \tag{14}$$

where ET is the total ET per unit area in mm yr^{-1}, GPP is the total GPP in g C m^{-2} yr^{-1}.

The water-use efficiency, reflecting the water stress under the different water conditions of the ecosystem, is the ratio of the *GPP* to the *ET*, and can be calculated as:

$$WUE = \frac{GPP}{ET} = a * ET + b + c * ET^{-1} \tag{15}$$

where *WUE* is the water use efficiency in g C m^{-2} per mm water or kg C m^{-3} H$_2$O. The parameters a, b, and c were determined with the observed *GPP* from 2001 to 2014 and the varying ET products.

3. Results

3.1. ET Comparison among the Five ET Products

The fluctuations of the annual ET are obvious, where the minimum values appeared in 2006, and the maximum values appeared in 2011 (except for the TerraClimate ET) (Figure 3a). Although the change trends of all ET products are similar, the differences between their extreme annual values are large. For example, the largest is up to 455.3 mm for the TerraClimate ET while the minimal is only 214.7 mm for the MODIS ET. The minimum and maximum annual ET values are 324.7 mm and 617.8 mm (AWAP); 364.3 mm and 611.2 mm (CSIRO); 366.7 mm and 732.5 mm (GLDAS); 249.6 mm and 464.3 mm (MODIS); and 257.8 mm and 713.1 mm (TerraClimate), respectively. Their ratios of the maximum to the minimum annual ET values are 1.9, 1.7, 2.0, 1.9 and 2.8, respectively. Moreover, there are differences among their average values (Table 2). The average value of GLDAS ET is 14% more than the AWAP ET, while MODIS ET is 24% less than the AWAP ET. The CSIRO ET and TerraClimate ET are very close to the AWAP ET. Their RMSE and MAE were smaller as RSR's values were around 0.5, and the PBIAS values were close to 0.

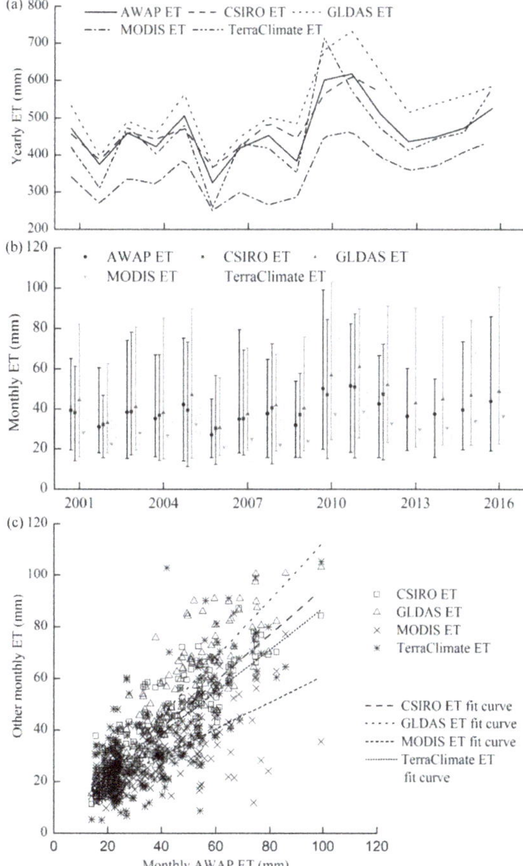

Figure 3. The comparison of annual and monthly ET products: (**a**) the annual ET, (**b**) mean monthly ET with the maximum and minimum values, and (**c**) the linear fit curves between AWAP ET and the other four ET products.

The mean, maximum, and minimum values of the monthly ET each year are shown in Figure 3b, and the variations of the mean monthly ET are the same as the yearly ET, but the changes of maximum and minimum values are very different. Compared with the AWAP ET, both the maximum and minimum values of GLDAS ET are larger, while the maximum values of MODIS ET are less, and its minimum values are less before 2010 and then almost the same since 2010. For the CSIRO and TerraClimate ET, their maximum values are similar to the AWAP ET, but the minimum values are less than the AWAP ET, and the TerraClimate ET is more obvious. The range of the monthly values are from 14.0 mm to 99.2 mm (AWAP), 11.2 mm to 87.2 mm (CSIRO), 15.0 mm to 103.1 mm (GLDAS), 9.8 mm to 77.3 mm (MODIS), and 5.1 mm to 105.3 mm (TerraClimate). Their ratios of the maximum to the minimum monthly ET values are up to 7.1, 7.8, 6.9, 7.9, and 20.6, respectively. Figure 3c shows the scatter diagram between the monthly AWAP ET with the other ET products from 2001 to 2016, and the performances of the fitting relationship between the GLDAS ET and CSIRO ET and AWAP ET are better than the other two ET products, which can be reflected by the slopes of fitting lines around 1 and correlation coefficients of more than 0.75 (Table 2). According to the other evaluation indicators, the CSIRO ET is the best among the four ET products, because its NSE is the largest, while the RMSE, RSR, and MAE are the smallest.

Table 2. Comparison of the four evapotranspiration (ET) products with the AWAP ET.

	A_ET	C_ET	G_ET	M_ET	T_ET	A_ET	C_ET	G_ET	M_ET	T_ET
			Yearly					Monthly		
Mean (mm)	464.14	473.53	529.91	352.32	450.37	38.68	39.46	44.16	29.36	37.53
N	16	12	16	16	16	192	144	192	192	192
NSE [h]	1	0.84	0.07	−1.42	0.61	1	0.74	0.61	0.21	0.34
R^2	1	0.87	0.93	0.82	0.87	1	0.77	0.82	0.49	0.52
RMSE [i] (mm)	0	33.45	71.90	116.25	46.43	0	9.31	11.21	15.84	14.50
RSR [j]	0	0.40	0.96	1.56	0.62	0	0.51	0.63	0.89	0.81
PBIAS [k]	0	0.03	0.14	−0.24	−0.03	0	0.03	0.14	−0.24	−0.03
MAE [l] (mm)	0	27.20	65.82	111.78	38.23	0	7.15	7.91	11.33	11.65

Note: [h] Nash–Sutcliffe efficiency (NSE), [i] root mean square error (RMSE), [j] root mean square error observation standard deviation ratio (RSR), [k] percent bias (PBIAS), and [l] mean absolute error (MAE).

3.2. The Effects of ET Variations on Estimation of Runoff and Water Storage Changes

3.2.1. The Water Balance in the Murrumbidgee River Catchment (MRC) Using Australian Water Availability Project (AWAP) Datasets

Figure 4 shows the annual and monthly water balance in the MRC using AWAP datasets. The mean annual precipitation, ET, runoff, and water storage changes are 518.0 mm, 464.1 mm, 49.5 mm, and 4.4 mm, respectively. The years of 2002, 2006, and 2009 were the dry years, while 2010, 2011, and 2016 were wet years. In most years, the precipitation is largely lost by ET, and in some years even the latter is larger than the former due to relatively less precipitation and consumption of the soil water. The runoff is low except in the very wet years, and the water storage is in a deficit condition. This situation was most serious in 2006.

The precipitation and ET in the MRC are concentrated in the period from December to February. The runoff only appeared in heavy rain years, for example, in 2010, 2011, and 2016. In some months, ET is more than precipitation, which leads to water storage changes: a recharge in heavy rain years and a deficit in drought years.

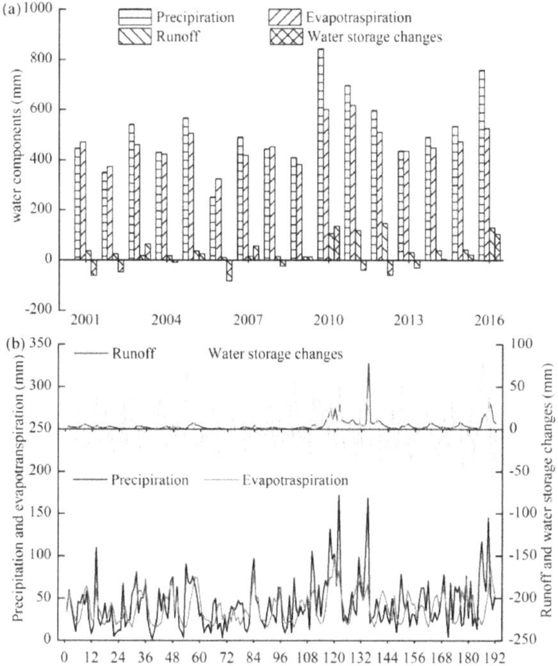

Figure 4. The annual (**a**) and monthly (**b**) water balance in MRC.

3.2.2. The Effects of ET Variations on Runoff Estimation

Figure 5 shows the annual runoff estimated using the four ET products combining the AWAP precipitation and water storage data. Compared with the AWAP runoff, the runoff estimated using the MODIS ET is much larger, while the runoff estimated using the TerraClimate ET is larger in some years and less in other years and even is negative in 2010. The runoff estimated using the GLDAS ET is much less than the AWAP runoff and even becomes negative. The runoff estimated using the CSIRO ET is complex, because it is larger or less than the AWAP runoff in some years but is negative in other years. The mean annual runoff estimated using the ET products from AWAP, CSIRO, GLDAS, MODIS, and TerraClimate are 49.5 mm, 34.4 mm, −16.3 mm, 161.3 mm, and 63.2 mm respectively. Application of the CRISO and GLADS ET products results in the underestimation of runoff, on the contrary, the MODIS and TerraClimate ET products result in the overestimation of runoff. The runoff estimated using the GLDAS ET is negative, while the runoff estimated using the MODIS ET is more than triple that of the AWAP runoff.

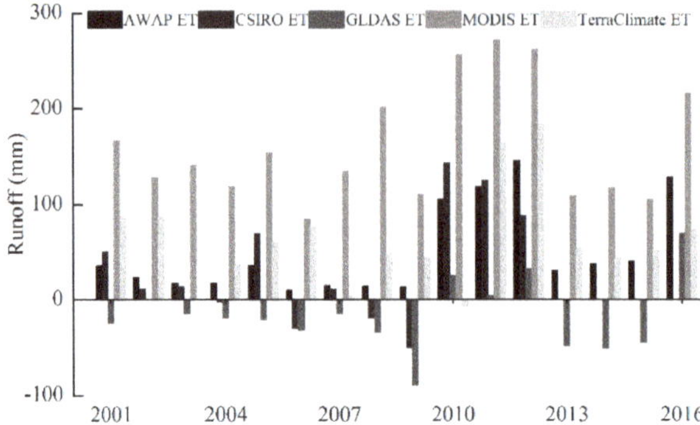

Figure 5. The yearly runoff estimations using different ET products.

3.2.3. The Effects of ET Variations on Water Storage Estimation

Figure 6 shows the annual water storage estimations using the four ET products combining the AWAP precipitation and runoff data. Compared with the AWAP water storage, the water storage estimated using the MODIS ET is much larger and positive except for the negative value in 2006 which was an extremely dry year, while the water storage estimated using the TerraClimate ET is larger in most years and less in a few years. The water storage estimated using the GLDAS ET is much less than the AWAP water storage and was negative in most years except in 2003, 2007, 2010, and 2016. The water storage estimated using the CSIRO ET is similar to the AWAP's estimations, but the former was negative while the latter was positive in 2009. The mean annual water storage estimated using the ET products from AWAP, CSIRO, GLDAS, MODIS, and TerraClimate are 4.4 mm, −14.1 mm, −61.5 mm, 116.1 mm, and 18.1 mm, respectively. The CSIRO and GLDAS ET products result in the underestimation of water storage, which indicates that the water storage is in deficit. By contrast, the MODIS and TerraClimate ET products result in the overestimation of water storage. The water storage estimated using the MODIS ET is even up to 110 mm and more than the AWAP water storage.

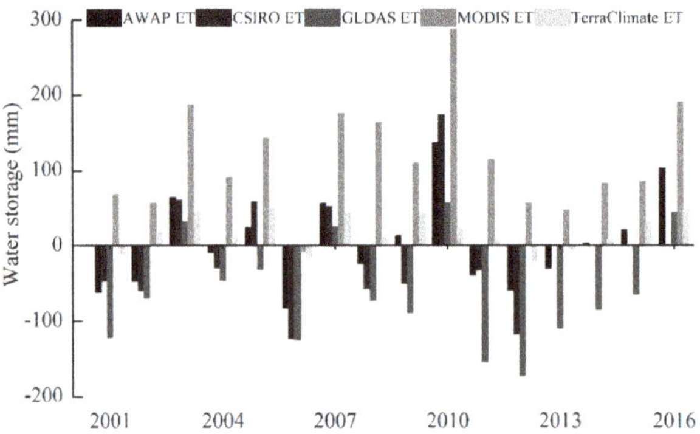

Figure 6. The yearly water storage changes estimated using different ET products.

3.3. The Effects of ET Variations on Irrigation Estimation

Figure 7 shows the monthly and annual irrigation water estimated using the five ET products combining the AWAP precipitation and runoff data in irrigated area in the MRC. There are significant differences between them, and the average values of annual irrigation water estimated using the AWAP, CSIRO, GLDAS, MODIS, and TerraClimate ET products are 87.0 mm, 114.2 mm, 154.6 mm, 115.7 mm, and 49.1 mm, respectively. The maximal one, estimated using MODIS ET, is more than three times of the minimal one estimated using TerraClimate ET. Compared with the annual irrigation water estimated using the AWAP, the CSIRO's are larger except in 2001, the GLDAS's are larger, the MODIS's are larger in most years, and the TerraClimate's are less except in 2010 and 2011. Although the irrigation water estimated using CSIRO and MODIS ET are close, they are different both over monthly and annual scales. The maximum values of the annual irrigation water using the AWAP, CSIRO, and MODIS ET products appeared in 2006 and were 118.8 mm, 171.8 mm, and 196.6 mm, respectively, while the maximum values using the GLDAS and TerraClimate ET products appeared in 2011 and were 256.5 mm and 110.9 mm, respectively.

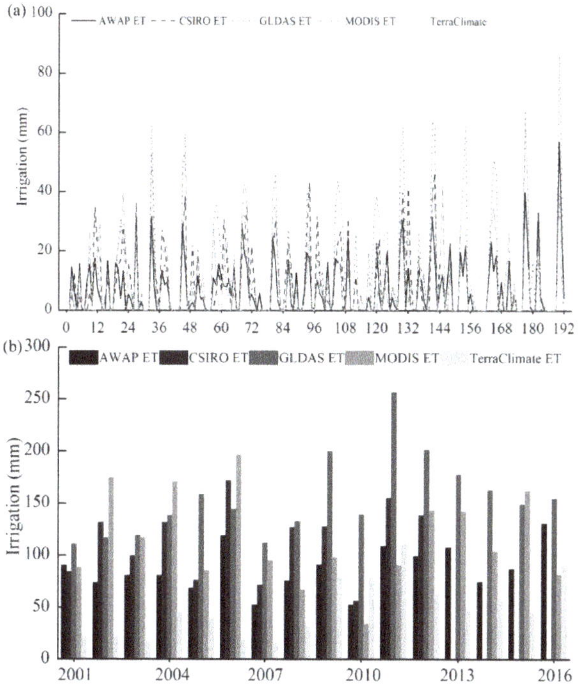

Figure 7. The monthly (**a**) and annual (**b**) irrigation water estimated using different ET products.

3.4. The Effects of ET Variations on the Gross Primary Productivity (GPP) and Water-Use Efficiency of the Grassland and Forest

The annual GPPs range is from 976.8 g C m^{-2} to 1279.5 g C m^{-2} for forest and 397.0 g C m^{-2} to 719.6 g C m^{-2} for grassland, and the annual ETs range is from 384.8 mm to 766.6 mm for forest and 118.0 to 618.8 mm for grassland (Figure 8a,c). The points representing different ET products overlap each other except for the MODIS ET for forest (Figure 8a), which indicates the ET products are close. However, the points representing GLDAS ET and MODIS ET are located at the left and right respectively, and the other three ET products are located in the middle and have some overlap for grassland, which reflects the obvious differences among the ET products (Figure 8c). There are significant quadratic relationships between the yearly GPP and all ET for forest and grassland, but the quadratic functions are different from each other (Figure 8a,c). For forest, the mean water use efficiency estimated using AWAP, CSIRO, GLDAS, and TerraClimate ET are close and about 1.9 kg C m^{-3} H$_2$O, while the water-use efficiency estimated using MODIS ET is up to 2.3 kg C m^{-3} H$_2$O (Figure 8b). For grassland, the mean water use efficiency estimated using AWAP, CSIRO, GLDAS, MODIS, and TerraClimate ET are 1.5 kg C m^{-3} H$_2$O, 1.5 kg C m^{-3} H$_2$O, 1.2 kg C m^{-3} H$_2$O, 2.4 kg C m^{-3} H$_2$O, and

1.6 kg C m^{-3} H$_2$O (Figure 8d). The ranges of the annual water-use efficiency estimated using AWAP, CSIRO, MODIS, and TerraClimate ET for forest are less than those for grassland, and the ranges using GLDAS ET are the same.

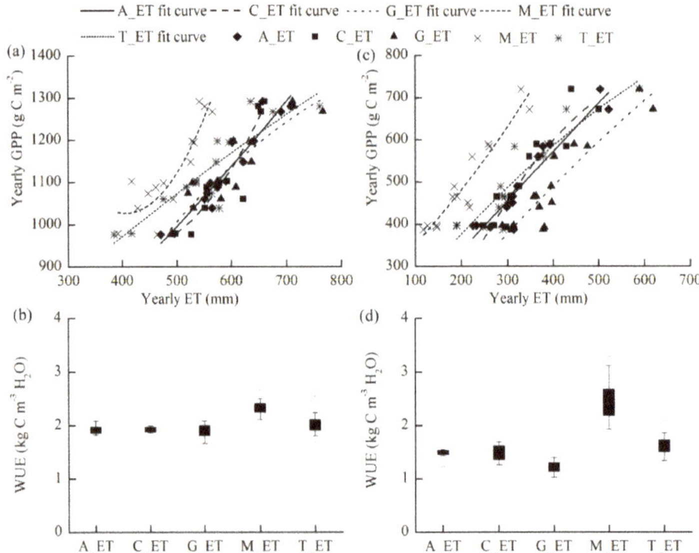

Figure 8. The relationships between yearly GPP and ET and the water-use efficiency of forest and grassland: (**a,b**) forest, and (**c,d**) grassland.

4. Discussion

4.1. The ET Variations among the Five Products

The results show that there are obvious differences in the annual and monthly ET values among these five ET products. The possible reasons for these ET variations could be explained as follows.

(1) Different estimation approach. GLDAS ET is the simulation from a land surface model which is upscaling of space- and ground-based observations with data assimilation techniques [27]. MODIS ET is the terrestrial ET based on the logic of the Penman–Monteith equation, which includes inputs of daily meteorological reanalysis data from the Global Modelling and Assimilation Office of NASA along with MODIS remotely sensed data products [28]. TerraClimate ET is developed using a water balance model that incorporates reference evapotranspiration, precipitation, temperature, and interpolated plant extractable soil water capacity, while the AWAP ET is also developed using a water balance model, and their results are close except that there is a larger difference in grassland. This is because the TerraClimate ET uses a simplified one-dimensional Thornthwaite–Mather water balance model while AWAP employs a two-layer model to simulate the changes in the shallow (thickness 0–0.7 m) and deep (0.2–1.5 m) soil layers, and the latter can depict the loss of the deep soil water [24,29]. Compared with the above three ET products, the CSIRO ET is similar to the AWAP ET, which is because they share many basic input data in this catchment [24,26]. CSIRO ET uses an observation-driven Penman–Monteith–Leuning (PML) model, supported with meteorological forcing data and along with satellite derived vegetation forcing data, land cover data, emissivity, and albedo. This is similar with the MODIS ET using the Penman–Monteith method and remotely sensed data, but their results are different in the interannual change, annual change, average value of the ET, and the ET of each land use type. This might be explained as the input datasets applied in CSIRO ET better represent the characteristics of Australian territory [26,28].

(2) Different inputs. There are obvious differences among inputs, such as meteorological forcing data, vegetation forcing data, and land surface data. For example, the precipitation, a key factor determining water availability and a basic component of the hydrological cycle, is different from each other in these ET products. GLDAS ET uses the space- and ground-based precipitation observations with data assimilation techniques, MODIS ET uses the daily meteorological reanalysis data, while AWAP ET uses a gridded daily rainfall dataset, which can directly influence the ET [24,27,28].

(3) Different spatial-temporal scales. The spatial scale mismatch between finer vegetation data and coarser meteorological forcing could result in large variations in ET products [11,35]. The spatial resolutions of these five ET products are different from each other and range from 500 m of MODIS ET to 2.5° of the TerraClimate ET. Generalisation of surface features (such as land use/cover and terrain) leads to the decreases of the resolution, thus the loss of details of surface features. The temporal resolutions of these five ET products vary from 3 h to one month, which could result in the representation of different meteorological processes and vegetation physiological processes in these ET products [14].

(4) The different expression of dynamics in vegetation index-based data could also cause differences among ET products. Some ET estimation methods, based on remotely sensed data like MODIS, can timely reflect the vegetation changes, but others cannot. If the vegetation growth process cannot be truly expressed, the evapotranspiration of vegetation, as well as its photosynthesis and other related hydrological processes (such as canopy interception and infiltration) will be affected [14].

4.2. The Effects of the Variations among the ET Products on Water Resources Management

ET is a significant climate variable that uniquely links the water cycle. It can help understand both sides of the water supply-demand equation [5,36]. The model calibration, using ET products combined with gaged runoff data, have effectively improved the performances of the hydrological models; moreover, ET products are also attractively used to estimate the runoff in the ungauged basins [5,13,14]. Thus, the value variations of ET could have a large influence on the other water balance components. Compared with the gaged runoff data and the condition of the total water storage changes estimated with the GRACE satellites from 2002 to 2014 [37], the runoff and water storage change in the MRC provided by the AWAP are reasonable respectively. The results in this study show the obvious influences of different ET products on the estimation of runoff and water storage undoubtedly creates a challenge for the calibration of hydrological models with such large variations of water balance components [13,14,19]. More importantly, such inaccurate information would seriously influence the precision and reliability of the decision-making on regional water resources allocation and regulation, which may artificially aggravate the unbalance of water allocation between different reaches in catchments, and between human resource use and ecosystem protection. In addition, such large variations of ET cast a shadow over the objective assessment of the effects of land use and cover change and climate change on basin water cycles, which are attracting increasing global attention [38,39].

As the predominant variable for water management in agricultural food production, the atmospheric demand for ET is a reference to irrigation [40,41]. In the MRC, the uncertainty of water allocation and low water availability were the main impediments to irrigated farms including dairy, broadacre and horticultural industries [42]. It was found in this study that the estimations of annual irrigation with these five ET products range from 49.1 mm using TerraClimate ET to 154.6 mm using GLDAS ET. Such a large range of estimated irrigation could lead to over-irrigation with water loss via infiltration of groundwater or it could result in unnecessarily returning water to rivers for environmental protection at the expense of necessary irrigation, resulting in lower crop yields due to crops suffering the effects of insufficient water, especially during the growth stage of the crop.

4.3. The Influences of the Variations among the ET Products on Ecosystem Management

The latest research found that the global carbon sink anomaly was driven by growth of semi-arid vegetation in the Southern Hemisphere, with almost 60 per cent of carbon uptake attributed to Australian ecosystems. This finding suggested that the higher turnover rates of carbon pools in semi-arid biomes were an increasingly important driver of inter-annual variability of global carbon cycle [43]. ET is the loss of water when mitigating vegetation stress responses and becomes the indicator of ecosystem function and health, such as productivity and carbon dioxide regulation [5]. In this study, it was found that there are less differences between the five ET products when applied to forest than when applied to grassland, thus there are also different relationships between GPP and different ET products for forest and grassland. However, all of their relationships are in significantly quadratic correlations. This means that we likely think that it is credible to predict the GPP using any of these relationships and the related ET products, which will probably result in different GPP estimations with different ET products. Water-use efficiency reflects the water stress of forest and grassland and therefore is a very important indicator of ecosystem health. It is found in this study that the mean water use efficiency of the forest using AWAP, CSIRO, GLDAS, and TerraClimate ET are about 1.9 kg C m^{-3} H$_2$O and are all greater than the grasslands, from the 1.2 kg C m^{-3} H$_2$O of GLDAS to 1.6 kg C m^{-3} H$_2$O of TerraClimate ET. However, the water-use efficiency estimated using MODIS ET is 2.3 kg C m^{-3} H$_2$O for forest, while grassland's is up to 2.4 kg C m^{-3} H$_2$O showing greater water use efficiency than forest. These values are reasonable according to the previous findings reporting C3 species range between 1.4 and 3.6 kg C m^{-3} H$_2$O [44], but the large range of GPP estimation functions and WUE among the five ET products could influence the precision in the assessment of the ecosystem function (e.g., carbon dioxide regulation) and how terrestrial ecosystems respond to environmental changes (e.g., a drier climate). Furthermore, the value variations of ET can seriously affect the allocation of environmental flows, especially in arid regions where the water resource is even scarcer. It also can affect the migration of water, salt and nutrient in the basin water quality modeling, then lead to the inaccurate prediction and evaluation of salinization, eutrophication, and ecological degradation.

4.4. Limitations

It should be noted that there are limitations in the methods and data in this study. We used the conventional water balance equation with fewer data requirements to assess the effects of ET variations on each water balance component. It should be known that the physical hydrological models reflecting the complex hydrological processes of the catchment could have more precise results. Meanwhile, we estimated the irrigation water amount according to the ET without consideration of the influence of different agricultural management practices, especially with the assumption of the unchanged water mass storage under irrigation practices. Nevertheless, as the focus of this study was to compare the effects of different ET products on runoff, water storage, and irrigation water amount, this simplified method was good enough to address the research questions in this study.

5. Conclusions

In this study, we provided quantitative effects of the value variations of several widely used ET products (AWAP, CSIRO, GLDAS, MODIS, and TerraClimate) on water resources management and ecosystem assessment using a water balance equation and its transformations, the relationships between GPP, WUE and ET. There are obvious differences of ET value among the five ET products in the MRC at annual and monthly scales. This leads to huge differences in the estimations of annual runoff (−16.3 to 161.3 mm) and annual water storage (−61.5 to 116.1 mm) and also for the annual irrigation water per area requirements ranging from 49.1 mm to 154.6 mm. The ET products for forest are relatively consistent among each other, but those for grassland present a large varying range. There are different productivity predictions of the ecosystem with the same water use based on the relationships between the annual GPP and varying ET for both forest and grassland, however, the forest has a better capacity of dealing with water stress than the grassland according to the values and the change amplitudes of their WUE. The effects of the variations among the ET products on water resources and ecosystem management are remarkable and need to attract more attention.

Author Contributions: Y.W. and Q.F. designed the case study. Z.L. and Y.Z. processed the data. Z.L. and Y.W. wrote the draft. Y.Z. and J.X. revised the draft. All of the authors contributed to the discussion of the results.

Funding: This work was funded by the National Natural Science Foundation of China (Project No: 41601036, 41571031), "Light of West China" Program of CAS, the International Postdoctoral Exchange Fellowship Program, the Australian Research Council (Project No: FT130100274), and the Strategic Priority Research Program of the Chinese Academy of Sciences (Project No: Y82CG11001).

Acknowledgments: The data are available from the corresponding data sites and providers listed in Table 1. We thank the two referees as well as the editors for providing valuable comments.

Conflicts of Interest: The authors declare no conflict of interest.

References

1. Elimelech, M.; Phillip, W.A. The future of seawater desalination: Energy, technology, and the environment. *Science* **2011**, *333*, 712–717. [CrossRef] [PubMed]
2. Jia, Z.; Liu, S.; Xu, Z.; Chen, Y.; Zhu, M. Validation of remotely sensed evapotranspiration over the Hai River Basin, China. *J. Geophys. Res. Atmos.* **2012**, *117*, 1–21. [CrossRef]
3. Vörösmarty, C.J.; McIntyre, P.B.; Gessner, M.O.; Dudgeon, D.; Prusevich, A.; Green, P.; Glidden, S.; Bunn, S.E.; Sullivan, C.A.; Liermann, C.R. Global threats to human water security and river biodiversity. *Nature* **2010**, *467*, 555. [CrossRef]
4. Bakker, K. Water security: Research challenges and opportunities. *Science* **2012**, *337*, 914–915. [CrossRef]
5. Fisher, J.B.; Melton, F.; Middleton, E.; Hain, C.; Anderson, M.; Allen, R.; McCabe, M.F.; Hook, S.; Baldocchi, D.; Townsend, P.A. The future of evapotranspiration: Global requirements for ecosystem functioning, carbon and climate feedbacks, agricultural management, and water resources. *Water Resour. Res.* **2017**, *53*, 2618–2626. [CrossRef]
6. Owen, C.R. Water budget and flow patterns in an urban wetland. *J. Hydrol.* **1995**, *169*, 171–187. [CrossRef]
7. Fahle, M.; Dietrich, O. Estimation of evapotranspiration using diurnal groundwater level fluctuations: Comparison of different approaches with groundwater lysimeter data. *Water Resour. Res.* **2014**, *50*, 273–286. [CrossRef]
8. Or, D.; Hanks, R. Spatial and temporal soil water estimation considering soil variability and evapotranspiration uncertainty. *Water Resour. Res.* **1992**, *28*, 803–814. [CrossRef]
9. Fatichi, S.; Ivanov, V.Y. Interannual variability of evapotranspiration and vegetation productivity. *Water Resour. Res.* **2014**, *50*, 3275–3294. [CrossRef]
10. Zhou, S.; Huang, Y.; Wei, Y.; Wang, G. Socio-hydrological water balance for water allocation between human and environmental purposes in catchments. *Hydrol. Earth Syst. Sci.* **2015**, *19*, 3715–3726. [CrossRef]
11. Long, D.; Longuevergne, L.; Scanlon, B.R. Uncertainty in evapotranspiration from land surface modeling, remote sensing, and GRACE satellites. *Water Resour. Res.* **2014**, *50*, 1131–1151. [CrossRef]
12. Drexler, J.Z.; Snyder, R.L.; Spano, D.; Paw U, K.T. A review of models and micrometeorological methods used to estimate wetland evapotranspiration. *Hydrol. Process.* **2004**, *18*, 2071–2101. [CrossRef]
13. Rajib, A.; Evenson, G.R.; Golden, H.E.; Lane, C.R. Hydrologic model predictability improves with spatially explicit calibration using remotely sensed evapotranspiration and biophysical parameters. *J. Hydrol.* **2018**, *567*, 668–683. [CrossRef]
14. Rajib, A.; Merwade, V.; Yu, Z. Rationale and efficacy of assimilating remotely sensed potential evapotranspiration for reduced uncertainty of hydrologic models. *Water Resour. Res.* **2018**, *54*, 4615–4637. [CrossRef]
15. Zhong, Y. Analysis of water rights system based on ET. *Water Resour. Dev. Res.* **2007**, *2*, 14–16.
16. Tang, W.; Zhong, Y.; Wu, D.; Deng, L. Analysis of water resource management mode based on ET. *China Rural Water Hydropow.* **2007**, *10*, 8–10.
17. Sörensson, A.A.; Ruscica, R.C. Intercomparison and uncertainty assessment of nine evapotranspiration estimates over South America. *Water Resour. Res.* **2018**, *54*, 2891–2908. [CrossRef]
18. Allen, R.G.; Pereira, L.S.; Howell, T.A.; Jensen, M.E. Evapotranspiration information reporting: I. Factors governing measurement accuracy. *Agric. Water Manag.* **2011**, *98*, 899–920. [CrossRef]
19. Lu, Z.; Zou, S.; Xiao, H.; Zheng, C.; Yin, Z.; Wang, W. Comprehensive hydrologic calibration of SWAT and water balance analysis in mountainous watersheds in northwest China. *Phys. Chem. Earth Parts A/B/C* **2015**, *79*, 76–85. [CrossRef]

20. Xing, W.; Wang, W.; Shao, Q.; Peng, S.; Yu, Z.; Yong, B.; Taylor, J. Changes of reference evapotranspiration in the Haihe River Basin: Present observations and future projection from climatic variables through multi-model ensemble. *Glob. Planet. Chang.* **2014**, *115*, 1–15. [CrossRef]

21. Lymburner, L.; Tan, P.; Mueller, N.; Thackway, R.; Lewis, A.; Thankappan, M.; Randall, L.; Islam, A.; Senarath, U. The Dynamic Land Cover Datset. *Geoscience* **2011**, 1–95.

22. Kingsford, R.; Thomas, R. Destruction of wetlands and waterbird populations by dams and irrigation on the Murrumbidgee River in arid Australia. *Environ. Manag.* **2004**, *34*, 383–396. [CrossRef] [PubMed]

23. Sivapalan, M.; Savenije, H.H.; Blöschl, G. Socio-hydrology: A new science of people and water. *Hydrol. Process.* **2012**, *26*, 1270–1276. [CrossRef]

24. Mr, R.; Briggs, P.; Haverd, V.; King, E.; Paget, M.; Trudinger, C. *Australian Water Availability Project, Data Release 26 m*; CSIRO Oceans and Atmosphere: Canberra, Australia, 2018.

25. Evans, J.P.; McCabe, M.F. Regional climate simulation over Australia's Murray-Darling basin: A multitemporal assessment. *J. Geophys. Res. Atmos.* **2010**, *115*, 1–15. [CrossRef]

26. Zhang, Y.; Pena Arancibia, J.; McVicar, T.; Chiew, F.; Vaze, J.; Zheng, H.; Wang, Y.P. *Monthly Global Observation-Driven Penman-Monteith-Leuning (PML) Evapotranspiration and Components*; Data Collection; CSIRO: Canberra, Australia, 2016.

27. Rodell, M.; Houser, P.R.; Jambor, U.; Gottschalck, J.; Mitchell, K.; Meng, C.-J.; Arsenault, K.; Cosgrove, B.; Radakovich, J.; Bosilovich, M.; et al. The Global Land Data Assimilation System. *Bull. Am. Meteorol. Soc.* **2004**, *85*, 381–394. [CrossRef]

28. NASA. *Evapotranspiration/Latent Heat Flux Product*; Version 6; NASA EOSDIS Land Processes DAAC, USGS Earth Resources Observation and Science (EROS) Center: Sioux Falls, SD, USA, 2018.

29. Abatzoglou, J.T.; Dobrowski, S.Z.; Parks, S.A.; Hegewisch, K.C. TerraClimate, a high-resolution global dataset of monthly climate and climatic water balance from 1958–2015. *Sci. Data* **2018**, *5*, 170191. [CrossRef] [PubMed]

30. NASA. *MOD17A3.055: Terra Net Primary Production Product*; Version 055; NASA EOSDIS Land Processes DAAC, USGS Earth Resources Observation and Science (EROS) Center: Sioux Falls, SD, USA, 2018.

31. Moriasi, D.N.; Arnold, J.G.; Van Liew, M.W.; Bingner, R.L.; Harmel, R.D.; Veith, T.L. Model evaluation guidelines for systematic quantification of accuracy in watershed simulations. *Trans. ASABE* **2007**, *50*, 885–900. [CrossRef]

32. Nash, J.E.; Sutcliffe, J.V. River flow forecasting through conceptual models part I—A discussion of principles. *J. Hydrol.* **1970**, *10*, 282–290. [CrossRef]

33. Lu, Z.; Wei, Y.; Xiao, H.; Zou, S.; Ren, J.; Lyle, C. Trade-offs between midstream agricultural production and downstream ecological sustainability in the Heihe River basin in the past half century. *Agric. Water Manag.* **2015**, *152*, 233–242. [CrossRef]

34. Beer, C.; Reichstein, M.; Ciais, P.; Farquhar, G.; Papale, D. Mean annual GPP of Europe derived from its water balance. *Geophys. Res. Lett.* **2007**, *34*, 1–4. [CrossRef]

35. Yang, Y.; Long, D.; Shang, S. Remote estimation of terrestrial evapotranspiration without using meteorological data. *Geophys. Res. Lett.* **2013**, *40*, 3026–3030. [CrossRef]

36. Wong, S.; Cowan, I.; Farquhar, G. Stomatal conductance correlates with photosynthetic capacity. *Nature* **1979**, *282*, 424. [CrossRef]

37. Xie, Z.; Huete, A.; Restrepo-Coupe, N.; Ma, X.; Devadas, R.; Caprarelli, G. Spatial partitioning and temporal evolution of Australia's total water storage under extreme hydroclimatic impacts. *Remote Sens. Environ.* **2016**, *183*, 43–52. [CrossRef]

38. Du, L.; Rajib, A.; Merwade, V. Large scale spatially explicit modeling of blue and green water dynamics in a temperate mid-latitude basin. *J. Hydrol.* **2018**, *562*, 84–102. [CrossRef]

39. Rajib, A.; Merwade, V. Hydrologic response to future land use change in the Upper Mississippi River Basin by the end of 21st century. *Hydrol. Process.* **2017**, *31*, 3645–3661. [CrossRef]

40. Anderson, M.; Kustas, W.; Norman, J.; Hain, C.; Mecikalski, J.; Schultz, L.; González-Dugo, M.; Cammalleri, C.; d'Urso, G.; Pimstein, A. Mapping daily evapotranspiration at field to continental scales using geostationary and polar orbiting satellite imagery. *Hydrol. Earth Syst. Sci.* **2011**, *15*, 223–239. [CrossRef]

41. Allen, R.G.; Pereira, L.S.; Raes, D.; Smith, M. Crop evapotranspiration-Guidelines for computing crop water requirements-FAO Irrigation and drainage paper 56. *Fao* **1998**, *300*, D05109.

42. Wei, Y.; Langford, J.; Willett, I.R.; Barlow, S.; Lyle, C. Is irrigated agriculture in the Murray Darling Basin well prepared to deal with reductions in water availability? *Glob. Environ. Chang.* **2011**, *21*, 906–916. [CrossRef]
43. Poulter, B.; Frank, D.; Ciais, P.; Myneni, R.B.; Andela, N.; Bi, J.; Broquet, G.; Canadell, J.G.; Chevallier, F.; Liu, Y.Y. Contribution of semi-arid ecosystems to interannual variability of the global carbon cycle. *Nature* **2014**, *509*, 600. [CrossRef]
44. Kocacinar, F.; Sage, R.F. Hydraulic properties of the xylem in plants of different photosynthetic pathways. In *Vascular Transport in Plants*; Elsevier: Amsterdam, The Netherlands, 2005; pp. 517–533.

MDPI

St. Alban-Anlage 66

4052 Basel

Switzerland

Tel. +41 61 683 77 34

Fax +41 61 302 89 18

www.mdpi.com

Remote Sensing Editorial Office

E-mail: remotesensing@mdpi.com

www.mdpi.com/journal/remotesensing